Time-Frequency Analysis in Biomedical Engineering

Time-frequency analysis is critical in biomedical signal analysis, which helps diagnose and monitor physiological conditions such as heart rate variability, seizure detection, and brain-computer interfacing. This edited book includes original theoretical, practical, and review chapters aimed at proposing advancements in time-frequency signal processing methods for biomedical healthcare applications. Exemplary themes of interest include biomedical signal processing challenges in complex physiological data, signals from remote sensors, wearables, and nearables such as ballistography-based sensors.

Features:

- Discusses detailed time-frequency signal processing applications for simple to complex biomedical research.
- Reports novel time-frequency techniques used for biomedical signals.
- Presents the theoretical basis of time-frequency analysis and state-of-the-art applications tailored for various biomedical problems.
- Provides a forum for presenting new and improved techniques and theories related to time-frequency analysis.
- Combines the primary knowledge of time-frequency signal analysis and processing, from theory and applications.

This book is aimed at graduate students and researchers in bioengineering and signal processing.

Time-Frequency Analysis in Biomedical Engineering

Contemporary Methods and Applications

Edited by

Ganesh R. Naik

CRC Press
Taylor & Francis Group
Boca Raton London New York

CRC Press is an imprint of the
Taylor & Francis Group, an **informa** business

First edition published 2026
by CRC Press
2385 NW Executive Center Drive, Suite 320, Boca Raton FL 33431

and by CRC Press
4 Park Square, Milton Park, Abingdon, Oxon, OX14 4RN

CRC Press is an imprint of Taylor & Francis Group, LLC

ISBN: 9781041007456 (hbk)
ISBN: 9781041007449 (pbk)
ISBN: 9781003611448 (ebk)

DOI: 10.1201/9781003611448

Typeset in Times
by codeMantra

Contents

SECTION V Advanced Techniques and Machine Learning Applications

Preface

The field of biomedical engineering has long relied on signal processing to extract meaningful information from the complex data generated by the human body. Among the many techniques available, time-frequency analysis has emerged as a potent tool, capable of revealing dynamic changes in both time and frequency domains that are often obscured by traditional methods. This edited volume, *Time-Frequency Analysis in Biomedical Engineering*, is a curated collection of research that showcases the breadth and depth of this crucial methodology. It serves as a comprehensive resource for understanding how time-frequency techniques, such as the wavelet transform and its variants, are being applied to solve significant challenges in medicine and healthcare.

This book is organized into five thematic sections, each presenting a focused view on a key area of application. The initial chapters introduce the fundamental concepts and their application to various physiological signals, including the electrocardiogram (ECG) and other time-series data with variable spectral content. Subsequent sections delve into the rapidly evolving field of neurological signal processing, providing insights into the denoising of electroencephalogram (EEG) signals and the application of time-frequency features for tasks such as driver drowsiness detection and motor imagery classification. A significant portion of this book is dedicated to the critical area of seizure detection and classification, presenting a range of innovative approaches from automated systems to deep learning models and advanced classifier fusion techniques. The final chapters broaden the scope to include advanced techniques in machine learning and biometrics, demonstrating the versatility of time-frequency analysis in applications like arrhythmia detection and EEG-based authentication.

Time-Frequency Analysis in Biomedical Engineering is intended for a diverse readership, including researchers, academics, graduate students, and professionals in biomedical engineering, signal processing, and medical informatics. I hope that this volume will not only provide valuable insights into the current state of the art but also inspire new research and foster further innovation in the application of time-frequency methods to enhance diagnostic capabilities and improve patient care. I am deeply grateful to all the contributing authors for their scholarly work and dedication to advancing this important field.

Ganesh R. Naik
August 2025

About the Editor

Ganesh R. Naik is a globally recognized biomedical engineer and signal processing expert, ranked in the top 2% of researchers by Stanford University. He holds a PhD from RMIT University and is currently a senior academic at Torrens University Australia. A prolific researcher, he has edited 16 books and authored over 150 papers. Dr. Naik is an associate editor for several prestigious journals, including IEEE ACCESS. His career includes significant research roles at Flinders University, Western Sydney University, and the University of Technology Sydney, where he contributed to advancements in sleep health and wearable technologies. He has received numerous fellowships, including from the Royal Academy of Engineering UK, the Government of Australia, and Germany's Baden–Württemberg Scholarship.

Contributors

Saif Al-Jumaili
Electrical and Computer Engineering
School of Engineering and Natural Sciences
Altınbaş University
Istanbul 34218,
Türkiye

Salam Alyassri
Department of Electrical and Computer Engineering
Altinbas University
Istanbul, Turkey

Durga Siva Teja Behara
Department of Electronics and Communication Engineering
Rajiv Gandhi University of Knowledge Technologies
Hyderabad, India

Manvir Bhatia
Neurology and Sleep Centre
New Delhi, India

Hasan Birol Çotuk
Neuroscience and Psychology Research in Sports Lab
Faculty of Sport Science
Marmara University
Istanbul, 34722,
Türkiye

Rodrigo Capobianco Guido
Instituto de Biociências, Letras e Ciências Exatas
Universidade Estadual Paulista (São Paulo State University),
São Paulo, Brazil

Gabriel José Pellisser Dalalana
Department of Electrical and Computer Engineering
School of Engineering of São Carlos
University of São Paulo
São Paulo, Brazil

Adil Deniz Duru
Neuroscience and Psychology Research in Sports Lab
Faculty of Sport Science
Department of Sports and Health Sciences
Marmara University
Istanbul, 34722, Türkiye

Muhammed Enes Subasi
Faculty of Medicine
Izmir Katip Celebi University
Izmir, Turkey

Natalia Filimonova
Institute of Mathematics
University of Kassel
Kassel, Germany

Elfriede Friedmann
Department of Mathematics
Heidelberg University
Heidelberg, Germany

Tapan Kumar Gandhi
Department of Electrical Engineering
Indian Institute of Technology
New Delhi, India

Alex Marino Gonçalves de Almeida
Ourinhos College of Technology
Brazil

Md. Anik Hasan
Department of Electrical & Electronic Engineering
Rajshahi University of Engineering and Technology
Station Rd, Rajshahi 6204,
Bangladesh

G. K. M. Hasanuzzaman
Department of Electrical and Electronic Engineering
Rajshahi University of Engineering & Technology
Rajshahi, Bangladesh

Roopa B. Hegde
Nitte (Deemed to be University) Department of ECE
NMAM Institute of Technology
Nitte-574110, Udupi, India

Divya Jain
Oriental College of Technology
Bhopal, Madhya Pradesh, India

Sumaiya Kabir
Department of Electrical & Electronic Engineering
Rajshahi University of Engineering and Technology
Station Rd, Rajshahi 6204,
Bangladesh

Suyash Khare
Centre for Artificial Intelligence
Madhav Institute of Technology and Science
Deemed University, Gwalior, India

Anirudh Kumar
Department of Electrical Engineering
Malaviya National Institute of Technology
Jaipur, India

Mykola Makarchuk
Department of Physiology
Taras Shevchenko National University of Kyiv
Ukraine

Saeed Mian Qaisar
College of Engineering and Technology
American University of the Middle East
Hadiya, Kuwait

Ahmed A. Mohammed
Department of Medical Physics
College of Applied Science
University of Fallujah
Al-Anbar, Iraq
and
Electronics Materials Lab, College of Science and Engineering
James Cook University
Townsville, Australia

Mohammod Abdul Motin
Department of Electrical & Electronic Engineering
Rajshahi University of Engineering and Technology
Station Rd, Rajshahi 6204, Bangladesh

Osman Nuri Uçan
Department of Electrical and Computer Engineering
Altinbas University
Istanbul, Turkey

Bijaya Ketan Panigrahi
Department of Electrical Engineering
Indian Institute of Technology
New Delhi, India

G. R. Pradyumna
Department of Electronics and Communication Engineering
Nitte (Deemed to be University)
Karnataka, India

Abhishek Sathyendran
Nitte (Deemed to be University)
Department of ECE
NMAM Institute of Technology
Nitte-574110, Udupi, India

Ankith P. Shetty
Department of Electronics and Communication Engineering
Nitte (Deemed to be University)
Karnataka, India

Priya Shree
Department of Microsoft Ads
Microsoft
Bengaluru, Karnataka, India
and
Department of Electrical Engineering
Indian Institute of Technology
New Delhi, India

Abdulhamit Subasi
Institute of Biomedicine
Faculty of Medicine
University of Turku
Turku, Finland

Muhammad Sudipto Siam Dip
Department of Electrical & Electronic Engineering
Rajshahi University of Engineering and Technology
Station Rd, Rajshahi 6204, Bangladesh

Piyush Swami
Télécom SudParis
Institut Polytechnique de Paris
Palaiseau, France
and
Centre for Biomedical Engineering
Indian Institute of Technology
Delhi, India

Durga Siva Teja Behara
Department of Electronics and Communication Engineering
Rajiv Gandhi University of Knowledge Technologies
Hyderabad, India

Akshra Tiwari
School of Sciences
Christ University
Delhi, India

Vibha Tiwari
Centre for Artificial Intelligence
Madhav Institute of Technology and Science
Deemed University
Gwalior, Madhya Pradesh, India

Section I

Introduction to Time–Frequency Analysis for Biomedical Engineering

1 Wavelet-Based Biomedical Signal Analysis

A Tutorial Approach for Pathological Assessment

Gabriel José Pellisser Dalalana, Alex Marino Gonçalves de Almeida, and Rodrigo Capobianco Guido

1.1 INTRODUCTION

Biomedical signals, such as electroencephalography (EEG), electrocardiography (ECG), and disordered speech, exhibit strongly non-stationary dynamics driven by physiological processes that operate on widely disparate time scales. Classical spectral techniques, most notably the short-time Fourier transform (STFT), become inadequate when short transients or rapidly evolving spectral components must be characterized because they rely on a fixed analysis window whose time-frequency resolution trade-off cannot adapt to local signal properties [1]. In contrast, the wavelet transform (WT) supplies a multi-resolution decomposition in which the analysis window automatically matches each scale, offering simultaneous time- and frequency-domain descriptions that enable pathological biomarkers, artifacts, and noise to be isolated with greater precision.

The WT is obtained by convolving the signal with scaled and translated replicas of a single *mother wavelet*. This construction yields fine temporal resolution at high scales (high frequencies) and fine spectral resolution at low scales (low frequencies). Consequently, events such as epileptic discharges, QRS complexes, and motor unit potentials can be detected even when embedded in background noise or obscured by spectral overlap [2]. Moreover, the multi-level representation facilitates thresholding, entropy-based feature extraction, and coefficient-based analyses that feed statistical classifiers and deep learning models.

Given the practical importance of the WT, this chapter first reviews the theoretical foundations underlying the transform and consolidates them through step-by-step, reproducible examples. We then illustrate how the WT can be applied to a range of biomedical problems, emphasizing algorithmic choices, parameter selection, and implementation pitfalls. By integrating fundamental theory, a critical survey of

DOI: 10.1201/9781003611448-2

recent literature, and hands-on tutorials, the chapter aims to serve as a comprehensive reference for biomedical engineers, clinicians, and data scientists seeking to apply the WT in real clinical scenarios, thereby enhancing early pathology detection and enabling personalized therapeutic interventions.

Beyond the foundational concepts and theoretical exposition of WTs, the remainder of this chapter systematically explores their practical applications in biomedical signal analysis. It begins by tracing the evolution from classical Fourier-based methods to advanced multi-resolution techniques, culminating in detailed coverage of continuous and discrete wavelet transforms (DWTs) and practical implementation strategies such as filter banks and matrix operations. Subsequently, the chapter delves into numerous case studies demonstrating the utility of wavelets across various biomedical contexts, including epilepsy detection using EEG, cardiac signal analysis, EEG artifact removal, and acoustic signal processing for voice pathology assessment. These applications are critically analyzed concerning their methodological frameworks, computational efficiency, and clinical relevance. Step-by-step tutorials and numerical examples support them to facilitate practical adoption by researchers and practitioners.

1.2 FROM FOURIER TO WAVELETS: THE EVOLUTION OF TIME-FREQUENCY ANALYSIS

Signal analysis is fundamental in numerous scientific and engineering disciplines, from audio processing and medical diagnostics to financial forecasting and geophysical exploration. At its core lies a critical challenge: effectively characterizing signals exhibiting temporal and spectral features. This challenge forms the foundation of time-frequency analysis, a field dedicated to simultaneously capturing when events occur in a signal and what frequency components constitute those events. Traditional signal analysis methods often focus exclusively on the time or frequency domains. Time-domain analysis examines how a signal's amplitude varies over time, providing excellent temporal resolution but offering limited insight into its frequency composition. Conversely, frequency-domain analysis, primarily through Fourier techniques, reveals the spectral content of a signal but typically sacrifices information about when specific frequencies occur. This inherent limitation of conventional approaches becomes particularly problematic when analyzing non-stationary signals—those whose frequency content changes over time. Such signals are ubiquitous in real-world applications: speech and music, electrocardiograms, seismic recordings, and financial time series exhibit time-varying spectral characteristics that pure time- or frequency-domain methods cannot adequately capture.

Time-frequency analysis addresses this limitation by providing a framework for joint representation of a signal's temporal and spectral properties. This section traces the evolution of time-frequency analysis techniques, beginning with the foundational Fourier transform and its limitations. It progresses through intermediate approaches such as the STFT and culminates in wavelet-based methods that offer adaptive resolution across time and frequency dimensions.

As we will explore, the development of WTs represents a significant paradigm shift in signal processing—one that overcomes the fundamental constraints of earlier methods and enables more nuanced analysis of complex, real-world signals.

1.2.1 Fourier Analysis and Its Limitations

Fourier analysis, a foundational tool in signal processing, is instrumental in transforming signals from the time domain to the frequency domain. This transformation, achieved through the Fourier transform, reveals the frequency content of a signal by decomposing it into a sum of sinusoidal components. Its digital counterpart, the discrete Fourier transform, is a key player in modern signal processing applications, from audio analysis to image compression, demonstrating Fourier analysis's practical relevance and applicability in your field [3]. Despite its mathematical elegance and computational efficiency, especially when implemented as the Fast Fourier Transform, the Fourier transform has a critical limitation: it provides a global frequency representation that completely discards temporal information. This global perspective, while broad, lacks specific temporal information about when frequency components occur. When we transform a signal using the discrete Fourier transform, we obtain information about which frequency components are present but no insight into when these components arise. This global averaging across the entire signal duration makes the Fourier transform particularly ill-suited for analyzing non-stationary signals, whose spectral characteristics evolve.

Consider, for example, a musical piece containing notes of different pitches played in sequence. The Fourier transform would identify all the frequency components throughout the piece but would fail to reveal which notes were played at which moments. Similarly, in electrocardiogram analysis, transient abnormalities might be obscured in the global frequency spectrum, despite their critical diagnostic importance.

This fundamental limitation stems from the nature of the basis functions employed in the Fourier analysis: infinitely extending sinusoids with perfect frequency localization but no time localization. When we project a signal onto these basis functions, we inevitably sacrifice all temporal information for precise frequency resolution.

Furthermore, this limitation is not merely a practical inconvenience but reflects a deeper theoretical constraint known as Heisenberg's uncertainty principle in signal processing: it is mathematically impossible to achieve arbitrarily high resolution simultaneously in both time and frequency. Any attempt to improve resolution in one domain necessarily degrades resolution in the other.

These limitations of the Fourier analysis have motivated the development of alternative approaches that can better capture the time-varying spectral characteristics of real-world signals, as we will explore in the following sections.

1.2.2 Short-Time Fourier Transform: A First Approach

The limitations of the standard Fourier transform led to the development of the STFT, the first attempt to simultaneously capture both time and frequency information. The STFT's fundamental insight is straightforward. Rather than analyzing an entire signal at once, we can apply the Fourier transform to short segments or windows of the signal, thereby localizing the spectral analysis in time [4].

The STFT involves sliding a window function across the signal and computing the Fourier transform at each position. Mathematically, for a signal $x(t)$ and a window function $w(t)$, the STFT is defined as follows:

$$\text{STFT}\{x(t)\}(\tau,\omega) = \int_{-\infty}^{\infty} x(t)\cdot w(t-\tau)\cdot e^{-j\omega t}\,dt$$

where τ represents the time the spectrum is evaluated and ω is the angular frequency. The window function $w(t-\tau)$ effectively isolates a portion of the signal centered at time τ, allowing the Fourier transform to capture the frequency content specifically within that temporal neighborhood.

This approach yields a two-dimensional representation—often visualized as a spectrogram—that shows how the signal's frequency content evolves. The STFT thus provides a rudimentary form of joint time-frequency analysis, offering significant advantages over the standard Fourier transform for non-stationary signals.

Despite this improvement, the STFT introduces a fundamental trade-off between time and frequency resolution governed by the width of the window function. This trade-off becomes especially problematic when analyzing signals that contain both low- and high-frequency components, a common characteristic of real-world signals.

Consider the dilemma: narrow windows provide sound temporal localization but poor frequency resolution, as they capture fewer oscillation cycles. This is particularly problematic for low-frequency components, requiring more extended observation periods to characterize accurately. Conversely, expansive windows offer improved frequency resolution but at the cost of temporal precision, potentially blurring transient events and violating the assumption of stationarity within the window.

This unavoidable compromise is not merely a practical limitation but reflects a fundamental constraint in signal processing. It is formalized as Heisenberg's uncertainty principle: it is impossible to simultaneously achieve arbitrarily high resolution in both time and frequency domains. Mathematically, if Δt represents time resolution and $\Delta\omega$ represents frequency resolution, then

$$\Delta t \cdot \Delta\omega \geq \frac{1}{2}$$

The critical weakness of the STFT lies in its use of a fixed-size window for all frequencies. Once the window size is chosen, it remains constant across the entire time-frequency plane, imposing the same resolution trade-off for all frequency bands. This "one-size-fits-all" approach is inherently suboptimal for signals with diverse spectral characteristics, motivating the search for more adaptive time-frequency analysis methods.

1.2.3 Wavelet Transform: Foundations

The limitations of the STFT—particularly its fixed resolution across all frequencies—motivated the development of a more adaptive approach to time-frequency analysis: the WT. Rather than decomposing a signal into sinusoidal components of infinite duration, the WT projects the signal onto a family of localized waveforms called wavelets, which are inherently limited in time and frequency.

The fundamental innovation of wavelet analysis lies in its multi-resolution capability. Unlike the STFT, which applies a fixed-size window across all frequencies, the WT automatically adjusts its resolution to match the analyzed frequency. This adaptive approach provides acceptable temporal resolution for high-frequency components (where rapid changes often occur) and better frequency resolution for low-frequency components (which require more extended observation periods)—precisely the behavior needed to analyze real-world signals effectively.

At the core of wavelet analysis is the concept of a *wavelet*—a small, oscillatory waveform with finite duration and zero mean. It is essential to distinguish between the WT, which is the mathematical transformation process, and a *wavelet*, which is the specific function used to perform this analysis. These terms are not interchangeable, just as we distinguish between the Fourier transform operation and the sinusoidal functions it employs.

The wavelet function, typically denoted as $\psi(t)$, serves as a prototype or template from which an entire family of analysis functions is generated through two fundamental operations: translation (shifting in time) and dilation (scaling). $\psi(t-b)$ gives a shifted wavelet version, where the parameter $b \in \mathbb{R}$ controls the temporal position. This shifted function effectively acts as a sliding window centered at time b.

To introduce control over the scale—and consequently the frequency resolution—we define a dilation parameter $a \in \mathbb{R}$, leading to the scaled and shifted function:

$$\psi\left(\frac{t-b}{a}\right)$$

The dilation parameter a controls the width of the wavelet: when $|a| > 1$, the wavelet stretches in time, becoming more suitable for analyzing low-frequency components; when $|a| < 1$, the wavelet compresses, enhancing sensitivity to high-frequency, transient features. This dual parametrization by both position and scale enables the WT to achieve localized analysis with adaptive resolution—a significant advancement over the fixed resolution of the STFT.

Two primary formulations of the WT are the continuous Wavelet Transform (CWT) and the DWT. In the CWT, the parameters a and b vary continuously, producing a highly redundant representation that offers excellent localization but at a high computational cost. The DWT, by contrast, restricts these parameters to discrete values (typically dyadic scales), yielding a more efficient representation while preserving the essential multi-resolution property [5].

Despite the terminology—"continuous" versus "discrete"—both the CWT and DWT can be applied to continuous-time signals. When the input signal itself is discrete (as is common in digital applications), a third variant called the discrete-time Wavelet Transform (DTWT) is employed. In practice, the distinction between DWT and DTWT is often blurred, and it is common to refer to the DTWT simply as the DWT when applied to sampled signals.

The wavelet framework provides a powerful and flexible approach to signal analysis, overcoming the fundamental limitations of Fourier-based methods while maintaining mathematical rigor. In the following sections, we will explore the specific formulations of the CWT and DWT, their mathematical properties, and their practical implementation through filter banks and multi-resolution analysis (MRA).

The framework for analyzing signals with optimal time-frequency resolution at each scale offers exceptional flexibility at the cost of computational intensity.

1.2.4 Continuous Wavelet Transform (CWT)

Having established the foundational concepts of wavelet analysis, we now turn to the most direct implementation of these ideas: the CWT. The CWT provides a comprehensive framework for analyzing signals with optimal time-frequency resolution at each scale, offering exceptional flexibility at the cost of computational intensity.

The CWT systematically compares a signal with scaled and shifted versions of a prototype function—the mother wavelet. By translating the wavelet function via the parameter b, we ensure complete coverage of the time domain. Simultaneously, by dilating it via the parameter a, we adjust the effective frequency band captured at each point. This dual parametrization enables the CWT to adapt its resolution to the scale being analyzed, overcoming the fixed-resolution limitation of the STFT.

In effect, the scaled and shifted wavelet $\psi\left(\frac{t-b}{a}\right)$ acts as a band-pass filter adapted to the specific temporal and spectral characteristics of the signal. Small scales (corresponding to high frequencies) yield short-duration wavelets that excel at detecting rapid, transient events. Conversely, large scales (corresponding to low frequencies) result in longer wavelets ideal for resolving slowly varying components. This scale-dependent resolution makes the WT effective for analyzing signals with diverse spectral content.

Formally, the CWT of a signal $x(t)$ is defined as follows:

$$C(a,b)=\frac{1}{\sqrt{a}}\int_{-\infty}^{\infty}x(t)\psi\left(\frac{t-b}{a}\right)dt,$$

where $\frac{1}{\sqrt{a}}$ is an energy normalization factor ensuring that the transform preserves signal energy across scales. This normalization is crucial: without it, the transform would artificially amplify high-frequency components (small a values) and attenuate low-frequency components (large a values).

The output coefficient $C(a,b)$ quantifies the degree of similarity—or correlation—between the input signal $x(t)$ and the wavelet ψ at scale a and position b. A large magnitude of $C(a,b)$ indicates a strong presence of signal features matching the wavelet at that particular scale and location. Thus, the CWT maps a one-dimensional time-domain signal to a two-dimensional time-frequency representation, with each point (a,b) encoding the presence of signal components at a given scale and time.

This representation is often visualized as a scalogram—a heat map where the horizontal axis represents time, the vertical axis represents a scale (or frequency), and the color intensity indicates the magnitude of the wavelet coefficients. Such visualizations reveal both the temporal localization of transient events and the spectral composition of the signal, providing insights that are impossible to obtain from either time- or frequency-domain analysis alone.

The CWT offers remarkable analytical power, particularly for detecting localized features, analyzing non-stationary signals, and characterizing scale-invariant processes. Its continuous nature ensures no information is lost during the transformation, making it ideal for theoretical analysis and applications requiring maximum precision [6].

However, this theoretical elegance comes with practical limitations. The CWT generates a highly redundant representation—infinitely many coefficients for a finite-length signal—making it computationally intensive and memory demanding. Moreover, it cannot be computed for all real values of a and b in practical implementations, especially when full spectral coverage is desired. These limitations motivate the development of more efficient, discretized formulations, such as the DWT, which we will explore in the next subsection.

1.2.5 Discrete Wavelet Transform (DWT)

While the CWT offers exceptional analytical power, its practical implementation faces significant computational challenges. The CWT generates a highly redundant representation with infinitely many coefficients for a finite-length signal, making it prohibitively expensive for many real-world applications. The DWT addresses these limitations by discretizing the scaling and translation parameters, yielding a more efficient yet robust framework for time-frequency analysis.

The DWT achieves this efficiency by strategically sampling the scale parameter a and the shift parameter b. Rather than allowing these parameters to vary continuously, the DWT restricts them to discrete values. Specifically, the scale parameter follows a dyadic sequence $a = 2^j$ (where j is an integer), and the shift parameter is sampled proportionally as $b = k \cdot 2^j$ (where k is also an integer). This dyadic sampling creates a logarithmic frequency scale that aligns well with how many natural signals behave, where lower frequencies typically require finer resolution than higher frequencies.

This discretization leads to an MRA in which the wavelet function is progressively stretched by powers of two, effectively halving its bandwidth at each scale level. However, this approach presents a critical limitation: as the scale increases, the wavelet functions become increasingly dilated, leaving gaps in the frequency coverage—particularly in the low-frequency region near 0 Hz. This gap would prevent complete signal reconstruction, a fundamental requirement for any viable transform.

To resolve this issue, the DWT introduces a second function: the *scaling function* $\phi(t)$, also known as the *father wavelet*. While the wavelet function $\psi(t)$ acts as a band-pass filter capturing details at specific scales, the scaling function serves as a complementary low-pass filter that captures the coarse approximation components of the signal. $\phi(t)$ and $\psi(t)$ form an orthogonal basis that spans the entire signal space, ensuring complete representation and perfect reconstruction.

The signal is then decomposed into a linear combination of shifted and scaled versions of both functions:

$$x(t) = \sum_k c_{J,k}\phi_{J,k}(t) + \sum_{j=1}^{J}\sum_k d_{j,k}\psi_{j,k}(t)$$

where $\phi_{J,k}(t) = 2^{-J/2}\phi(2^{-J}t - k)$ represents the scaling function at the coarsest scale J, $\psi_{j,k}(t) = 2^{-j/2}\psi(2^{-j}t - k)$ represents the wavelet function at scale j, and the coefficients $c_{J,k}$ and $d_{j,k}$ capture the approximation and detail components, respectively.

This decomposition creates a hierarchical, multi-resolution representation of the signal. At each scale level, the approximation coefficients capture the low-frequency content, while the detail coefficients capture the high-frequency features. As we move from finer to coarser scales, the approximation becomes increasingly smooth, retaining only the signal's most fundamental structures, while the details preserve the scale-specific features that would otherwise be lost.

There are two conceptually distinct but mathematically equivalent approaches to implementing the DWT. The first is grounded in linear algebra, viewing the transform as a projection of the signal onto an orthogonal wavelet basis. Alfred Haar pioneered this perspective in 1910 with the simplest wavelet family, providing a solid theoretical foundation for wavelet analysis.

The second approach, which dominates practical implementations, is based on digital filtering operations. Rather than explicitly constructing orthogonal basis functions, this method performs successive convolutions and downsampling operations using a pair of filters: a low-pass filter $h[\cdot]$ derived from the scaling function and a high-pass filter $g[\cdot]$ derived from the wavelet function. This filter-based formulation, formalized in Stéphane Mallat's MRA framework, transforms the abstract mathematical concept into an efficient computational algorithm.

For decades, these approaches—the algebraic and filtering perspectives—evolved separately in different academic communities. Mathematicians focused on the theoretical properties of wavelet bases, while signal processing engineers developed efficient filtering algorithms. The formal establishment of their equivalence marked a significant milestone in wavelet theory, unifying these perspectives and enabling cross-fertilization between the fields.

The DWT's discrete nature and efficient implementation make it well-suited for practical applications. Its hierarchical structure naturally supports progressive transmission and MRA, while its energy compaction properties enable effective compression. These characteristics have led to the widespread adoption of the DWT in diverse fields, from image and video compression standards to biomedical signal analysis and feature extraction algorithms.

In the next section, we will explore the DTWT, which adapts the DWT framework specifically for discrete-time signals. We will also examine its practical implementation through filter banks and the fast WT algorithm.

1.2.6 Discrete-Time Wavelet Transform

The DTWT adapts the wavelet framework specifically for discrete-time signals, providing a practical implementation methodology through filter banks and downsampling operations [7,8]. This section explores the computational aspects of the DTWT, its hierarchical decomposition structure, and the mathematical relationships that govern its implementation.

The DTWT maps a discrete-time signal $s[\cdot]$ of length M to another signal $y[\cdot]$ of the same length. The transformed signal encodes both the spectral content of the original signal and its temporal localization, enabling practical time-frequency analysis of discrete signals. While the theoretical foundation of wavelets can be approached from a functional analysis perspective, the practical implementation relies on digital filtering operations.

There are two equivalent but conceptually distinct approaches to explaining the DTWT: one grounded in linear algebra and the other based on filtering operations. The signal is decomposed in the linear algebraic formulation as a linear combination of orthogonal basis functions. In practice, however, the filter-based approach is universally adopted, performing successive convolutions and downsampling operations using complementary filters.

1.2.6.1 Multi-Level Decomposition Structure

The DTWT implementation follows a hierarchical binary tree structure, as illustrated in Figure 1.1. Let us consider an input signal $s[\cdot]$ sampled at $2T$ samples per second. By the Nyquist–Shannon sampling theorem, this signal contains frequency content up to T Hz.

The decomposition process begins by convolving the input signal $s[\cdot]$ with a low-pass filter $h[\cdot]$ and a high-pass filter $g[\cdot]$, producing two filtered signals. The output from the low-pass filter retains frequency content in the range $\left[0, T/2^j\right]$ Hz, while the high-pass filter captures the complementary range $\left[T/2^j, T\right]$ Hz. Since each output occupies only half the original bandwidth, their sampling rate can be

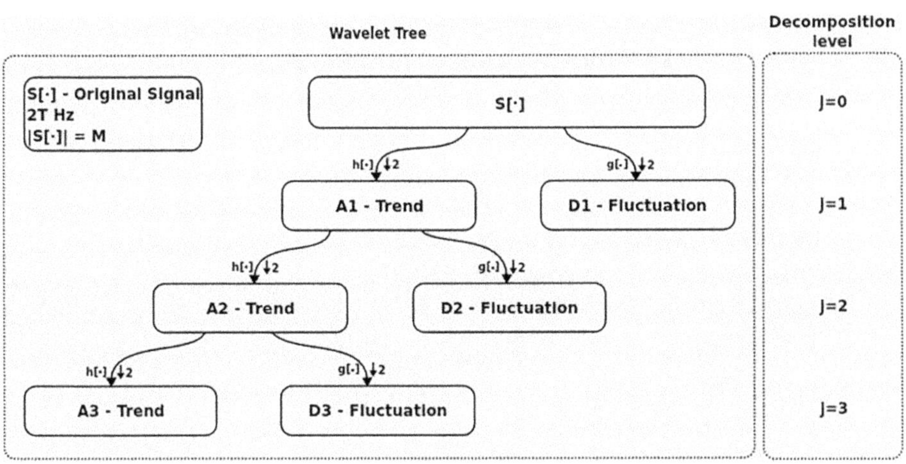

FIGURE 1.1 Multi-level wavelet decomposition tree showing the hierarchical structure of the DTWT. The original signal is decomposed into trend and fluctuation components at $j = 1$. The approximation component is then recursively decomposed into further trends at subsequent levels $j = 2$ and $j = 3$, and fluctuation components at levels $j = 2$ and $j = 3$, creating a binary tree structure.

reduced by a factor of two without aliasing—an operation known as *downsampling by 2,* denoted by "↓2".

After filtering and downsampling, we obtain two signals of length $M/2$:

$$A1[n]=\sum_k h[k]\cdot s[2n-k] \tag{1.1}$$

$$D1[n]=\sum_k g[k]\cdot s[2n-k] \tag{1.2}$$

The low-pass result is referred to as the *approximation* or *trend* component (denoted A1), and the high-pass result is called the *detail* or *fluctuation* component (denoted D1). These are depicted in the decomposition tree: the trend is shown as the left child of the root node, while the fluctuation is the right child.

This two-branch structure corresponds to the first level of the DTWT. Formally, the first-level DTWT of the signal $s[\cdot]$, denoted $\mathrm{DTWT}_1(s[\cdot])$, is the concatenation of the trend and fluctuation components:

$$\mathrm{DTWT}_1(s[\cdot])=\mathrm{A1}\,||\,\mathrm{D1}$$

This process can be recursively applied to the trend component (A1), generating further levels of approximation (A2, A3, ...) and detail (D2, D3, ...). At each level *j*, the approximation and detail components are computed as follows:

$$Aj[n]=\sum_k h[k]\cdot A(j-1)[2n-k] \tag{1.3}$$

$$Dj[n]=\sum_k g[k]\cdot A(j-1)[2n-k] \tag{1.4}$$

Each successive level halves both the signal length and bandwidth. For instance, if the original signal has length M and bandwidth $[0,T]$ Hz, then

- A1 has length $M/2$ and covers $[0,T/2]$ Hz
- D1 has length $M/2$ and covers $[T/2,T]$ Hz
- A2 has length $M/4$ and covers $[0,T/4]$ Hz
- D2 has length $M/4$ and covers $[T/4,T/2]$ Hz
- A3 has length $M/8$ and covers $[0,T/8]$ Hz
- D3 has length $M/8$ and covers $[T/8,T/4]$ Hz

1.2.6.2 Maximum Decomposition Level

A critical question in DTWT implementation is determining the maximum possible decomposition level. Since each level halves the signal length, there is a natural limit to how many decomposition levels can be applied. For a signal of length M, the maximum decomposition level $j_{\max}$ is given by

$$j_{\max} = \lfloor \log_2 M \rfloor$$

This formula ensures that the approximation component at the deepest level contains at least one sample. For example, a signal with $M = 256$ samples can undergo a maximum of $\lfloor \log_2 256 \rfloor = 8$ levels of decomposition, while a signal with $M = 1{,}000$ samples is limited to $\lfloor \log_2 1{,}000 \rfloor = 9$ levels.

In practice, the filter length may further constrain the maximum decomposition level. If the filters $h[\cdot]$ and $g[\cdot]$ have length L, the effective maximum level might be reduced to ensure that the approximation signal at each level remains longer than the filter length, avoiding boundary issues in the convolution operation.

1.2.6.3 Number of Components in Each Subband

The number of coefficients in each subband follows a systematic pattern determined by the decomposition level. For a signal of length M, the number of coefficients in the approximation and detail components at level j is given by

$$|A_j| = |D_j| = \frac{M}{2^j}$$

This relationship reflects the dyadic downsampling at each level. For example, in a three-level decomposition of a signal with $M = 64$ samples:

$$|A_1| == \frac{64}{2^1} = 32 \text{ coefficients} \tag{1.5}$$

$$|A_2| = |D_2| = \frac{64}{2^2} = 16 \text{ coefficients} \tag{1.6}$$

$$|A_3| = |D_3| = \frac{64}{2^3} = 8 \text{ coefficients} \tag{1.7}$$

The total wavelet coefficients across all subbands equal the original signal length M. This can be verified by summing the coefficients at the leaf nodes of the decomposition tree:

$$|A_{j_{\max}}| + \sum_{j=1}^{j_{\max}} |D_j| = \frac{M}{2^{j_{\max}}} + \sum_{j=1}^{j_{\max}} \frac{M}{2^j} = M$$

This property ensures that the DTWT completely represents the original signal, preserving all information in a more structured form.

1.2.6.4 Generalizing the Decomposition Rule

For a j-level DTWT, the transformed signal consists of the approximation coefficients at level j, followed by the detail coefficients from levels j down to 1:

$$\text{DTWT}_j(s[\cdot]) = A_j \,||\, D_j \,||\, D_{j-1} \,||\, \ldots \,||\, D_1$$

This ordering places the coarsest approximation first, followed by details of increasing resolution. The resulting transform preserves the original signal length while reorganizing its content into a hierarchical, multi-resolution structure that facilitates analysis at different scales.

To illustrate this process, consider a simple discrete-time signal:

$$s[\cdot] = \{1, 2, 3, 4, 5, 6, 7, 8\}$$

Assuming the first-level decomposition yields

$$A_1 = \left\{\frac{3}{2}, \frac{7}{2}, \frac{11}{2}, \frac{15}{2}\right\}, \quad D_1 = \left\{-\frac{1}{2}, -\frac{1}{2}, -\frac{1}{2}, -\frac{1}{2}\right\}$$

Then the first-level DTWT is

$$\text{DTWT}_1(s[\cdot]) = \left\{\frac{3}{2}, \frac{7}{2}, \frac{11}{2}, \frac{15}{2}, -\frac{1}{2}, -\frac{1}{2}, -\frac{1}{2}, -\frac{1}{2}\right\}$$

For a second-level decomposition, we apply the same process to A_1, obtaining

$$A_2 = \left\{\frac{5}{2}, \frac{13}{2}\right\}, \quad D_2 = \{-1, -1\}$$

The second-level DTWT is then

$$\text{DTWT}_2(s[\cdot]) = \left\{\frac{5}{2}, \frac{13}{2}, -1, -1, -\frac{1}{2}, -\frac{1}{2}, -\frac{1}{2}, -\frac{1}{2}\right\}$$

This example demonstrates how the DTWT organizes signal information into a hierarchical structure, with each level capturing features at different resolutions while maintaining the original signal length.

In practical implementations, the DTWT is computed efficiently using Mallat's algorithm, which formalizes the filter bank approach through matrix operations. This algorithm ensures that the transform remains computationally tractable even for large signals, making wavelets a powerful tool for various signal processing applications.

1.2.7 Filter Bank Implementation

While the previous section introduced the DTWT's conceptual framework, this section delves into its practical implementation through filter banks—a computational approach formalized by Stéphane Mallat that has become the standard method for wavelet analysis in digital signal processing.

1.2.7.1 Mallat's Algorithm and Matrix Representation

Mallat's algorithm provides an efficient mechanism for computing the DTWT using complementary filter pairs and structured matrix operations. The algorithm addresses several practical challenges, including boundary effects and computational efficiency, while ensuring perfect reconstruction capabilities.

In theory, convolving an M-sample input signal with an N-sample filter yields a length signal $M+N-1$. This expansion creates boundary effects that complicate the implementation of multi-level decomposition. Furthermore, while downsampling reduces the data length, the exact output length depends on the filter size, potentially leading to inconsistent dimensions across decomposition levels.

To overcome these challenges, Mallat's approach ensures fixed-length outputs through a wrapping scheme and structured matrix multiplication. The DTWT of a signal $s[\cdot]$ can be expressed as a matrix–vector product:

$$\underbrace{\text{Matrix A[.][.]}}_{M\times M}.\underbrace{\begin{pmatrix} s_0 \\ s_1 \\ . \\ . \\ . \\ s_{M-1} \end{pmatrix}}_{\text{Input } S[.]} = \underbrace{\begin{pmatrix} r_0 \\ r_1 \\ . \\ . \\ . \\ r_{M-1} \end{pmatrix}}_{\text{Output } R[.]}$$

The matrix $A[\cdot][\cdot]$ is constructed using the coefficients of the low-pass filter $h[\cdot]$ and high-pass filter $g[\cdot]$. For a first-level decomposition, the matrix has a specific structure:

- The even-indexed rows (0, 2, 4, ...) contain the coefficients of the low-pass filter $h[\cdot]$.
- The odd-indexed rows (1, 3, 5, ...) contain the coefficients of the high-pass filter $g[\cdot]$.
- Each pair of rows (low-pass and high-pass) is shifted by two positions relative to the previous pair.

This structure effectively implements filtering and downsampling in a single matrix multiplication. When filter coefficients extend beyond the signal bounds, they "wrap around" to the beginning of the row, implementing what is known as *circular convolution with wraparound.*

1.2.7.2 Circular Convolution and Boundary Handling

The circular convolution approach is a key aspect of Mallat's algorithm, addressing the boundary effects that would otherwise complicate the transform. For a signal $s[\cdot]$ of length M and a filter $f[\cdot]$ of length N, the circular convolution is defined as follows:

$$(s \circledast f)[n] = \sum_{k=0}^{N-1} f[k]\cdot s[(n-k) \bmod M]$$

This operation treats the signal as if it were periodically extended, wrapping around the boundaries. The resulting convolution has the same length M as the original signal, eliminating the expansion effect of standard convolution.

This circular behavior is implemented in the matrix representation by wrapping filter coefficients to the beginning of the row when they would otherwise extend beyond the signal length. For example, in an 8×8 matrix with a filter of length 4, the first row might contain the pattern $[h_0,h_1,h_2,h_3,0,0,0,0]$, while a later row might have the pattern $[h_2,h_3,0,0,0,0,h_0,h_1]$ due to the wraparound effect.

1.2.7.3 Reordering for the DTWT Output

After computing the matrix–vector product $R[\cdot]=A[\cdot][\cdot]\cdot S[\cdot]$, the first-level DTWT is obtained by reordering the result. Specifically,

- The trend (approximation) component consists of the even-indexed samples: $\{r_0,r_2,r_4,\ldots,r_{M-2}\}$
- The fluctuation (detail) component consists of the odd-indexed samples: $\{r_1,r_3,r_5,\ldots,r_{M-1}\}$

The first-level DTWT is then defined as the concatenation of these components:

$$\mathrm{DTWT}_1(s[\cdot])=\{r_0,r_2,r_4,\ldots,r_{M-2},r_1,r_3,r_5,\ldots,r_{M-1}\}$$

This reordering prioritizes all approximation coefficients, followed by all detail coefficients, while maintaining the original signal length M.

1.2.7.4 Handling Short Signals: Periodic Extension

As the decomposition progresses to deeper levels, the approximation signal used as input becomes progressively shorter—halving in length at each stage. Eventually, this signal may become shorter than the support length of the filters used in the convolution. For example, if $M=16$ and the filters have $N=4$ taps, then at level $j=4$, the input segment has only 2 samples, which is insufficient for a standard convolution with a four-tap filter.

A periodic extension strategy is employed in such cases: the input signal is repeated until it reaches the minimum required length for valid matrix–vector multiplication. Mathematically, if the approximation signal at level j has length $M_j=M/2^j$ and the filter has length N, with $M_j<N$, then the signal is periodically extended to length N before convolution:

$$s_{\text{extended}}[n]=s\left[n \bmod M_j\right]$$

This approach preserves the algorithm's computational structure and ensures that deeper decomposition levels remain operable, even with very short approximation signals.

1.2.7.5 Numerical Example: Matrix Implementation

To illustrate the matrix-based implementation, consider a discrete-time signal and wavelet filters:

$$s[\cdot]=\{4,6,10,12,8,6,5,5\},\quad h[\cdot]=\left\{\frac{1}{\sqrt{2}},\frac{1}{\sqrt{2}}\right\},\quad g[\cdot]=\left\{\frac{1}{\sqrt{2}},-\frac{1}{\sqrt{2}}\right\}.$$

For the first-level decomposition, we construct the matrix $A[\cdot][\cdot]$ using the filter coefficients:

$$A[\cdot][\cdot]=\begin{pmatrix} h_0 & h_1 & 0 & 0 & 0 & 0 & 0 & 0 \\ g_0 & g_1 & 0 & 0 & 0 & 0 & 0 & 0 \\ 0 & 0 & h_0 & h_1 & 0 & 0 & 0 & 0 \\ 0 & 0 & g_0 & g_1 & 0 & 0 & 0 & 0 \\ 0 & 0 & 0 & 0 & h_0 & h_1 & 0 & 0 \\ 0 & 0 & 0 & 0 & g_0 & g_1 & 0 & 0 \\ 0 & 0 & 0 & 0 & 0 & 0 & h_0 & h_1 \\ 0 & 0 & 0 & 0 & 0 & 0 & g_0 & g_1 \end{pmatrix}$$

Substituting the filter values,

$$A[\cdot][\cdot]=\begin{pmatrix} \frac{1}{\sqrt{2}} & \frac{1}{\sqrt{2}} & 0 & 0 & 0 & 0 & 0 & 0 \\ \frac{1}{\sqrt{2}} & -\frac{1}{\sqrt{2}} & 0 & 0 & 0 & 0 & 0 & 0 \\ 0 & 0 & \frac{1}{\sqrt{2}} & \frac{1}{\sqrt{2}} & 0 & 0 & 0 & 0 \\ 0 & 0 & \frac{1}{\sqrt{2}} & -\frac{1}{\sqrt{2}} & 0 & 0 & 0 & 0 \\ 0 & 0 & 0 & 0 & \frac{1}{\sqrt{2}} & \frac{1}{\sqrt{2}} & 0 & 0 \\ 0 & 0 & 0 & 0 & \frac{1}{\sqrt{2}} & -\frac{1}{\sqrt{2}} & 0 & 0 \\ 0 & 0 & 0 & 0 & 0 & 0 & \frac{1}{\sqrt{2}} & \frac{1}{\sqrt{2}} \\ 0 & 0 & 0 & 0 & 0 & 0 & \frac{1}{\sqrt{2}} & -\frac{1}{\sqrt{2}} \end{pmatrix}$$

Computing the matrix–vector product,

$$A[\cdot][\cdot]\cdot s[\cdot]=\begin{pmatrix} \frac{1}{\sqrt{2}} & \frac{1}{\sqrt{2}} & 0 & 0 & 0 & 0 & 0 & 0 \\ \frac{1}{\sqrt{2}} & -\frac{1}{\sqrt{2}} & 0 & 0 & 0 & 0 & 0 & 0 \\ 0 & 0 & \frac{1}{\sqrt{2}} & \frac{1}{\sqrt{2}} & 0 & 0 & 0 & 0 \\ 0 & 0 & \frac{1}{\sqrt{2}} & -\frac{1}{\sqrt{2}} & 0 & 0 & 0 & 0 \\ 0 & 0 & 0 & 0 & \frac{1}{\sqrt{2}} & \frac{1}{\sqrt{2}} & 0 & 0 \\ 0 & 0 & 0 & 0 & \frac{1}{\sqrt{2}} & -\frac{1}{\sqrt{2}} & 0 & 0 \\ 0 & 0 & 0 & 0 & 0 & 0 & \frac{1}{\sqrt{2}} & \frac{1}{\sqrt{2}} \\ 0 & 0 & 0 & 0 & 0 & 0 & \frac{1}{\sqrt{2}} & \frac{1}{\sqrt{2}} \end{pmatrix}\cdot\begin{pmatrix} 4 \\ 6 \\ 10 \\ 12 \\ 8 \\ 6 \\ 5 \\ 5 \end{pmatrix}=\begin{pmatrix} \frac{10}{\sqrt{2}} \\ -\frac{2}{\sqrt{2}} \\ \frac{22}{\sqrt{2}} \\ -\frac{2}{\sqrt{2}} \\ \frac{14}{\sqrt{2}} \\ \frac{2}{\sqrt{2}} \\ \frac{10}{\sqrt{2}} \\ 0 \end{pmatrix}$$

Reordering the output to place even-indexed samples first (trend) and odd-indexed samples later (fluctuation),

$$\mathrm{DTWT}_1(s[\cdot])=\left\{\frac{10}{\sqrt{2}},\frac{22}{\sqrt{2}},\frac{14}{\sqrt{2}},\frac{10}{\sqrt{2}},-\frac{2}{\sqrt{2}},-\frac{2}{\sqrt{2}},\frac{2}{\sqrt{2}},0\right\}$$

For the second-level decomposition, we take the first four values (the trend) from $\mathrm{DTWT}_1(s[\cdot])$ and apply the same process with an appropriately sized matrix:

$$\begin{pmatrix} \frac{1}{\sqrt{2}} & \frac{1}{\sqrt{2}} & 0 & 0 \\ \frac{1}{\sqrt{2}} & -\frac{1}{\sqrt{2}} & 0 & 0 \\ 0 & 0 & \frac{1}{\sqrt{2}} & \frac{1}{\sqrt{2}} \\ 0 & 0 & \frac{1}{\sqrt{2}} & -\frac{1}{\sqrt{2}} \end{pmatrix}\cdot\begin{pmatrix} \frac{10}{\sqrt{2}} \\ \frac{22}{\sqrt{2}} \\ \frac{14}{\sqrt{2}} \\ \frac{10}{\sqrt{2}} \end{pmatrix}=\begin{pmatrix} 16 \\ -6 \\ 12 \\ 2 \end{pmatrix}$$

Reordering and combining with the unchanged detail coefficients from the first level,

$$\mathrm{DTWT}_2(s[\cdot])=\left\{16,12,-6,2,-\frac{2}{\sqrt{2}},-\frac{2}{\sqrt{2}},\frac{2}{\sqrt{2}},0\right\}$$

1.2.7.6 Filter Bank Structure

The matrix-based implementation can be equivalently represented as a cascade of filter banks. In this representation, each decomposition level consists of

1. A pair of analysis filters: low-pass $h[\cdot]$ and high-pass $g[\cdot]$
2. Downsampling operations that retain only even-indexed samples
3. Storage of detail coefficients and further processing of approximation coefficients

This filter bank structure is particularly efficient for hardware implementation and forms the basis for many wavelet-based signal processing systems. The cascade structure naturally implements the MRA, with each level operating at half the sampling rate of the previous level.

The filter bank implementation also facilitates the inverse transform, which follows a mirror-image structure: upsampling operations (inserting zeros between samples), filtering with synthesis filters, and summation to reconstruct the original signal. Wavelets' perfect reconstruction property ensures that this process exactly recovers the input signal, provided appropriate synthesis filters are used.

1.2.7.7 Relationship between Filters and Wavelet Functions

The filter coefficients used in the DTWT implementation are not arbitrary but are directly related to the underlying wavelet and scaling functions. For orthogonal wavelets, the relationship between the scaling function $\phi(t)$ and the low-pass filter $h[\cdot]$ is given by the dilation equation:

$$\phi(t) = \sum_k h[k]\sqrt{2}\phi(2t-k)$$

Similarly, the wavelet function $\psi(t)$ is related to the high-pass filter $g[\cdot]$ and the scaling function:

$$\psi(t) = \sum_k g[k]\sqrt{2}\phi(2t-k)$$

These relationships establish the theoretical connection between the functional analysis perspective of wavelets and their practical implementation through filter banks. The filter coefficients effectively encode the shape and properties of the wavelet, determining characteristics such as smoothness, symmetry, and number of vanishing moments.

In summary, Mallat's algorithm provides an elegant and computationally efficient framework for implementing the DTWT through filter banks and matrix operations. Addressing practical issues such as boundary effects and signal length constraints enables the widespread application of wavelet analysis in digital signal processing systems.

1.2.8 Numerical Examples

This section consolidates and expands upon the earlier numerical examples, providing step-by-step illustrations of the DTWT computation process. These examples aim to bridge the gap between theoretical concepts and practical implementation, demonstrating how abstract mathematical formulations translate into concrete computational procedures.

Example 1: Basic DTWT Computation

Let us begin with a simple discrete-time signal and examine its wavelet decomposition in detail:

$$s[\cdot]=\{4,6,10,12,8,6,5,5\}$$

For this example, we will use the simplest orthogonal wavelet filters, known as the Haar filters:

$$h[\cdot]=\left\{\frac{1}{\sqrt{2}},\frac{1}{\sqrt{2}}\right\},\quad g[\cdot]=\left\{\frac{1}{\sqrt{2}},-\frac{1}{\sqrt{2}}\right\}$$

These filters have a clear interpretation: the low-pass filter $h[\cdot]$ computes the average of adjacent samples, while the high-pass filter $g[\cdot]$ computes their difference. Both are normalized by $\frac{1}{\sqrt{2}}$ to preserve energy.

1.2.8.1 First-Level Decomposition

The first step in computing the DTWT is applying the filters to the input signal and downsizing the results. Let us compute this explicitly:

$$A1[0]=\frac{1}{\sqrt{2}}(s[0]+s[1])=\frac{1}{\sqrt{2}}(4+6)=\frac{10}{\sqrt{2}} \tag{1.8}$$

$$A1[1]=\frac{1}{\sqrt{2}}(s[2]+s[3])=\frac{1}{\sqrt{2}}(10+12)=\frac{22}{\sqrt{2}} \tag{1.9}$$

$$A1[2]=\frac{1}{\sqrt{2}}(s[4]+s[5])=\frac{1}{\sqrt{2}}(8+6)=\frac{14}{\sqrt{2}} \tag{1.10}$$

$$A1[3]=\frac{1}{\sqrt{2}}(s[6]+s[7])=\frac{1}{\sqrt{2}}(5+5)=\frac{10}{\sqrt{2}} \tag{1.11}$$

Similarly, for the detail coefficients,

$$D1[0] = \frac{1}{\sqrt{2}}\left(s[0] - s[1]\right) = \frac{1}{\sqrt{2}}(4-6) = -\frac{2}{\sqrt{2}} \tag{1.12}$$

$$D1[1] = \frac{1}{\sqrt{2}}\left(s[2] - s[3]\right) = \frac{1}{\sqrt{2}}(10-12) = -\frac{2}{\sqrt{2}} \tag{1.13}$$

$$D1[2] = \frac{1}{\sqrt{2}}\left(s[4] - s[5]\right) = \frac{1}{\sqrt{2}}(8-6) = \frac{2}{\sqrt{2}} \tag{1.14}$$

$$D1[3] = \frac{1}{\sqrt{2}}\left(s[6] - s[7]\right) = \frac{1}{\sqrt{2}}(5-5) = 0 \tag{1.15}$$

The first-level DTWT is the concatenation of these components:

$$\mathrm{DTWT}_1\left(s[\cdot]\right) = \left\{\frac{10}{\sqrt{2}}, \frac{22}{\sqrt{2}}, \frac{14}{\sqrt{2}}, \frac{10}{\sqrt{2}}, -\frac{2}{\sqrt{2}}, -\frac{2}{\sqrt{2}}, \frac{2}{\sqrt{2}}, 0\right\}$$

This result can also be obtained using the matrix formulation described in the previous section. The matrix $A[\cdot][\cdot]$ is constructed using the filter coefficients:

$$A[\cdot][\cdot] = \begin{pmatrix} \frac{1}{\sqrt{2}} & \frac{1}{\sqrt{2}} & 0 & 0 & 0 & 0 & 0 & 0 \\ \frac{1}{\sqrt{2}} & -\frac{1}{\sqrt{2}} & 0 & 0 & 0 & 0 & 0 & 0 \\ 0 & 0 & \frac{1}{\sqrt{2}} & \frac{1}{\sqrt{2}} & 0 & 0 & 0 & 0 \\ 0 & 0 & \frac{1}{\sqrt{2}} & -\frac{1}{\sqrt{2}} & 0 & 0 & 0 & 0 \\ 0 & 0 & 0 & 0 & \frac{1}{\sqrt{2}} & \frac{1}{\sqrt{2}} & 0 & 0 \\ 0 & 0 & 0 & 0 & \frac{1}{\sqrt{2}} & -\frac{1}{\sqrt{2}} & 0 & 0 \\ 0 & 0 & 0 & 0 & 0 & 0 & \frac{1}{\sqrt{2}} & \frac{1}{\sqrt{2}} \\ 0 & 0 & 0 & 0 & 0 & 0 & \frac{1}{\sqrt{2}} & -\frac{1}{\sqrt{2}} \end{pmatrix}$$

Multiplying this matrix by the input signal vector yields the intermediate result:

$$R[\cdot] = \begin{pmatrix} \frac{10}{\sqrt{2}} \\ -\frac{2}{\sqrt{2}} \\ \frac{22}{\sqrt{2}} \\ -\frac{2}{\sqrt{2}} \\ \frac{14}{\sqrt{2}} \\ \frac{2}{\sqrt{2}} \\ \frac{10}{\sqrt{2}} \\ 0 \end{pmatrix}$$

Reordering to place even-indexed samples first (trend) and odd-indexed samples later (fluctuation) gives the same DTWT result as the direct computation.

1.2.8.2 Second-Level Decomposition

For the second-level decomposition, we apply the same process to the approximation coefficients from the first level:

$$A1[\cdot] = \left\{ \frac{10}{\sqrt{2}}, \frac{22}{\sqrt{2}}, \frac{14}{\sqrt{2}}, \frac{10}{\sqrt{2}} \right\}$$

Computing the second-level approximation coefficients,

$$A2[0] = \frac{1}{\sqrt{2}}\left(A1[0] + A1[1]\right) = \frac{1}{\sqrt{2}}\left(\frac{10}{\sqrt{2}} + \frac{22}{\sqrt{2}}\right) = \frac{32}{\sqrt{2} \cdot \sqrt{2}} = 16 \tag{1.16}$$

$$A2[1] = \frac{1}{\sqrt{2}}\left(A1[2] + A1[3]\right) = \frac{1}{\sqrt{2}}\left(\frac{14}{\sqrt{2}} + \frac{10}{\sqrt{2}}\right) = \frac{24}{\sqrt{2} \cdot \sqrt{2}} = 12 \tag{1.17}$$

And the second-level detail coefficients

$$D2[0] = \frac{1}{\sqrt{2}}\left(A1[0] - A1[1]\right) = \frac{1}{\sqrt{2}}\left(\frac{10}{\sqrt{2}} - \frac{22}{\sqrt{2}}\right) = \frac{-12}{\sqrt{2} \cdot \sqrt{2}} = -6 \tag{1.18}$$

$$D2[1] = \frac{1}{\sqrt{2}}\left(A1[2] - A1[3]\right) = \frac{1}{\sqrt{2}}\left(\frac{14}{\sqrt{2}} - \frac{10}{\sqrt{2}}\right) = \frac{4}{\sqrt{2} \cdot \sqrt{2}} = 2 \tag{1.19}$$

The second-level DTWT is then

$$\text{DTWT}_2(s[\cdot]) = \left\{16, 12, -6, 2, -\frac{2}{\sqrt{2}}, -\frac{2}{\sqrt{2}}, \frac{2}{\sqrt{2}}, 0\right\}$$

1.2.8.3 Third-Level Decomposition

For the third-level decomposition, we apply the process to the second-level approximation coefficients:

$$A2[\cdot] = \{16, 12\}$$

Computing the third-level coefficients,

$$A3[0] = \frac{1}{\sqrt{2}}\left(A2[0] + A2[1]\right) = \frac{1}{\sqrt{2}}(16+12) = \frac{28}{\sqrt{2}} \tag{1.20}$$

$$D3[0] = \frac{1}{\sqrt{2}}\left(A2[0] - A2[1]\right) = \frac{1}{\sqrt{2}}(16-12) = \frac{4}{\sqrt{2}} \tag{1.21}$$

The third-level DTWT is

$$\text{DTWT}_3(s[\cdot]) = \left\{\frac{28}{\sqrt{2}}, \frac{4}{\sqrt{2}}, -6, 2, -\frac{2}{\sqrt{2}}, -\frac{2}{\sqrt{2}}, \frac{2}{\sqrt{2}}, 0\right\}$$

This example illustrates the recursive nature of the DTWT computation, where each level operates on the approximation coefficients from the previous level, progressively capturing coarser features of the signal.

Example 2: Handling Boundary Effects with Circular Convolution

When the filter length exceeds two samples, boundary effects become more significant. Consider a signal and a pair of longer filters:

$$s[\cdot] = \left\{\frac{1+\sqrt{3}}{4\sqrt{2}}, 0, 0, \frac{3-\sqrt{3}}{4\sqrt{2}}\right\}$$

$$h[\cdot] = \left\{\frac{1+\sqrt{3}}{4\sqrt{2}}, \frac{3+\sqrt{3}}{4\sqrt{2}}, \frac{3-\sqrt{3}}{4\sqrt{2}}, \frac{1-\sqrt{3}}{4\sqrt{2}}\right\}$$

$$g[\cdot] = \left\{\frac{1-\sqrt{3}}{4\sqrt{2}}, \frac{\sqrt{3}-3}{4\sqrt{2}}, \frac{3+\sqrt{3}}{4\sqrt{2}}, -\frac{1+\sqrt{3}}{4\sqrt{2}}\right\}$$

The Daubechies D4 wavelet coefficients have two vanishing moments and provide better frequency localization than the Haar wavelet.

1.2.8.4 First-Level Decomposition with Circular Convolution

Since the filters have length $N = 4$ and the signal $s[\cdot]$ also has four samples, we must apply circular convolution. The analysis matrix is constructed as follows:

$$A[\cdot][\cdot] = \begin{pmatrix} h_0 & h_1 & h_2 & h_3 \\ g_0 & g_1 & g_2 & g_3 \\ h_2 & h_3 & h_0 & h_1 \\ g_2 & g_3 & g_0 & g_1 \end{pmatrix}$$

Substituting the filter values,

$$A[\cdot][\cdot] = \begin{pmatrix} \frac{1+\sqrt{3}}{4\sqrt{2}} & \frac{3+\sqrt{3}}{4\sqrt{2}} & \frac{3-\sqrt{3}}{4\sqrt{2}} & \frac{1-\sqrt{3}}{4\sqrt{2}} \\ \frac{1-\sqrt{3}}{4\sqrt{2}} & \frac{\sqrt{3}-3}{4\sqrt{2}} & \frac{3+\sqrt{3}}{4\sqrt{2}} & -\frac{1+\sqrt{3}}{4\sqrt{2}} \\ \frac{3-\sqrt{3}}{4\sqrt{2}} & \frac{1-\sqrt{3}}{4\sqrt{2}} & \frac{1+\sqrt{3}}{4\sqrt{2}} & \frac{3+\sqrt{3}}{4\sqrt{2}} \\ \frac{3+\sqrt{3}}{4\sqrt{2}} & -\frac{1+\sqrt{3}}{4\sqrt{2}} & \frac{1-\sqrt{3}}{4\sqrt{2}} & \frac{\sqrt{3}-3}{4\sqrt{2}} \end{pmatrix}$$

Note how the filter coefficients wrap around in the third and fourth rows, implementing circular convolution. Computing the matrix–vector product,

$$A[\cdot][\cdot] \cdot s[\cdot] = \begin{pmatrix} \frac{1+\sqrt{3}}{4\sqrt{2}} & \frac{3+\sqrt{3}}{4\sqrt{2}} & \frac{3-\sqrt{3}}{4\sqrt{2}} & \frac{1-\sqrt{3}}{4\sqrt{2}} \\ \frac{1-\sqrt{3}}{4\sqrt{2}} & \frac{\sqrt{3}-3}{4\sqrt{2}} & \frac{3+\sqrt{3}}{4\sqrt{2}} & -\frac{1+\sqrt{3}}{4\sqrt{2}} \\ \frac{3-\sqrt{3}}{4\sqrt{2}} & \frac{1-\sqrt{3}}{4\sqrt{2}} & \frac{1+\sqrt{3}}{4\sqrt{2}} & \frac{3+\sqrt{3}}{4\sqrt{2}} \\ \frac{3+\sqrt{3}}{4\sqrt{2}} & -\frac{1+\sqrt{3}}{4\sqrt{2}} & \frac{1-\sqrt{3}}{4\sqrt{2}} & \frac{\sqrt{3}-3}{4\sqrt{2}} \end{pmatrix} \cdot \begin{pmatrix} \frac{1+\sqrt{3}}{4\sqrt{2}} \\ 0 \\ 0 \\ \frac{3-\sqrt{3}}{4\sqrt{2}} \end{pmatrix}$$

This yields

$$R[\cdot] = \begin{pmatrix} \frac{5-\sqrt{3}}{16} \\ -\frac{1+\sqrt{3}}{16} \\ \frac{\sqrt{3}+3}{16} \\ -\frac{3-5\sqrt{3}}{16} \end{pmatrix}$$

Reordering to place even-indexed samples first (trend) and odd-indexed samples later (fluctuation),

$$\mathrm{DTWT}_1\left(s[\cdot]\right)=\left\{\frac{5-\sqrt{3}}{16},\frac{\sqrt{3}+3}{16},-\frac{1+\sqrt{3}}{16},-\frac{3-5\sqrt{3}}{16}\right\}$$

1.2.8.5 Second-Level Decomposition with Periodic Extension

We encounter a practical limitation for the second-level decomposition: the trend portion $\{r_0, r_2\}$ has only two samples, but the filters require four taps. This is where periodic extension becomes necessary.

We extend the trend vector by repeating it:

$$\begin{pmatrix} r_0 \\ r_2 \\ r_0 \\ r_2 \end{pmatrix}=\begin{pmatrix} \frac{5-\sqrt{3}}{16} \\ \frac{\sqrt{3}+3}{16} \\ \frac{5-\sqrt{3}}{16} \\ \frac{\sqrt{3}+3}{16} \end{pmatrix}$$

Applying the same analysis matrix,

$$A[\cdot][\cdot]\cdot\begin{pmatrix} \frac{5-\sqrt{3}}{16} \\ \frac{\sqrt{3}+3}{16} \\ \frac{5-\sqrt{3}}{16} \\ \frac{\sqrt{3}+3}{16} \end{pmatrix}=\begin{pmatrix} \frac{1}{2\sqrt{2}} \\ \frac{1-\sqrt{3}}{8\sqrt{2}} \\ \frac{1}{2\sqrt{2}} \\ \frac{1-\sqrt{3}}{8\sqrt{2}} \end{pmatrix}$$

Since the result contains duplicates due to the periodic extension, we retain only the first two coefficients as the second-level trend and fluctuation:

$$\mathrm{DTWT}_2\left(s[\cdot]\right)=\left\{\frac{1}{2\sqrt{2}},\frac{1-\sqrt{3}}{8\sqrt{2}},-\frac{1+\sqrt{3}}{16},-\frac{3-5\sqrt{3}}{16}\right\}$$

This example demonstrates how circular convolution and periodic extension handle boundary effects in the DTWT computation, ensuring that the transform remains well defined even when the signal length becomes comparable to or smaller than the filter length.

Example 3: Complete Decomposition and Reconstruction

To illustrate the complete cycle of decomposition and reconstruction, let us return to our first example with the Haar wavelet. We computed the third-level DTWT as follows:

$$\mathrm{DTWT}_3\left(s[\cdot]\right)=\left\{\frac{28}{\sqrt{2}},\frac{4}{\sqrt{2}},-6,2,-\frac{2}{\sqrt{2}},-\frac{2}{\sqrt{2}},\frac{2}{\sqrt{2}},0\right\}$$

To reconstruct the original signal, we apply the inverse DTWT, which reverses the decomposition process. The inverse DTWT uses synthesis filters derived from the analysis filters:

$$\bar{h}[\cdot]=\{h[1],h[0]\}=\left\{\frac{1}{\sqrt{2}},\frac{1}{\sqrt{2}}\right\},\quad \bar{g}[\cdot]=\{g[1],g[0]\}=\left\{-\frac{1}{\sqrt{2}},\frac{1}{\sqrt{2}}\right\}$$

For the Haar wavelet, the synthesis filters are identical to the analysis filters except for $\bar{g}[\cdot]$, which has a sign change in the first coefficient.

1.2.8.6 Third-Level Reconstruction

Starting from the deepest level, we reconstruct the second-level approximation coefficients:

$$A2[0]=\bar{h}[0]\cdot A3[0]+\bar{g}[0]\cdot D3[0]=\frac{1}{\sqrt{2}}\cdot\frac{28}{\sqrt{2}}+\frac{1}{\sqrt{2}}\cdot\frac{4}{\sqrt{2}}=\frac{32}{2}=16 \quad (1.22)$$

$$A2[1]=\bar{h}[1]\cdot A3[0]+\bar{g}[1]\cdot D3[0]=\frac{1}{\sqrt{2}}\cdot\frac{28}{\sqrt{2}}-\frac{1}{\sqrt{2}}\cdot\frac{4}{\sqrt{2}}=\frac{24}{2}=12 \quad (1.23)$$

1.2.8.7 Second-Level Reconstruction

Next, we reconstruct the first-level approximation coefficients:

$$A1[0]=\bar{h}[0]\cdot A2[0]+\bar{g}[0]\cdot D2[0]=\frac{1}{\sqrt{2}}\cdot 16+\frac{1}{\sqrt{2}}\cdot(-6)=\frac{10}{\sqrt{2}} \quad (1.24)$$

$$A1[1]=\bar{h}[1]\cdot A2[0]+\bar{g}[1]\cdot D2[0]=\frac{1}{\sqrt{2}}\cdot 16-\frac{1}{\sqrt{2}}\cdot(-6)=\frac{22}{\sqrt{2}} \quad (1.25)$$

$$A1[2]=\bar{h}[0]\cdot A2[1]+\bar{g}[0]\cdot D2[1]=\frac{1}{\sqrt{2}}\cdot 12+\frac{1}{\sqrt{2}}\cdot 2=\frac{14}{\sqrt{2}} \quad (1.26)$$

$$A1[3]=\bar{h}[1]\cdot A2[1]+\bar{g}[1]\cdot D2[1]=\frac{1}{\sqrt{2}}\cdot 12-\frac{1}{\sqrt{2}}\cdot 2=\frac{10}{\sqrt{2}} \quad (1.27)$$

1.2.8.8 First-Level Reconstruction

Finally, we reconstruct the original signal:

$$s[0] = \bar{h}[0] \cdot A1[0] + \bar{g}[0] \cdot D1[0] = \frac{1}{\sqrt{2}} \cdot \frac{10}{\sqrt{2}} + \frac{1}{\sqrt{2}} \cdot \left(-\frac{2}{\sqrt{2}}\right) = \frac{8}{2} = 4 \quad (1.28)$$

$$s[1] = \bar{h}[1] \cdot A1[0] + \bar{g}[1] \cdot D1[0] = \frac{1}{\sqrt{2}} \cdot \frac{10}{\sqrt{2}} - \frac{1}{\sqrt{2}} \cdot \left(-\frac{2}{\sqrt{2}}\right) = \frac{12}{2} = 6 \quad (1.29)$$

$$s[2] = \bar{h}[0] \cdot A1[1] + \bar{g}[0] \cdot D1[1] = \frac{1}{\sqrt{2}} \cdot \frac{22}{\sqrt{2}} + \frac{1}{\sqrt{2}} \cdot \left(-\frac{2}{\sqrt{2}}\right) = \frac{20}{2} = 10 \quad (1.30)$$

$$s[3] = \bar{h}[1] \cdot A1[1] + \bar{g}[1] \cdot D1[1] = \frac{1}{\sqrt{2}} \cdot \frac{22}{\sqrt{2}} - \frac{1}{\sqrt{2}} \cdot \left(-\frac{2}{\sqrt{2}}\right) = \frac{24}{2} = 12 \quad (1.31)$$

$$s[4] = \bar{h}[0] \cdot A1[2] + \bar{g}[0] \cdot D1[2] = \frac{1}{\sqrt{2}} \cdot \frac{14}{\sqrt{2}} + \frac{1}{\sqrt{2}} \cdot \frac{2}{\sqrt{2}} = \frac{16}{2} = 8 \quad (1.32)$$

$$s[5] = \bar{h}[1] \cdot A1[2] + \bar{g}[1] \cdot D1[2] = \frac{1}{\sqrt{2}} \cdot \frac{14}{\sqrt{2}} - \frac{1}{\sqrt{2}} \cdot \frac{2}{\sqrt{2}} = \frac{12}{2} = 6 \quad (1.33)$$

$$s[6] = \bar{h}[0] \cdot A1[3] + \bar{g}[0] \cdot D1[3] = \frac{1}{\sqrt{2}} \cdot \frac{10}{\sqrt{2}} + \frac{1}{\sqrt{2}} \cdot 0 = \frac{10}{2} = 5 \quad (1.34)$$

$$s[7] = \bar{h}[1] \cdot A1[3] + \bar{g}[1] \cdot D1[3] = \frac{1}{\sqrt{2}} \cdot \frac{10}{\sqrt{2}} - \frac{1}{\sqrt{2}} \cdot 0 = \frac{10}{2} = 5 \quad (1.35)$$

The reconstructed signal is

$$s[\cdot] = \{4, 6, 10, 12, 8, 6, 5, 5\}$$

which exactly matches our original input, demonstrating the perfect reconstruction property of the WTs.

1.2.8.9 Interpretation of Wavelet Coefficients

The wavelet coefficients obtained through the DTWT provide valuable insights into the signal's structure at different scales:

- The approximation coefficient at the deepest level ($A3[0] = \frac{28}{\sqrt{2}}$) represents the overall average or DC component of the signal, scaled by a normalization factor.

- The detail coefficient at the deepest level ($D3[0] = \frac{4}{\sqrt{2}}$) captures the coarsest fluctuation in the signal, indicating a slight upward trend in the first half compared to the second half.
- The detail coefficients at the second level ($D2 = \{-6, 2\}$) reveal more localized variations, with the negative coefficient indicating a significant dip in the middle of the first half and the positive coefficient showing a slight rise in the middle of the second half.
- The detail coefficients at the first level $\left(D1 = \left\{-\frac{2}{\sqrt{2}}, -\frac{2}{\sqrt{2}}, \frac{2}{\sqrt{2}}, 0\right\}\right)$ capture the finest fluctuations, with the first two negative coefficients indicating local decreases, the third positive coefficient showing a local increase, and the fourth zero coefficient indicating no change in the last pair of samples.

This MRA is particularly valuable for signals with features at different scales, allowing separate examination and processing of each scale component.

In summary, these numerical examples demonstrate the practical implementation of the DTWT, illustrating the step-by-step computation process, the handling of boundary effects through circular convolution and periodic extension, and the WT's perfect reconstruction capability. These concrete illustrations bridge the gap between abstract mathematical formulations and their practical application in signal processing.

1.3 WAVELET TRANSFORM APPLICATIONS IN BIOELECTRICAL AND ACOUSTIC SIGNALS

This section reviews the applications of WTs in bioelectrical and acoustic signal processing. The discussion is divided into detailed subsections covering epilepsy detection via EEG, cardiac signal analysis, EEG artifact removal, and voice pathology classification. Each subsection explores methodologies, highlights key studies, and emphasizes WT's crucial role in enhancing analysis accuracy and computational efficiency.

1.3.1 Epilepsy (EEG: Detection, Classification, and Seizure Prediction)

WTs have demonstrated significant advantages in epilepsy studies, particularly for the detection, classification, and prediction of epileptic seizures using EEG data. This subsection is divided into two categories: classical entropy- and statistics-based methods and contemporary deep learning and hybrid approaches.

1.3.1.1 Classical Entropy- and Statistics-Based Approaches

Mursalin et al. [9] demonstrated state-of-the-art performance for automatic seizure detection using the DTWT in EEG. Their methodology employed a four-level decomposition using Daubechies-4 wavelets, isolating critical physiological frequency bands (0–60 Hz). They extracted hybrid features from time, frequency, and entropy domains and applied an improved correlation-based feature selection, significantly reducing the feature set. A random forest classifier achieved an accuracy of 98.45%, with maximum sensitivity and specificity of 100%, underscoring the role of wavelets in capturing spectral–temporal features essential for robust classification.

Similarly, Subasi et al. [10] leveraged a five-level Daubechies-4 wavelet decomposition of EEG signals, extracting absolute mean, power, standard deviation, and inter-band ratios from subbands D_3–D_5 and A_5. A support vector machine with an ANOVA kernel, optimized using genetic algorithms and particle swarm optimization, attained an accuracy of up to 99.38% with superior sensitivity and specificity, highlighting the effectiveness of evolutionary tuning strategies combined with wavelets.

Zarei and Mohammadzadeh Asl [11] proposed a hybrid approach combining DTWT and orthogonal matching pursuit for EEG decomposition. Metrics such as fuzzy entropy, sample entropy, and recurrence quantification were derived from these decompositions. Sequential feature selection fed into an Radial Basis Function kernel support vector machine achieved an average sensitivity of 99.39%, specificity of 99.72%, and accuracy of 99.63%, outperforming contemporary techniques and confirming WT's efficacy in EEG-based seizure detection.

Savadkoohi et al. [12] introduced a multi-scale feature engineering approach where DTWT (level-4, Daubechies-6) significantly improved EEG classification accuracy. Employing statistical features and rigorous feature selection (t-test and sequential feature selection), the approach achieved perfect accuracy, sensitivity, and specificity in distinguishing healthy from ictal states, reinforcing DTWT's capability for hierarchical signal analysis and efficient prediction.

1.3.1.2 Deep Learning and Hybrid Architectures

Deep learning methodologies incorporating WTs have recently emerged as highly effective for seizure prediction and detection. Hussein et al. [13] developed a semi-dilated convolutional network using CWT-generated scalograms to achieve a sensitivity of 98.90% and an accuracy of 98.82% on scalp EEG data, demonstrating reduced computational complexity compared to traditional convolutional neural networks (CNNs).

Liu et al. [14] utilized S-transform spectrograms from intracranial EEG pre-filtered with DTWT–Db4 as inputs to a 15-layer CNN. This system achieved high sensitivity (97.01%) and accuracy (98.12%), demonstrating real-time operational feasibility even with minimal training.

Rashed-Al-Mahfuz et al. [15] compared spectrograms and scalograms using CNN architectures, achieving peak performance with CWT-generated scalograms (99.21% accuracy) and full fine-tuning of VGG16, confirming wavelets' superior discriminative power.

Zhang et al. [16] combined wavelet packet decomposition and common spatial patterns with CNNs for EEG-based seizure prediction, achieving an average sensitivity of 92.2%, demonstrating the efficacy of wavelet-enhanced CNNs in handling imbalanced data.

1.3.2 ECG and Other Cardiac Signals

WTs have also been pivotal in cardiac signal analysis, specifically in beat classification, continuous monitoring, and signal preprocessing.

1.3.2.1 Beat Classification and Continuous Monitoring

Saadatnejad et al. [17] integrated wavelet features with lightweight long short-term memory networks for continuous ECG monitoring. Wavelet-based multi-resolution features enabled efficient real-time processing with high classification accuracy in wearable devices.

Sudarshan et al. [18] applied dual tree complex WT to ECG signals for rapid congestive heart failure detection, achieving exceptional accuracy and clinical suitability for primary care settings. In particular, the dual-tree complex WT provides an MRA that is nearly shift-invariant and directionally selective by employing two parallel filter banks that produce complex coefficients with reduced aliasing, rendering it especially effective for image processing and biomedical signal analysis.

1.3.2.2 Denoising, Compression, and Preprocessing

Rakshit and Das [19] demonstrated the synergy between empirical mode decomposition and wavelet-based denoising, significantly improving ECG signal quality while preserving critical morphological information.

Amri et al. [20] developed an IoT-based ECG system using DTWT (sym5) for noise reduction and signal enhancement, showing robust performance in mobile tele-cardiology.

1.3.3 EEG: Artifact Removal, Denoising, and Preprocessing

WTs have also been instrumental in EEG preprocessing and artifact removal. Maddirala and Veluvolu [21] combined independent component analysis, CWT, and singular spectrum analysis to effectively remove ocular artifacts while preserving neural information, outperforming traditional artifact removal techniques.

Tibdewal et al. [22] employed a combination of DTWT Haar and adaptive thresholding using stationary WT Daubechies-3, efficiently eliminating ocular artifacts and network interference without significant loss of EEG data quality. Interestingly, the stationary WT preserves shift invariance by eliminating the decimation stage present in the DTWT, yielding redundant coefficients that enhance robustness in denoising tasks and the detection of transient events in biomedical and image signals.

1.3.4 Voice Pathology (Acoustic Speech Signals)

In speech pathology classification, wavelet packet transforms have shown significant promise. Kassim et al. [23] used dual-tree complex wavelet packet transform combined with entropy metrics for fine-grained pathology classification, significantly outperforming traditional methods.

Zhang et al. [24] applied wavelet packet decomposition alongside nonlinear complexity metrics, achieving high accuracy in voice disorder differentiation.

Dalalana et al. [25] proposed a robust classifier combining discrete wavelet packet transform and paraconsistent logic for distinguishing between healthy and pathological vocal conditions, demonstrating high accuracy and computational efficiency suitable for clinical use.

1.4 CONCLUSION

This chapter has charted the conceptual and practical journey from classical Fourier analysis to modern wavelet-based methodologies, demonstrating how multi-resolution principles resolve the inherent limitations imposed by Heisenberg's uncertainty when fixed-window techniques are applied to non-stationary biomedical signals.

By formally deriving the CWT, the DWT, and their discrete-time counterparts, and by explicating Mallat's algorithm in both matrix and filter bank form, we have shown that the WT furnishes a complete, perfectly reconstructible representation whose time-frequency granularity is naturally matched to physiological events of interest.

These findings highlight three practical insights:

1. **Resolution Adaptivity**: Proper selection of mother wavelet and decomposition depth aligns spectral resolution with the temporal scales of biomedical phenomena.
2. **Algorithmic Robustness**: Careful treatment of boundary conditions via circular convolution and periodic extension safeguards coefficient integrity even for short or irregularly sampled recordings.
3. **Synergistic Integration**: Wavelet coefficients form an informative front end for both classical statistical learning and contemporary deep learning architectures, enabling end-to-end pipelines that balance interpretability and accuracy.

Looking ahead, three avenues merit concerted investigation. First, *adaptive or learned wavelets* generated through lifting schemes or neural optimization promise application-specific performance gains while retaining interpretability. Second, *hybrid time-frequency networks* that embed wavelet layers within deep architectures could reduce data requirements and improve robustness to artifacts. Third, *hardware-efficient implementations*, whether on low-power edge devices or neuromorphic accelerators, are essential for translating wavelet analytics into point-of-care and wearable technologies.

In summary, the WT is not merely a legacy tool but a continually evolving framework whose multi-resolution ethos aligns uniquely with the complex, multi-scale nature of biomedical signals. By synthesizing theoretical foundations, implementation strategies, and domain-specific applications, this chapter provides researchers and practitioners with a rigorous yet accessible roadmap for harnessing wavelets in next-generation biomedical signal analysis.

REFERENCES

1. Emanuel Guariglia, Rodrigo C. Guido, and Gabriel J. P. Dalalana. "From wavelet analysis to fractional calculus: A review". In: *Mathematics* 11.7 (2023), p. 1606. DOI: 10.3390/math11071606.
2. Shazia Baloch, Sunil Srivatsav Samsani, and Mannan Saeed Muhammad. "Fault protection in microgrid using wavelet multiresolution analysis and data mining". In: *IEEE Access* 9 (2021), pp. 86382–86391. DOI: 10.1109/AC-CESS.2021.3088900.
3. Hao Zhou et al. "Interpretations of frequency domain analyses of neu-ral entrainment: Periodicity, fundamental frequency, and harmonics". In: *Frontiers in Human Neuroscience* 10 (2016), p. 274. DOI: 10.3389/fnhum.2016.00274.
4. Samuel Pearce-Davies. *Sonic Analysis for Machine Learning: Multi-Layer Perceptron Training Using Spectrograms*. Thesis. University of Plymouth, 2019. DOI: 10.24382/628.
5. Chunxu Xia and Chao Chunguang. "Identification and representation of multi-pulse near-fault strong ground motion using adaptive wavelet trans-form". In: *Applied Sciences* 9.2 (2019), p. 275. DOI: 10.3390/app9020259.
6. Tuomas P. Rossi et al. "Strong plasmon–molecule coupling at the nanoscale revealed by first-principles modeling". In: *Nature Communications* 10.1 (2019), p. 3336. DOI: 10.1038/s41467-019-11315-5.
7. Rodrigo Capobianco Guido. "Wavelets behind the scenes: Practical as-pects, insights, and perspectives". In: *Physics Reports* 985 (2022), pp. 1–23. DOI: 10.1016/j.physrep.2022.08.001.
8. Rodrigo Capobianco Guido et al. "CWT×DWT×DTWT×SDTWT: Clar-ifying terminologies and roles of different types of wavelet transforms". In: *International Journal of Wavelets, Multiresolution and Information Processing* 18.06 (2020), p. 2030001. DOI: 10.1142/S0219691320300017.
9. Md Mursalin et al. "Automated epileptic seizure detection using improved correlation-based feature selection with random forest classifier". In: *Neu-Rocomputing* 241 (2017), pp. 204–214. DOI: 10.1016/j.neucom.2017.02.053.
10. Abdulhamit Subasi, Jasmin Kevric, and M. Abdullah Canbaz. "Epilep-tic seizure detection using hybrid machine learning methods". In: *Neural Computing and Applications* 31 (2019), pp. 317–325. DOI: 10.1007/s00521-017-3003-y.
11. Asghar Zarei and Babak Mohammadzadeh Asl. "Automatic seizure detec-tion using orthogonal matching pursuit, discrete wavelet transform, and entropy-based features of EEG signals". In: *Computers in Biology and Medicine* 131 (2021), p. 104250. DOI: 10.1016/j.compbiomed.2021.104250.
12. Marzieh Savadkoohi, Timothy Oladunni, and Lara Thompson. "A ma-chine learning approach to epileptic seizure prediction using electroencephalogram (EEG) signal". In: *Biocybernetics and Biomedical Engineer-ing* 40.3 (2020), pp. 1328–1341. DOI: 10.1016/j.bbe.2020.07.004.
13. Ramy Hussein et al. "Semi-dilated convolutional neural networks for epilep-tic seizure prediction". In: *Neural Networks* 139 (2021), pp. 212–222. DOI: 10.1016/j.neunet.2021.03.008.
14. Guoyang Liu, Weidong Zhou, and Minxing Geng. "Automatic seizure de-tection based on S-transform and deep convolutional neural network". In: *International Journal of Neural Systems* 30.04 (2020), p. 1950024. DOI: 10.1142/S0129065719500242.
15. Md Rashed-Al-Mahfuz et al. "A deep convolutional neural network method to detect seizures and characteristic frequencies using epileptic electroencephalogram (EEG) data". In: *IEEE Journal of Translational Engineering in Health and Medicine* 9 (2021), pp. 1–12. DOI: 10.1109/JTEHM.2021.3050925.

16. Yuan Zhang et al. "Epilepsy seizure prediction on EEG using common spatial pattern and convolutional neural network". In: *IEEE Journal of Biomedical and Health Informatics* 24.2 (2020), pp. 465–474. DOI: 10.1109/JBHI.2019.2933046.
17. Saeed Saadatnejad, Mohammadhosein Oveisi, and Matin Hashemi. "LSTM-based ECG classification for continuous monitoring on personal wearable devices". In: *IEEE Journal of Biomedical and Health Informatics* 24.2 (2020), pp. 515–523. DOI: 10.1109/JBHI.2019.2911367.
18. Vidya K. Sudarshan et al. "Automated diagnosis of congestive heart fail-ure using dual tree complex wavelet transform and statistical features ex-tracted from 2 s of ECG signals". In: *Computers in Biology and Medicine* 83 (2017), pp. 48–58. DOI: 10.1016/j.compbiomed.2017.01.019.
19. Manas Rakshit and Susmita Das. "An efficient ECG denoising method-ology using empirical mode decomposition and adaptive switching mean filter". In: *Biomedical Signal Processing and Control* 40 (2018), pp. 140–148. DOI: 10.1016/j.bspc.2017.09.020.
20. M. Faizal Amri, Muhammad I. Rizqyawan, and Arjon Turnip. "ECG sig-nal processing using offline-wavelet transform method based on ECG-IoT device". In: *Proceedings of the 2016 3rd International Conference on Information Technology, Computer, and Electrical Engineering (ICITACEE)*. IEEE, New York, 2016, pp. 1–6. DOI: 10.1109/ICITACEE.2016.7892404.
21. Ajay Kumar Maddirala and Kalyana C. Veluvolu. "ICA with CWT and k-means for eye-blink artifact removal from fewer channel EEG". In: *IEEE Transactions on Neural Systems and Rehabilitation Engineering* 30 (2022), pp. 1361–1373. DOI: 10.1109/TNSRE.2022.3176575.
22. Manish N. Tibdewal et al. "Power line and ocular artifact denoising from EEG using notch filter and wavelet transform". In: *Proceedings of the 2016 3rd International Conference on Computing for Sustainable Global Development (INDIACom)*. IEEE, New York, 2016, pp. 1654–1659.
23. Farah Nazlia Che Kassim et al. "Optimization of dual-tree complex wavelet packet-based entropy features for voice pathologies detection". In: *Jurnal Teknologi* 82.6 (2020), pp. 21–28. DOI: 10.11113/jurnalteknologi.v82.14748.
24. Xuehui Zhang and Weiping Hu. "Pathological voice classification based on wavelet packet multiscale analysis". In: *Proceedings of the 2018 Inter-National Conference on Algorithms, Computing and Artificial Intelligence*. IEEE, New York, 2018, pp. 1–6. DOI: 10.1145/3302425.3302450.
25. Gabriel José Pellisser Dalalana et al. "Application of wavelet analysis and paraconsistent feature extraction in the classification of voice pathologies". In: *Journal of Voice* (2025). 10.1016/j.jvoice.2025.01.021.

Section II

Time-Frequency Analysis of Specific Biomedical Signals

2 Time-Frequency Analysis of ECG Signal

Divya Jain, Vibha Tiwari, Akshra Tiwari, and Suyash Khare

2.1 INTRODUCTION

Traditional frequency analysis techniques are used for analyzing a signal's spectral content and the distribution of frequency components of stationary signals, whose frequency content does not change over time. However, in biomedical applications, signals such as electrocardiograms (ECGs), electroencephalograms (EEGs), and other physiological recordings are typically non-stationary, meaning their statistical properties—particularly frequency components—change over time due to varying physiological conditions, movement, or external influences. This dynamic nature challenges traditional frequency analysis techniques for biomedical signals [1].

Time-frequency analysis (TFA) provides a solution to this challenge; it allows for the study of signals simultaneously in time and frequency domains. TFA provides a more extensive understanding of signal dynamics, making it particularly well suited for analyzing non-stationary and transient signals.

In TFA, the constituent frequency components of a signal are decomposed across different time intervals. With this decomposition, we can analyze how the energy or power of each frequency component varies over time. It can uncover the transient phenomena and temporal patterns obscured by traditional frequency analysis methods.

Different techniques have been developed to study and analyze diverse signal characteristics. These techniques include the short-time Fourier transform (STFT), which gives a time-frequency space representation organized into equally sized grids, and applies the Fourier transform to short, overlapping segments of the signal to capture its time-varying frequency content. Additionally, the continuous wavelet transform (CWT) and discrete wavelet transform (DWT) utilize wavelet functions to provide both time and frequency localization, allowing for efficient analysis of transient signals.

For TFA, alternative approaches such as the Hilbert–Huang transform (HHT) and the Stockwell transform have their unique strengths and applications in telecommunications, geophysics, and notably, biomedical signal processing. In biomedical applications, there is a crucial role for TFA in understanding complex physiological phenomena.

Many biomedical phenomena, such as heartbeats, muscle contractions, or neural spikes, manifest as transient events in the signal. TFA can localize and characterize

DOI: 10.1201/9781003611448-4

these transient events by revealing their time-varying frequency content. TFA of EEG signals is useful in neuroscience and neuroengineering to extract information about brain activity across both time and frequency domains. EEG signals are the brain's electrical activity recorded through electrodes placed on the scalp. TFA of ECG signals is essential for understanding the dynamic changes in cardiac activity over time and frequency.

2.1.1 Basics of Time-Frequency Analysis Techniques

A TFA technique gives a representation of how the frequency content of a signal changes over time. The following are the techniques used in TFA.

2.1.1.1 Windowed Fourier Transform (STFT)

The Fourier transform converts a time-domain signal into a frequency-domain signal, providing information about the frequency content of the entire signal. STFT breaks the signal into short segments using a sliding window and computes the Fourier transform for each segment. By applying STFT, we can observe how the signal's frequency content changes over time. However, the trade-off is that the window size limits the resolution in both time and frequency.

2.1.1.2 Wavelet Transform

Wavelet transform is a method to analyze a signal in the time and frequency domain. It uses a mathematical function called wavelet function, which consists of small waves localized in both time and frequency. There are two types of wavelet transforms [2]:

- CWT
- DWT

Wavelet transform is particularly useful for analyzing non-stationary signals with varying frequency components over time.

2.1.1.3 Time-Frequency Distributions

Time-frequency distributions (TFDs) are the mathematical tools used to analyze non-stationary signals. TFDs give the ability to analyze signal frequency changes over time. Common types of TFD techniques are STFT, Wigner–Ville distribution, and spectrogram.

TFDs find applications in various fields, including image processing, audio and speech processing, and biomedical signal analysis. They are essential tools for extracting meaningful information from non-stationary signals.

TFDs provide a representation of signal energy in both the time and frequency domains.

2.1.1.4 Hilbert–Huang Transform

The HHT is mainly used for analyzing non-stationary and non-linear signals. The two main components of HHT are given below:

Empirical Mode Decomposition: It is a data-driven adaptive decomposition technique that breaks the signal into a finite set of intrinsic mode functions (IMFs).

Hilbert Spectral Analysis: After decomposing the signal into IMFs, the instantaneous frequency and amplitude of each IMF are calculated using the Hilbert transform. HHT gives an understanding of the time-varying spectral characteristics of the signal by analyzing the instantaneous frequency and amplitude of each IMF, making it suitable for analyzing non-linear and non-stationary signals.

2.2 TIME- FREQUENCY ANALYSIS IN ELECTROCARDIOGRAPHY

2.2.1 Overview of ECG Signals and Their Significance

Analyzing the electrical activities of heart ECG signals is an important tool for providing valuable information. ECG signals are essential for diagnosing cardiac conditions, monitoring patients, and assessing overall cardiac health. They can be recorded by placing electrodes on the skin. During each heartbeat, heart muscle depolarization causes electrical changes to arise on the skin, which are recorded by the placed electrodes [3] (Figure 2.1).

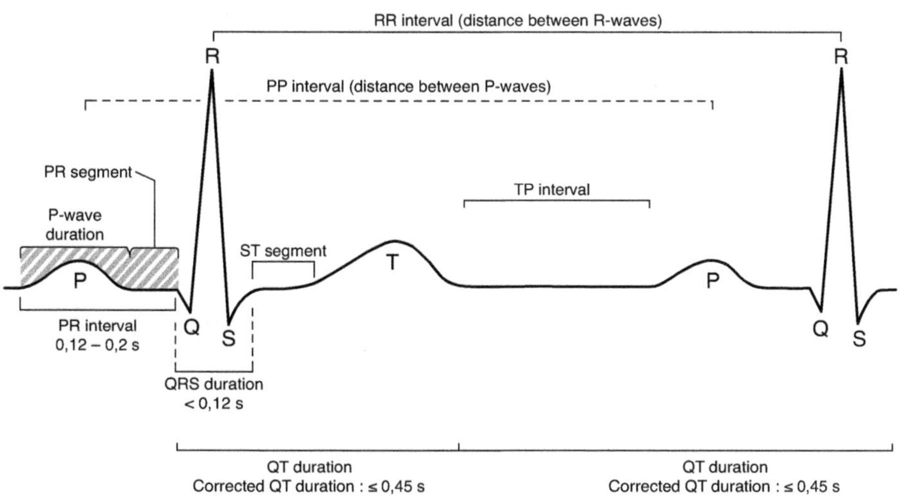

FIGURE 2.1 ECG components representation during the trace of cardiac cycle.

2.2.2 ECG Signal

Components: It consists of several components.

P Wave: It represents atrial depolarization, or the contraction of the atria. Abnormalities in the P wave may indicate atrial enlargement or conditions affecting atrial conduction.

QRS Complex: It represents ventricular depolarization, or the contraction of the ventricles.

T Wave: It represents ventricular repolarization, or the relaxation of the ventricles. Changes in the T wave can indicate electrolyte disturbances, myocardial ischemia, or drug effects.

2.2.3 ECG Peak Detection

For ECG signal analysis, peak detection is an important activity; before applying a peak detection algorithm, pre-processing of the ECG signal is mandatory [2,4]. After pre-processing, the symbolic aggregate approximation (SAX) method can be used along with a digital differentiator for ECG peaks. For the pre-processing of ECG signals, we can apply a low-pass filter, a baseline wander filter, followed by a notch filter. The low-pass filter is applied to remove high-frequency noisy components and limit the ECG signal within a specified cut-off frequency range. Baseline wander, which is unwanted drift in the ECG signal around the baseline due to breathing, body movements, behavior of electrodes, and other reasons, can be filtered using a baseline wander filter. Power line interference, which can disturb the weak ECG signal, can be filtered out by using a notch filter. The pre-processed ECG signal, passed through three filtering cascaded schemes, can be approximated by the SAX method. After approximation, a digital differentiator can be used to identify the location of peaks (Figure 2.2).

2.2.4 Low-Pass Filtering

Electronic and electrical disturbances, as well as muscle movements, corrupt the recorded ECG signal. Mathematically, the recorded ECG signal is the summation of

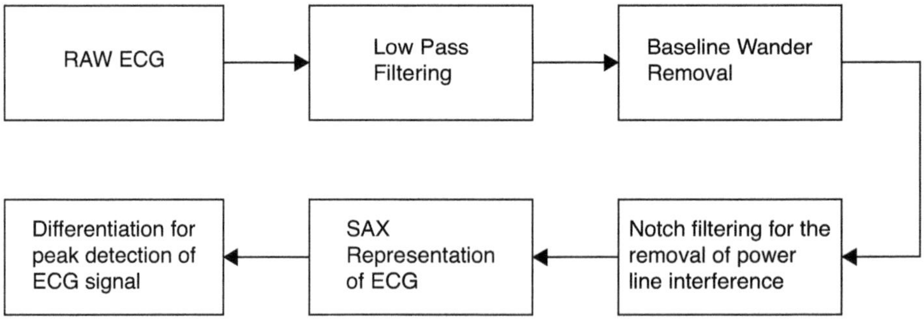

FIGURE 2.2 Block diagram for pre-processing and peak detection of ECG signal.

the ECG signal ($\alpha(n)$) and external noise ($\varphi(n)$), i.e., $y(n) = \alpha(n) + \varphi(n)$. A filtering mechanism is applied to filter out the external noise $\varphi(n)$, which is high-frequency content and random in nature [5]. For noise suppression, we use temporal statistics in place of ensemble statistics because it is treated as an ergodic process. In this process, we can obtain output at various points in time using moving averages based on a few samples.

A formulation for the moving average filter is given by

$$y(n) = \sum_{m=0}^{N} \rho_m \alpha(n-m) \tag{2.1}$$

where $y[n]$ is the output and $\alpha[n]$ is the input, and values of ρ_m are weight coefficients. The Z transform of the above equation is

$$H[z] = \rho_0 + \rho_1 z^{-1} + \rho_2 z^{-2} + \rho_3 z^{-3} + \ldots\ldots\ldots + \rho_N z^{-N} \tag{2.2}$$

The simple averaging filter derived from (2.1) is given by

$$y[n] = k\left(\alpha[n] + 2\alpha[n-1] + \alpha[n-2]\right) \tag{2.3}$$

where k is the gain. Taking the z transform, we get

$$Y[z] = k\left(\alpha[z] + 2z^{-1}\alpha[z] + z^{-2}\alpha[z]\right)$$

or

$H[z] = k(1 + 2z^{-1} + z^{-2})$ at $z = 1$, the normalized gain is unity; therefore, $k = ¼$.

Hence, $$H[z] = \frac{1}{4}\left(1 + 2z^{-1} + z^{-2}\right) \tag{2.4}$$

In the low-pass filtering result, the original and filtered signals are shown in Figure 2.3a. In Figure 2.3b, filter magnitude and phase plots are shown; the magnitude is one up to 40 Hz, and after that, it falls slowly. In the phase plot, phase and frequency have linear relations.

2.2.5 Baseline Wander Removal Filtering

Electrode behavior, breathing, body movement, etc., lead to baseline wander generation, which fluctuates the ECG signal; because of these fluctuations, it is difficult to correctly identify ST-elevated myocardial infarction and non-STEMI patients [6].

Mathematically, the baseline wander can be represented as the addition of sinusoids that vary randomly with amplitude (β_k) and phase (Φ_k).

$$k_w(t) = \sum_{j=0}^{k} \beta_k \cos\left(2\pi k \Delta f t + \phi_k\right)$$

$$\Delta f = \frac{f_s}{N} \tag{2.5}$$

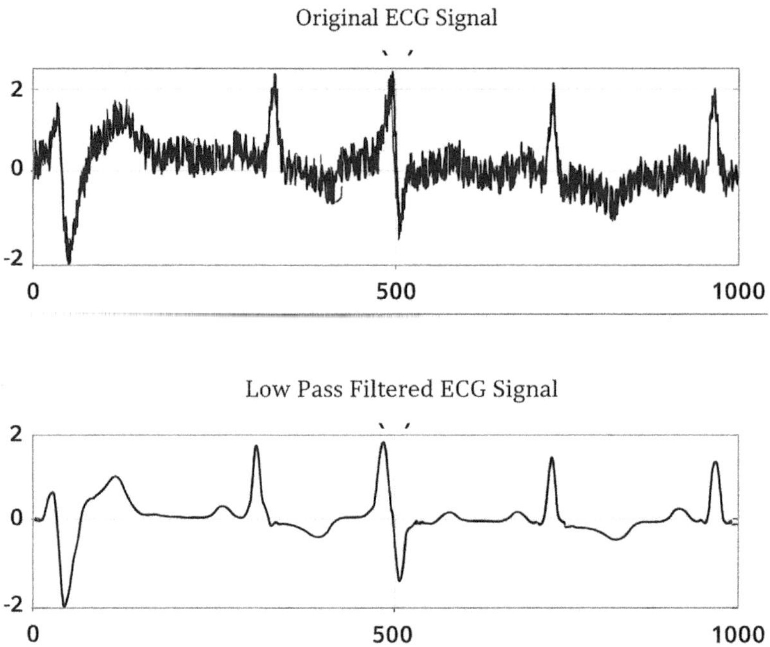

(a) Original ECG and low pass Filtered ECG

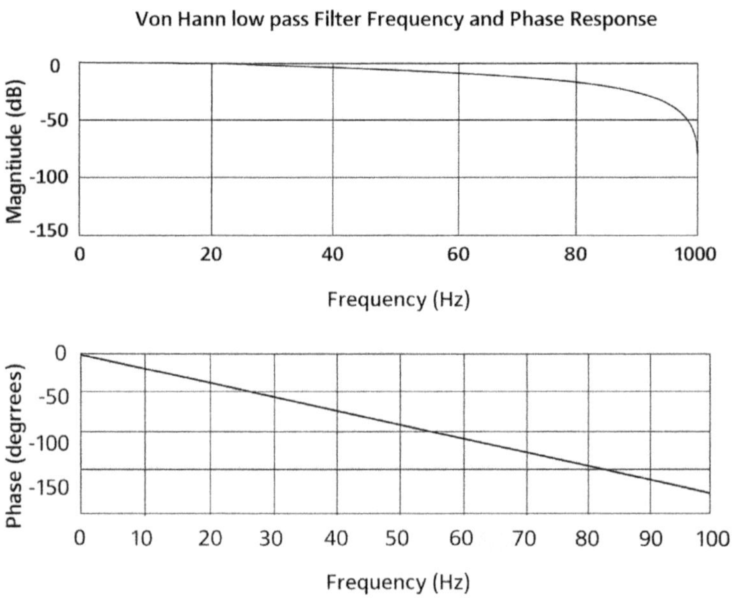

(b) Magnitude and Phase plot of low pass filtered ECG

FIGURE 2.3 Low-pass filtered ECG signal.

Here,

f_s = sampling frequency of ECG

N = total number of samples in the ECG

β_k = random variable uniformly distributed between [0, 1]

Φ_k = the random phase uniformly distributed between [0, 2π]. The function for the baseline wander added to the ECG signal is

$$y(t) = \alpha(t) + k_w(t) \tag{2.6}$$

where $y(t)$ is the recorded ECG signal *and* $\alpha(t)$ is the actual ECG signal free from baseline wander. DWT can be applied for the removal of baseline wander; with the help of DWT, the signal can be decomposed into many sets, in each set time series of coefficients, which gives the time evolution of the signal in the associated frequency band. The wavelet function for DWTs is [7]

$$\psi_{i,j}(t) = \frac{1}{\sqrt{a_0^j}}\psi\left(\frac{t - ib_0a_0^j}{a_0^j}\right) \tag{2.7}$$

The parameters i are used to measure dilation, and j is used to measure translation. The most preferred values for a_0 is 2 and for b_0 is 1.

$$\psi_{i,j}(t) = 2^{-j/2}\psi\left(2^{-j}t - i\right) \tag{2.8}$$

Under the condition that the two distinct wavelets are orthogonal to each other, i.e.,

$$\int_{-\infty}^{\infty} \psi_{i,j}(t)\psi_{p,q}(t)dt = \begin{cases} 1 & \text{if } i = p, j = q \\ 0 & \text{otherwise} \end{cases} \tag{2.9}$$

The wavelet transform coefficients for signals $y\ (t)$ are obtained as

$$T_{i,j} = \int_{-\infty}^{\infty} y(t)\psi_{i,j}(t)dt \tag{2.10}$$

Let the pure ECG signal discrete wavelet function be given by

$$\phi_{i,j}(t) = 2^{-j/2}\phi\left(2^{-j}t - i\right) \tag{2.11}$$

With the property that $\int_{-\infty}^{\infty} \phi_{0,0}(t)dt = 1$ (2.12)

Therefore, ideally the coefficient of the pure ECG signal can be expressed as

$$y_i(t) = \int_{j=-\infty}^{\infty} \zeta_{i,j}\phi_{i,j}(t)dt \tag{2.13}$$

The above equation in the presence of baseline wander can be written as

$$y_i(t) = \int_{j=-\infty}^{\infty} \zeta_{i0,j}\phi_{i0,j}(t)dt + \int_{i=-\infty}^{\infty}\int_{j=-\infty}^{\infty} T_{i,j}\psi_{i,j}(t)\,dt \tag{2.14}$$

In the above equation, the first term represents the slowly varying changes, and the second represents rapid signal changes.

Care should be taken when selecting the wavelet so that the baseline wander can be patterned as a slowly varying signal. Finally, baseline wander can be subtracted from y (t) and (t) can be obtained, which is the baseline wander-free actual ECG signal. Signal decomposition is shown in Figure 2.4.

Using (2.14), the decomposed signal is represented as the combination of an exact and approximate signal. The approximate signal is subtracted from the original signal to find the exact signal. Figure 2.5 shows the removal of baseline drift from the original signal.

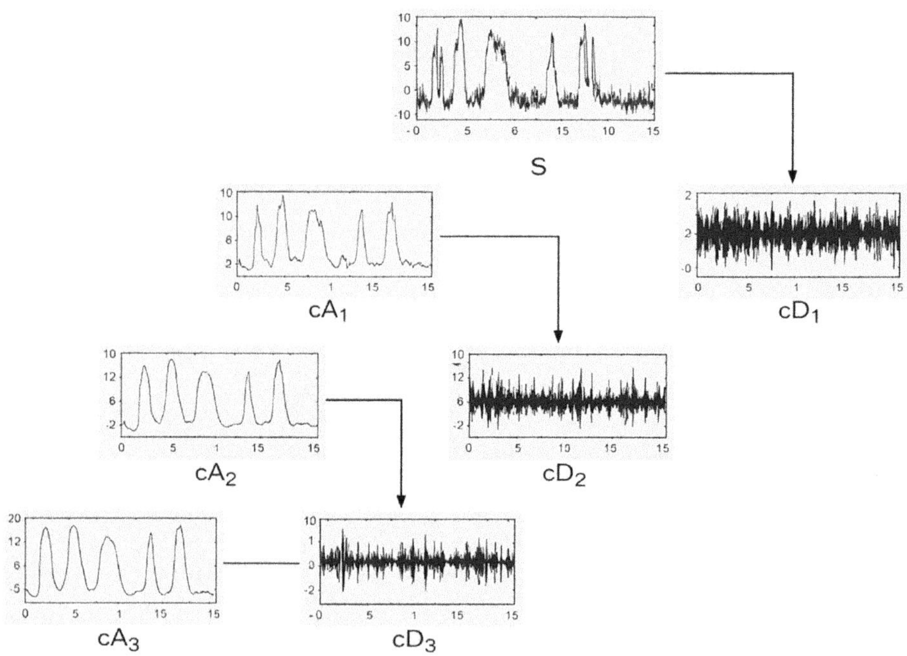

FIGURE 2.4 Process of signal decomposition in DWT.

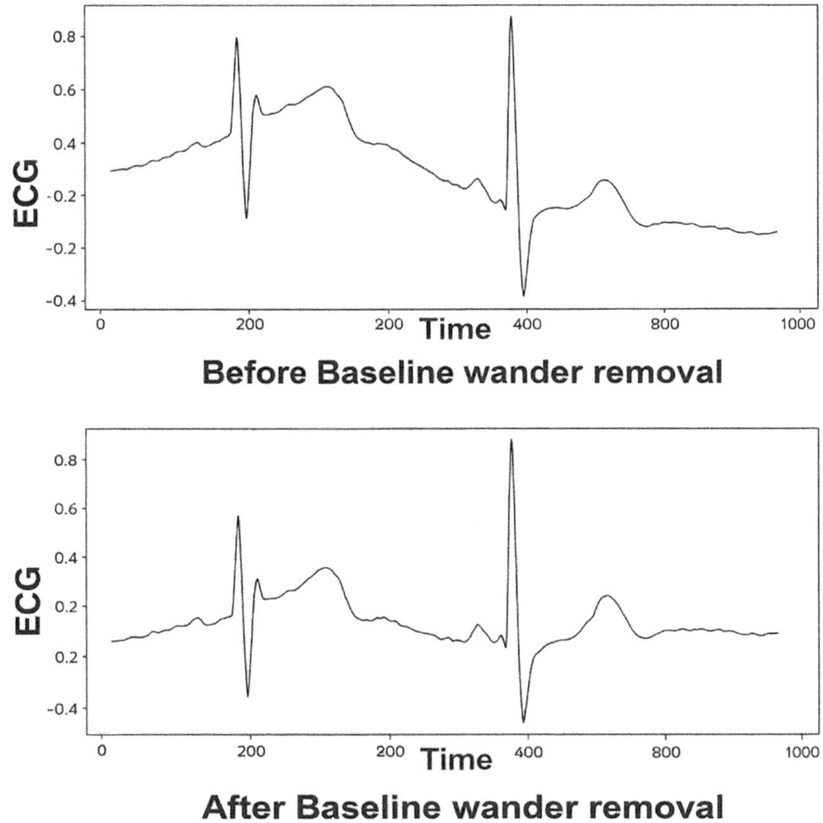

FIGURE 2.5 ECG signal after baseline wander filtering.

2.2.6 Notch Filtering for 60 Hz Power Supply Peak Removal

The noise created due to the power supply is sinusoidal. Its interference can be represented as $\eta\ (n) = A\ \cos(\Omega\ n)$ [8]. We must notch out the two frequencies in the cosine function to make our ECG signal $y\ (n)$ free from power supply interferences. The following transfer function can represent the notch filter:

$$H(\omega) = \frac{\left(e^{j\omega} - z_1\right)\left(e^{j\omega} - z_2\right)}{\left(e^{j\omega} - p_1\right)\left(e^{j\omega} - p_2\right)} \tag{2.15}$$

For notch filtering, we can add zeros to particular frequency components.

By adding zeros to undesirable frequency components, we can remove these components using a notch filter. The selected radius of the zeros should be one, and the angle should be

$$\Omega_0 = \pm \frac{f_0}{f_s}(2\pi)$$

The zeros can be defined as

$z_1 = e^{j\Omega_0}$ and $z_2 = e^{-j\Omega_0}$, where Ω_0 is defined by

$$\Omega_0 = 2\pi \frac{f_0}{f_s}$$

The filter transfer function is

$$H[z] = \frac{(z - z_1)(z - z_2)}{z^2} \tag{2.16}$$

Considering $f_s = 200$ Hz and the power supply frequency as f_0 −60 Hz; therefore,

$$H_N[z] = \frac{z^2 - 0.618z + 1}{z^2} \tag{2.17}$$

The magnitude and phase response of the notch filter is shown in Figure 2.6a; at the frequency of 60 Hz, we got suppression of −60 dB. Figure 2.6b shows the power spectral densities of the original and notch-filtered signals. Figure 2.7 shows the original ECG and notch-filtered ECG signals.

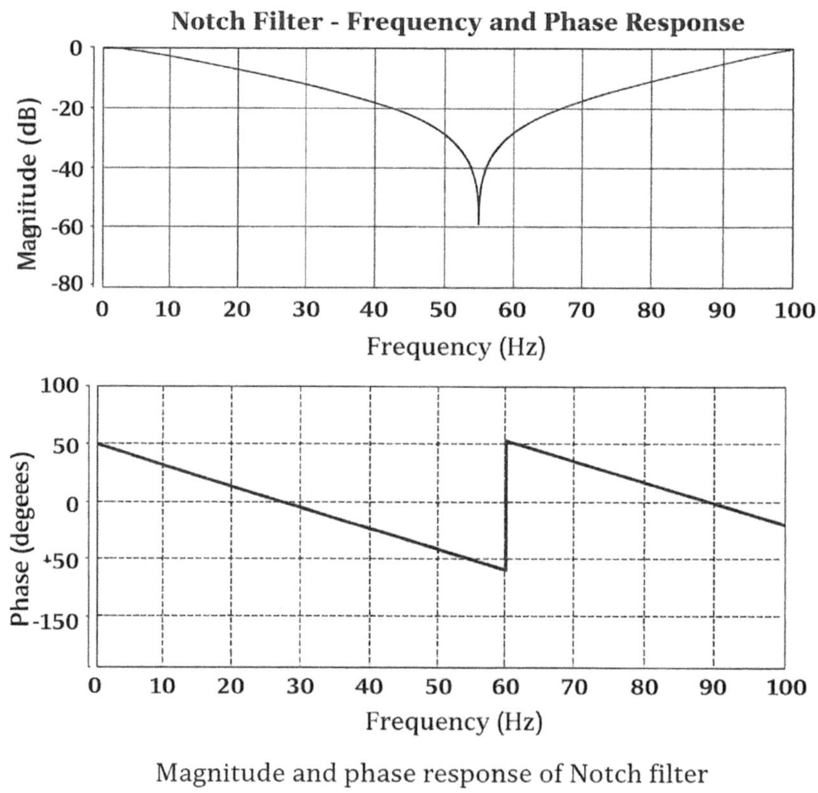

FIGURE 2.6 ECG signal representation after notch filtering.

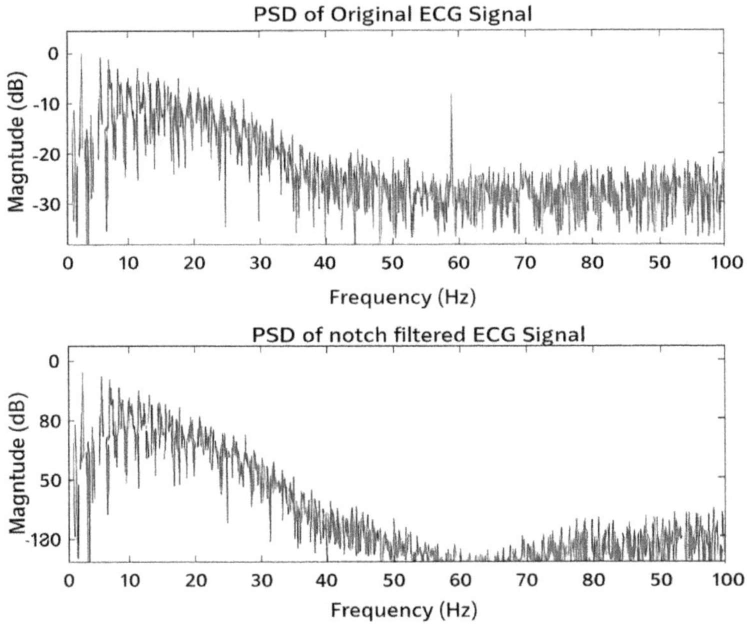

(b) Power spectral density of Original ECG and notch filtered ECG Signal

FIGURE 2.6 (*Continued*) ECG signal representation after notch filtering.

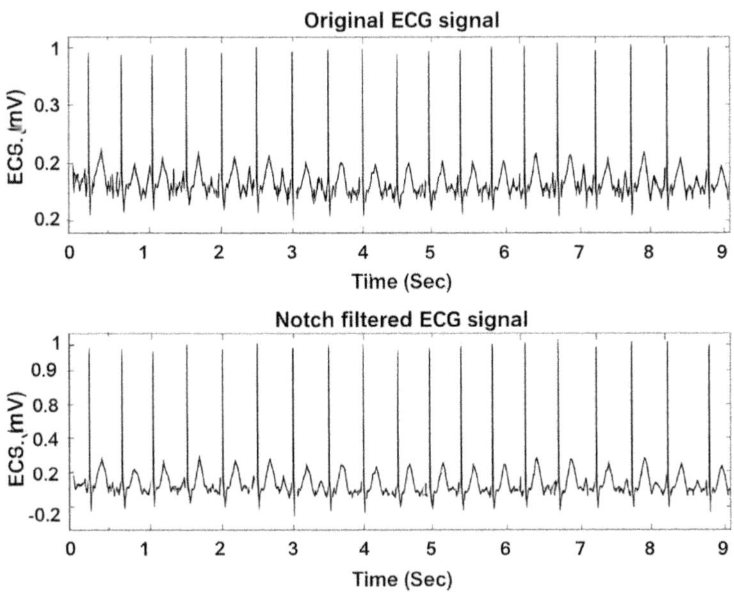

FIGURE 2.7 Original ECG and ECG signal after notch filtering.

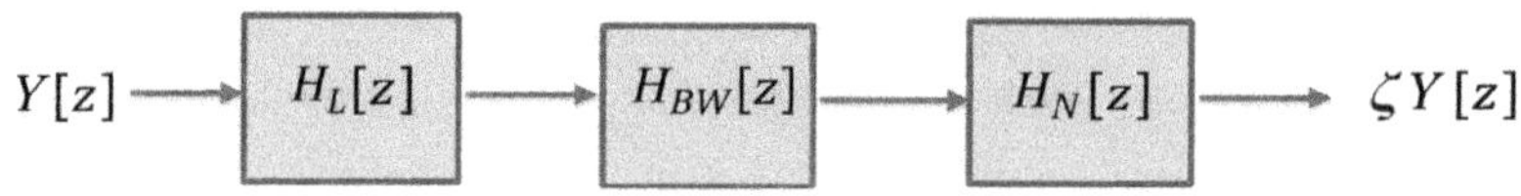

FIGURE 2.8 Combination of cascaded filters for pre-processing of the recorded ECG signal.

2.2.7 Composite Filtering

Finally, we can get the clean ECG signal after passing the recorded ECG signal through the cascaded filter system, where the first filter is the low-pass filter with the representation [*HL*(z)], the second is the baseline wander removal filter [*HBW*(z)] for baseline drift removal, and the third one is the notch filter *HN*[z] for power supply interference removal. The combined filtering scheme is shown in Figure 2.8.

$$\zeta[z] = H_L\,[z]H_{BW}\,[Z]H_N\,[z]Y[z]$$

where $Y[z]$ is the input ECG signal and $\zeta[z]$ is the cascaded filtered signal. Figure 2.7a shows a pre-processed clean ECG signal (Figures 2.9 and 2.10).

The clean ECG can be considered a quasi-periodic signal because its periodicity changes very slowly; thus, it is called a wide-sense stationary (WSS) signal [9]. This modeling has been discussed in the next section. We can model the ECG signal as a stochastic process, which is represented in the next section.

2.2.8 ECG Modeling as a Stochastic Process

In ECG, each sample can be modeled as an independent identically distributed process. Considering an ECG sample as $\alpha(n)$, noise as $\varphi(n)$, and then the noise-corrupted symbol can be defined as $y(n)$, represented as

$$y(n) = \alpha(n) + \varphi(n) \tag{2.18}$$

For the removal of noise, the recorded signal is allowed to pass through a series of filters with a combined impulse function as $h\,(n)$, which can be represented as

$$y_F(n) = [\alpha(n) + \varphi(n)] \otimes h(n) \tag{2.19}$$

$$y_F(n) = \alpha(n) \otimes h(n) + \varphi(n) \otimes h(n) \tag{2.20}$$

The received signal's time-delayed version is given by

$$y(n + n_0) = \alpha(n + n_0) + \varphi(n + n_0) \tag{2.21}$$

The auto-correlation function can be written in the following way:

$$E\big[y(n + n_0)(n)\big] = E\big[(\alpha(n + n_0) + \varphi(n + n_0))((n) + \varphi(n))\big]$$

or

$$R_{YY}(n_0) = R_{\alpha\alpha}(n_0) + \left[(n)\varphi(n+n_0)\right] + E\left[\alpha(n+n_0)\varphi(n)\right] + R_{\varphi\varphi}(n_0) \tag{2.22}$$

Because ECG and noise are not related to one another,

$$R_{YY}(n_0) = R_{\alpha\alpha}(n_0) + R_{\varphi\varphi}(n_0) \tag{2.23}$$

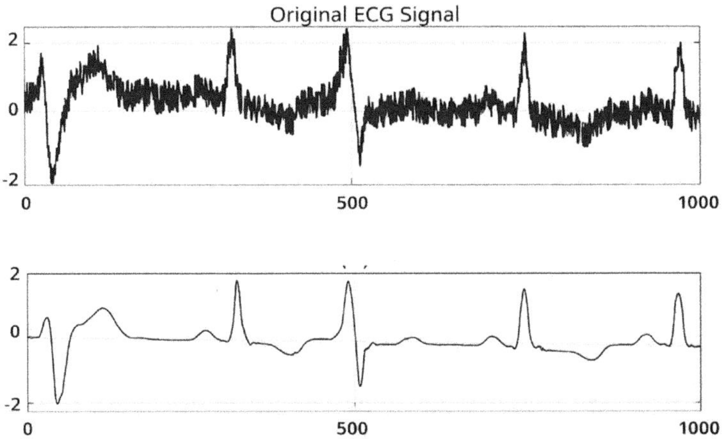

(a) Original ECG and ECG after cascaded filtering

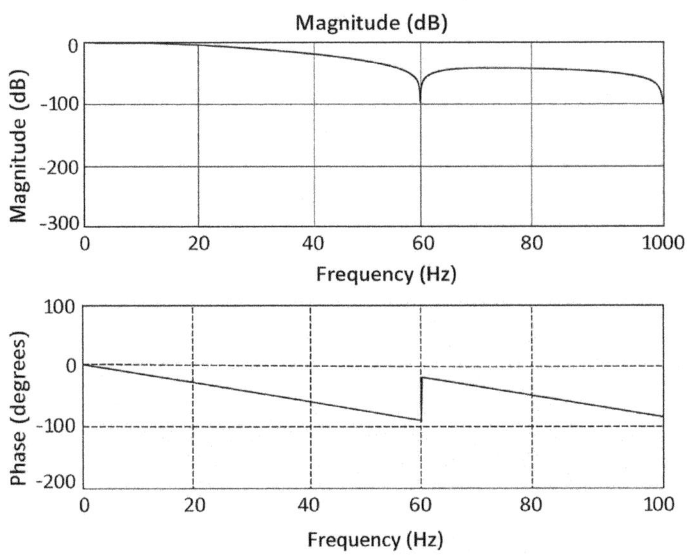

Magnitude and phase plot of cascaded filtered ECG

FIGURE 2.9 Pre-processing of ECG with cascaded filters.

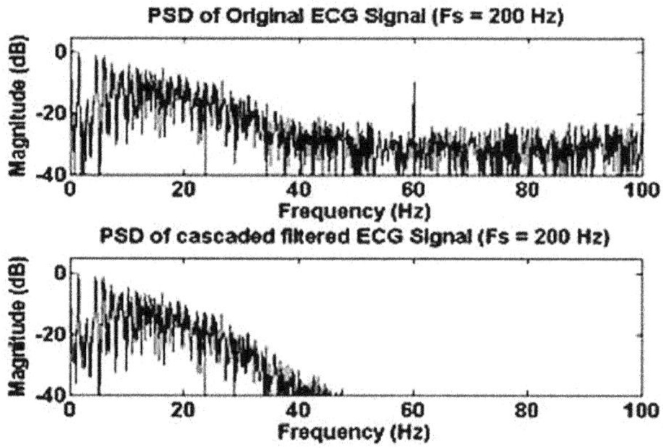

(c)Power spectral density of the original and the cascaded filtered ECG Signal

FIGURE 2.9 (*Continued*) Pre-processing of ECG with cascaded filters.

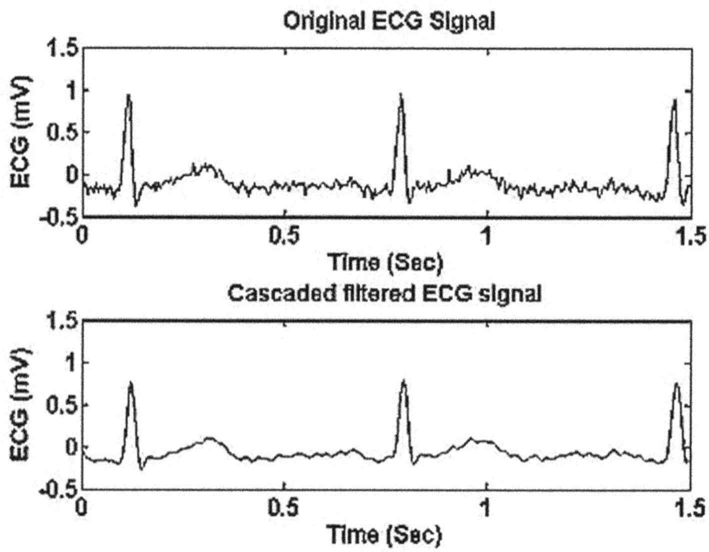

FIGURE 2.10 Original ECG signal and cascaded filtered ECG signal.

If noise is reduced through the use of a filtering method, the auto-correlation function can be written as

$$R_{YY}(n_0) = R_{\alpha\alpha}(n_0) = \sum_{n=-\infty}^{\infty} \alpha(n + n_0)\,\alpha(n) \tag{2.24}$$

If the original signal α (n) and the time-delayed signal y $(n+n_0)$ are identical, auto-correction is 1, but if they are not, auto-correction is less than 1. Therefore, for the accuracy calculation, auto-correlation can be directly used in place of the confusion matrix parameters (TP, TN, FP, and FN).

Due to the quasi-periodic nature of the ECG signal, some ECG data can be reduced by using the time series representation for the fast ECG signal processing. This data reduction and noise suppression can be achieved using SAX.

2.3 SYMBOLIC AGGREGATE APPROXIMATION

SAX is a process in which a time series representation of ECG signal is converted to a symbolic representation. In this process, storage, retrieval, and analysis of the ECG signal is more efficient.

For SAX representation of ECG signal, the following steps are followed.

2.3.1 Normalization

For normalizing the ECG signal, we set the mean value of the ECG to 0 and the standard deviation to 1, making the signal scale invariant.

2.3.2 Segmentation

The normalized ECG signal is divided into segments of a fixed length of time or a fixed number of data points.

2.3.3 PAA (Peace-Wise Aggregate Approximation)

In the peace-wise aggregate approximation (PAA) process, each segment is represented by its mean value so that we can reduce the dimensionality of data. A sequence of arbitrary length n can be minimized to a string of arbitrary length w ($w < n$, mostly $w << n$).

2.3.4 Discretization

After PAA, we got the mean values of segments mapped to discrete symbols. For mapping, we define breakpoints so that the range of possible values can be divided into intervals. A unique symbol represents each interval.

For symbolic representation, the alphabet size is an arbitrary integer '*a*', which is greater than 2, e.g., for the alphabet = {*a*, *b*, *c*}, '*a*' is 3.

	a	**b**	**c**
a	0	0	0.67
b	0	0	0
c	0.67	0	0

It is necessary that the symbols be equiprobable in discretization processes [10,11]. The Gaussian distribution has been applied to normalize the time series, making it easier to find the "breakpoints" that produce equal-sized areas under the Gaussian curve [11].

2.3.5 Breakpoints Calculation

To find the breakpoints, we have to find the points in the cumulative distribution function (CDF) of the standard normal distribution that distribute the data into equal-sized areas.

Steps to Calculate Breakpoints

Determine the CDF Values:

Calculate the cumulative probabilities that divide the normal distribution into $\frac{1}{a}, \frac{2}{a}, \frac{3}{a} \ldots\ldots\ldots\ldots \frac{(a-1)}{a}$

Use the Inverse CDF (z-scores):

Find the z-scores corresponding to these cumulative probabilities. These z-scores are the breakpoints.

Example

For a three-symbol alphabet ($a = 3$),

- We need two breakpoints to divide the data into three regions.
- The cumulative probabilities are $\frac{1}{3} = 0.333$ and $\frac{2}{3} = 0.667$.

Distribution table or a statistical function to find the z-scores:

$z\,(0.333) \approx -0.43$

$z\,(0.667) \approx 0.43$

So, the breakpoints are approximately −0.43 and 0.43.

Using these breakpoints, we can divide the range of the normalized data into three intervals. Table 2.1: The lookup table for alphabet size 3

	a	**b**	**c**
a	0	0	0.67
b	0	0	0
c	0.67	0	0

The PAA coefficients are mapped as "*a*" if they are present in the minimum breakpoint, the coefficient present in the second smallest breakpoint is mapped to the symbol "*b*," and so on. As shown in Figure 2.11, symbols "**a**," "**b**," and "**c**" are more or less equiprobable.

In SAX approximation, accuracy is dependent on window sizing, and the smaller the window size, the more accurate the results will be.

In our experiment, window sizes of 20, 10, 5, and 1 have been used with alphabet sizes of 4 and 10, as indicated in Figure 2.12a–d.

TABLE 2.1
A Lookup Table Showing the Breakpoints Which Give Equiprobable Regions (3–10) in the Gaussian Distribution

β_1	3	4	5	6	7	8	9	10
β_2	−0.43	−0.67	−0.84	−0.97	−1.07	−1.15	−1.22	−1.28
β_3	0.43	0	−0.25	−0.43	−0.57	−0.67	−0.76	−0.84
β_4		0.67	0.25	0	−0.18	−0.32	−0.43	−0.52
β_5			0.84	0.43	0.18	0	−0.14	−0.25
β_6				0.97	0.57	0.32	0.14	0
β_7					1.07	0.67	0.43	0.25
β_8						1.15	0.76	0.52
β_9							1.22	0.84
β_{10}								1.28

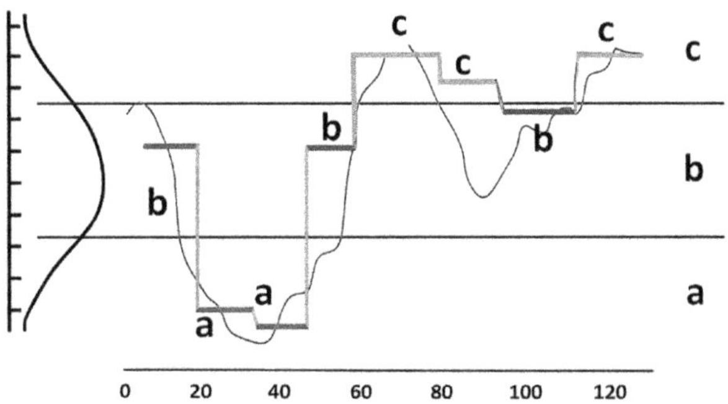

FIGURE 2.11 SAX representation using an alphabet size of 3.

In Figure 2.12a, ECG data after normalization is shown ECG using SAX representation with a window size of 20 and 4 alphabets is shown in Figure 2.12b, while in Figure 2.12c, ECG-SAX representation for the same window size of 20 and 10 alphabets is shown.

Therefore, it is clear that if we use more alphabets, it means the window size is low; SAX representation follows more closely to the actual ECG data, while for larger window sizes, SAX representation is far away from the actual data. To further improve accuracy, the window size has to be reduced; in Figure 2.12d, we reduced the window size to 5, while the alphabet size is 10. Here, most of the peaks can now be covered, except for larger peaks.

In Figure 2.13a, window size is further reduced to 2, with all the minor peaks covered, while some portions of the major peaks are still left. In Figure 2.13b, window size is reduced to 1, and now all the peaks are fully covered. The symbolic representation for the ECG signal for a window size of one is

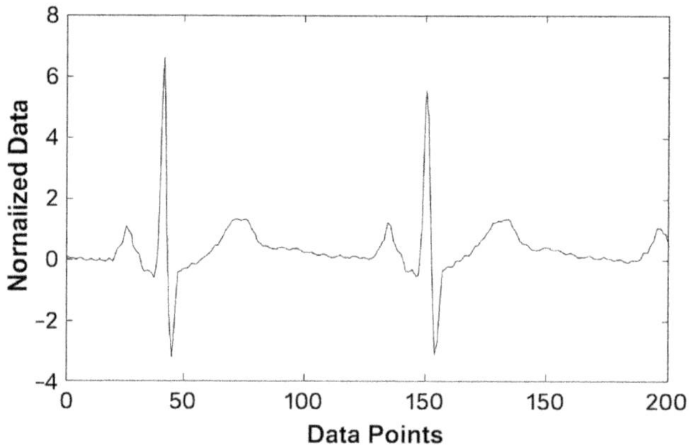

(a) ECG data Normaliztion

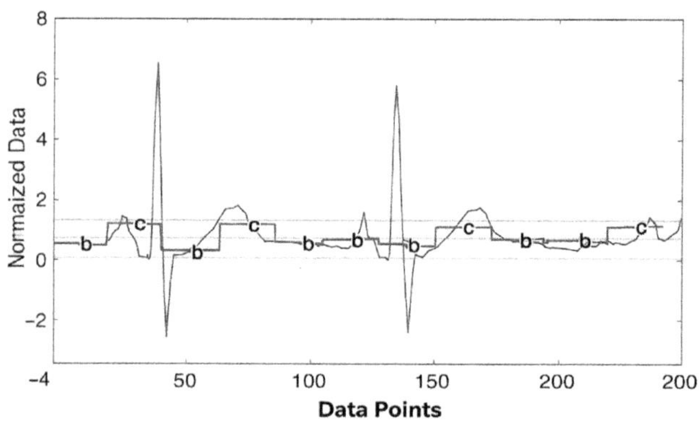

SAX representation of ECG (winsize-20, alphabet - 4)

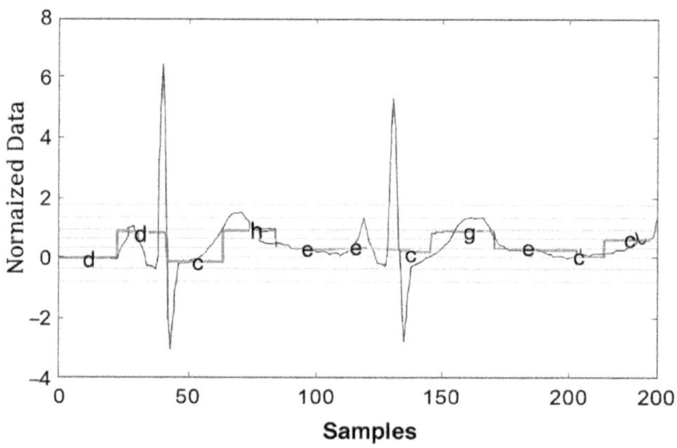

SAX representation of ECG (winsize-20, alphabet - 10)

FIGURE 2.12 ECG data generation using SAX under different window size and alphabet sizes.

(Continued)

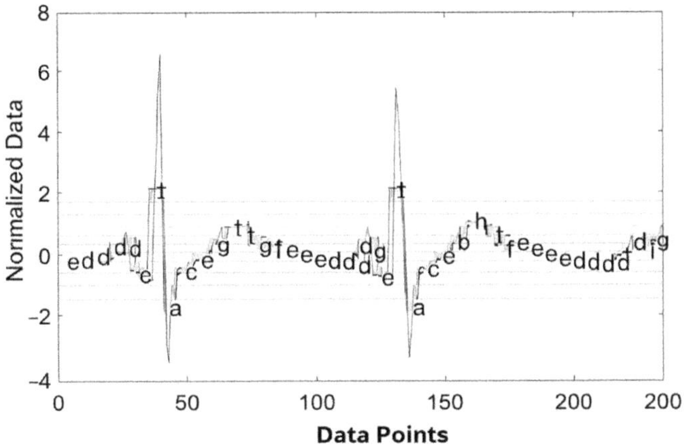

SAX representation of ECG (winsize-5, alphabet - 10)

FIGURE 2.12 (*Continued*) ECG data generation using SAX under different window size and alphabet sizes.

‘e’ ‘e’ ‘e’ ‘d’ ‘e’ ‘d’ ‘d’ ‘d’ ‘d’ ‘d’ ‘d’ ‘d’ ‘d’ ‘d’ ‘d’ ‘d’ ‘d’ ‘d’ ‘d’

‘d’ ‘d’ ‘e’ ‘f’ ‘f’ ‘f’ ‘h’ ‘h’ ‘f’ ‘f’ ‘e’ ‘d’ ‘c’ ‘c’ ‘c’ ‘c’ ‘b’ ‘d’ ‘j’ ‘j’

‘j’ ‘c’ ‘a’ ‘a’ ‘a’ ‘c’ ‘c’ ‘c’ ‘c’ ‘d’ ‘d’ ‘d’ ‘d’ ‘e’ ‘e’ ‘e’ ‘f’ ‘f’ ‘f’ ‘g’

‘h’ ‘h’ ‘h’ ‘i’ ‘i’ ‘i’ ‘i’ ‘i’ ‘i’ ‘i’ ‘h’ ‘h’ ‘g’ ‘f’ ‘f’ ‘f’ ‘f’ ‘f’ ‘f’ ‘e’

‘f’ ‘f’ ‘f’ ‘e’ ‘e’ ‘e’ ‘e’ ‘e’ ‘e’ ‘e’ ‘e’ ‘e’ ‘d’ ‘e’ ‘e’ ‘d’ ‘d’ ‘d’ ‘e’ ‘d’

‘d’ ‘d’ ‘d’ ‘d’ ‘d’ ‘d’ ‘d’ ‘d’ ‘d’ ‘d’ ‘d’ ‘e’ ‘f’ ‘f’ ‘g’ ‘i’ ‘h’ ‘g’ ‘e’

‘e’ ‘d’ ‘c’ ‘c’ ‘c’ ‘c’ ‘b’ ‘f’ ‘j’ ‘j’ ‘j’ ‘c’ ‘a’ ‘a’ ‘a’ ‘c’ ‘c’ ‘c’ ‘c’ ‘d’

‘d’ ‘d’ ‘d’ ‘d’ ‘e’ ‘e’ ‘f’ ‘f’ ‘f’ ‘g’ ‘h’ ‘h’ ‘h’ ‘i’ ‘i’ ‘i’ ‘i’ ‘i’ ‘i’ ‘h’

‘h’ ‘g’ ‘g’ ‘f’ ‘f’ ‘f’ ‘f’ ‘e’ ‘e’ ‘f’ ‘f’ ‘e’ ‘f’ ‘e’ ‘e’ ‘e’ ‘e’ ‘e’ ‘d’ ‘e’

‘d’ ‘d’ ‘d’ ‘d’ ‘d’ ‘d’ ‘d’ ‘d’ ‘d’ ‘d’ ‘d’ ‘d’ ‘d’ ‘d’ ‘d’ ‘d’ ‘d’ ‘f’ ‘f’ ‘g’ ‘i’ ‘h’

It is noticeable that for the highest and lowest peaks, the symbols “j” and “a” are repeatedly used. To address this issue, there are two solutions: firstly, more alphabets can be used (>10); secondly, higher positive peaks can be detected using a digital differentiator as (y $(n-) < y(n)$ and $y\ (n+1) < y(n)$ and $y(n) > \tau$), where $y(n)$ is the nth sample amplitude value and “τ” is the initially defined threshold, which in our case is greater than 2. Similarly, using the same analogy, negative peaks can be detected. The detected peaks are at indices 40, 42, 128, and 131, respectively (Figure 2.14).

A digital differentiator highlights rapid changes in the ECG signal, which correspond to the steep slopes of the QRS complex. The differentiator can be implemented using convolution with a finite impulse response filter.

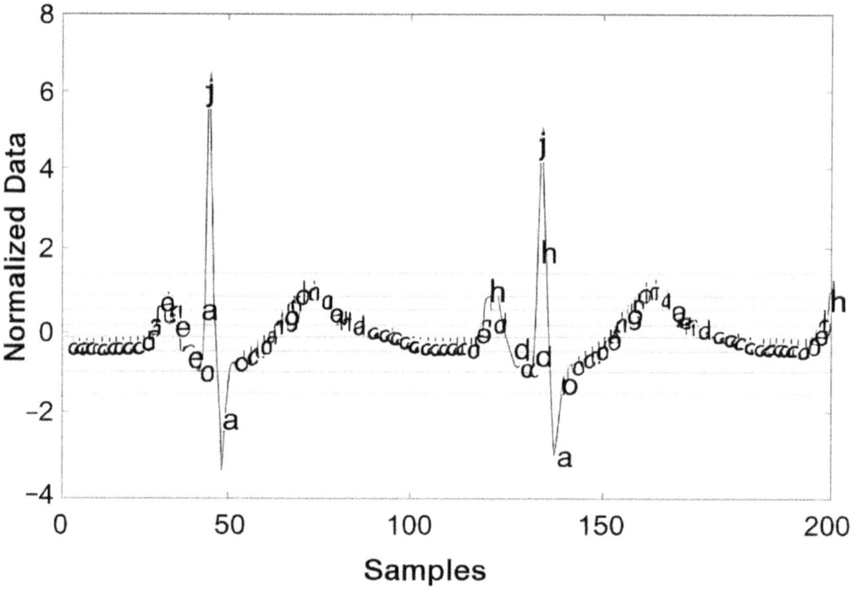

(a) SAX representation of ECG (winsize-2, alpabet – 10)

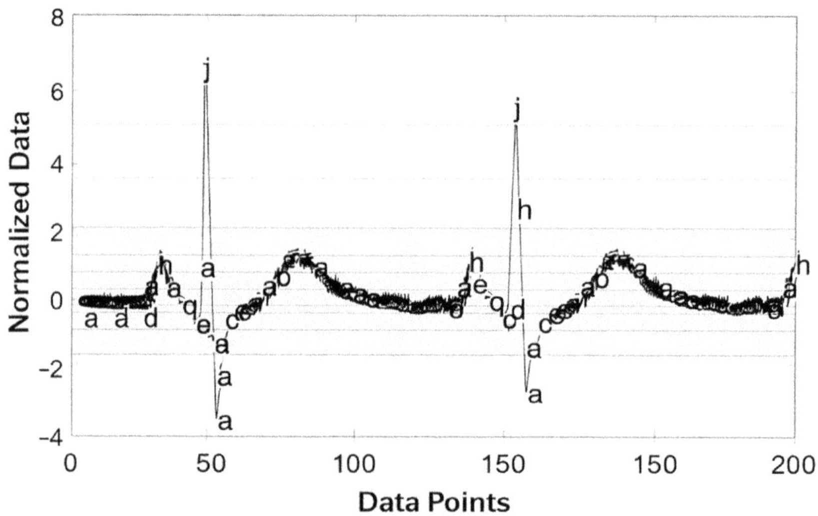

SAX representation of ECG (winsize-1, alphabet – 10)

FIGURE 2.13 SAX representation of ECG data with window sizes 1 and 2.

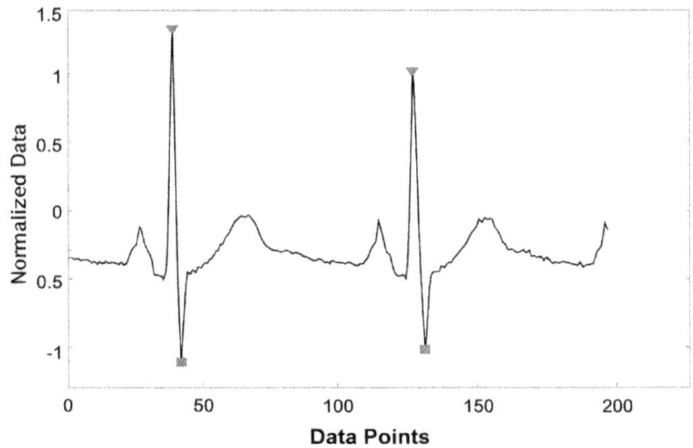

FIGURE 2.14 Detection of the highest and lowest peaks of ECG using a digital differentiator.

2.4 DIFFERENTIATOR FILTER DESIGN

A commonly used differentiator in ECG processing is a five-point

$$y[n]=\frac{1}{8}(-x\ [n-2]-2x[n-1]+2x[n+1]+x[n+2]$$

After differentiation, square the signal to highlight large values and suppress small ones, making it easier to detect the QRS complex

Therefore, using a combination of SAX representation and digital differentiation, every variation and all the peaks of ECG data can be detected error-free. Finally, to check the similarity between the original and SAX-generated ECG for a window size of one, the correlation coefficient is evaluated, which is found to be 0.9956, indicating that the original and the SAX-generated ECG are exactly the same. Therefore, the error is nearly zero.

2.5 CONCLUSION

In ECG signals, the accuracy of further analysis can be affected by various types of noise, such as baseline wander, power line interference, and motion artifacts. To improve this, pre-processing of ECG signals is required, which involves a combination of cascaded filters to effectively remove unwanted noise and enhance signal quality. These filters help preserve the ECG waveform's key characteristics while eliminating distortions that may interfere with peak detection and feature extraction.

Once the signal is pre-processed, the **SAX method** is applied to represent the ECG signal as a time series. SAX is an efficient technique that reduces dimensionality while maintaining essential trends within the signal. In the SAX-based representation, two crucial parameters—**window size** and **alphabet size**—directly impact the accuracy of peak detection. The selection of an appropriate window size ensures that significant variations in the ECG signal are captured without excessive loss of detail. Similarly, the alphabet size determines the level of granularity in symbolizing the signal, which in turn influences the detection of key events such as R-peaks in the ECG waveform.

Furthermore, a digital differentiator is employed to enhance peak identification when multiple peaks are present in the ECG signal. This approach allows for the precise detection of target peaks, ensuring that clinically significant features, such as R-peaks, are accurately distinguished from other fluctuations in the signal. Using a digital differentiator helps refine the results, leading to improved reliability in ECG signal analysis.

In conclusion, the proposed method integrates a well-structured pre-processing framework with SAX-based representation and digital differentiation techniques to improve ECG peak detection. This approach enhances signal clarity, optimizes feature extraction, and accurately identifies critical peaks, contributing to more reliable ECG analysis in biomedical applications.

REFERENCES

1. Akan A and Cura OK. Time–frequency signal processing: Today and future. *Digital Signal Processing*. 119, 103216, 2021.
2. Addison PS, Walker J, and Guido RC. Time–frequency analysis of biosignals. *IEEE Engineering in Medicine and Biology Magazine*. 28(5), 14–29, 2009.
3. Volna E, Kotyrba M, and Habiballa H. ECG prediction based on classification via neural networks and linguistic fuzzy logic forecaster. *The Scientific World Journal*. 2015, 512530, 2015.
4. Chauhan S and Vig L. Anomaly detection in ECG time signals via deep long short-term memory networks. In *Proceedings of the IEEE International Conference on Data Science and Advanced Analytics (DSAA)*, pp. 1–7. Paris, France, October 2015.
5. Holsinger WP, Kempner KM, and Miller MH. A QRS preprocessor based on digital differentiation. *IEEE Transactions on Biomedical Engineering*. 18(3), 212–217, 1971.
6. Zhang D, Zuo W, and Wang P. Baseline wander correction in pulse waveforms using wavelet-based cascaded adaptive filter. In *Computational Pulse Signal Analysis*, Ed. D Zheng, pp. 65–90. Springer, Singapore, 2018.
7. Daqrouq K. ECG baseline wander reduction using discrete wavelet transform. *Asian Journal of Information Technology*. 4(11), 989–995, 2005.
8. Tinati MA and Mozaffary B. A wavelet packets approach to electrocardiograph baseline drift cancellation. *International Journal of Biomedical Imaging*. 2006, 70624, 2006.
9. Scholkmann F, Boss J, and Wolf M. An efficient algorithm for automatic peak detection in noisy periodic and quasi-periodic signals. *Algorithms*. 5(4), 588–603, 2012.
10. Lin J, Keogh E, Wei L, and Lonardi S. Experiencing SAX: A novel symbolic representation of time series. *Data Mining and Knowledge Discovery*. 15(2), 107–144, 2007.
11. Lin J, Keogh E, Lonardi S, and Chiu B. A symbolic representation of time series, with implications for streaming algorithms. In *Proceedings of the 8th ACM SIGMOD Workshop on Research Issues in Data Mining and Knowledge Discovery*, pp. 2–11. ACM, San Diego, CA, 2003.

3 Application of Decomposition Techniques to Physiological Time Series with Variable Spectral Content

Adil Deniz Duru and Hasan Birol Çotuk

3.1 DECOMPOSITION TECHNIQUES FOR BIOMEDICAL TIME SERIES

Monitoring physiological data can facilitate a comprehensive and detailed assessment of a person's overall health and quality of life. As has been demonstrated in relation to heart rate variability, examining such data provides invaluable insight into understanding stress and health outcomes. It facilitates the development of effective treatment strategies when indicated (Thayer et al., 2012). In this context, time series analysis is the most appropriate method for gaining these relevant physiological insights. Monitoring time-dependent changes in physiological activity allows for the collection of information that is available in both the time and frequency domains as well as the time-frequency domain. Since almost all recorded time series represent an integration of numerous physiological processes and components, the appropriate decomposition of these series is the key procedure to extract relevant information from the complex content.

There are two different but basic approaches to decomposing time series components: data-driven and model-based. Model-based methods require a priori mathematical formulation to elucidate a physiological process or predict future values of the measured parameter. In contrast, data-driven methods for decomposing time series provide researchers with flexible and powerful analytical tools because they are not constrained by the assumptions of a priori models. The effectiveness of data-driven methods lies in their ability to exploit the structure inherent in the data itself, making them particularly suitable for analyzing complex and irregular time series. These techniques are used in a wide range of applications, from biomedical signals to financial data. The main methods used in this category are empirical mode

DOI: 10.1201/9781003611448-5

decomposition (EMD), variational mode decomposition (VMD), Hilbert–Huang transform, multivariate EMD, and ensemble EMD.

This section of the chapter provides an overview of the studies in biomedicine in which the principal decomposition methods have been used to analyze specific physiological time series measurements.

3.1.1 Electrocardiography and Photoplethysmography

The electrical activity of the heart can be monitored non-invasively with high temporal resolution using an electrocardiogram (ECG) to assess the rhythm and waveforms representing different phases of the cardiac cycle. In recent research, EMD has been applied to ECG time series for various other purposes. These include detection of ventricular fibrillation (Oh & Choi, 2024), feature extraction to calculate input parameters for cardiovascular disease classification (Li et al., 2023), detection of sudden cardiac death (Centeno-Bautista et al., 2023), and screening of patients with mitral regurgitation (Qi et al., 2023). In addition, EMD has been combined with deep learning techniques (Zhang et al., 2022) and a complex dual-tree wavelet transform (WT) (Lu et al., 2021) to solve ECG signal denoising problems. The method was also used to determine sleep apnea-related parameters from a single-lead ECG (Sharma & Sharma, 2020). Conventional EMD techniques are primarily hampered by mode aliasing between intrinsic mode functions (IMFs). To address this problem, Nazari and Sakhaei (2018) proposed a VMD-based method that minimizes the mode aliasing of the center frequency for the extracted components, thus facilitating the extraction of the respiratory signal from the ECG. Subsequently, Li et al. (2022) developed an improved version of the VMD algorithm for ECG signal denoising using the EMD method for filtering and an adaptive particle swarm optimization algorithm for determining the mode number and penalty factor of the VMD.

An alternative non-invasive method for measuring heart rate, peripheral blood flow, and oxygen saturation is based on the principle of light absorption by the blood associated with each heartbeat. This produces a measurable photoplethysmography (PPG) signal. The effectiveness of heart rate estimation from PPG signals by principal component analysis after VMD was demonstrated by Sharma (2019) for short intervals (30 s). The EMD method was also applied to extract respiratory rate (Motin et al., 2019) and heart rate (Pollreisz & TaheriNejad, 2022) from the PPG time series. In a more recent study, Pachori et al. (2024) demonstrated the effectiveness of VMD in identifying atrial fibrillation from the PPG signal. In addition, it has been used in conjunction with imaging PPG (Karlen et al., 2015) to determine respiratory rate.

3.1.2 Electromyography

Measuring the electrical impulses generated by muscle fibers during contraction provides important information for assessing muscle activity and diagnosing neuromuscular disorders. This activity can be quantified in two ways: non-invasively by surface electromyography (sEMG) or by needle electrodes inserted directly into the muscle tissue (needle electromyogram). Researchers have used EMD to classify muscle diseases based on electromyography (EMG) time series (Dubey et al., 2022) and for signal denoising in EMG (Koppolu & Chemmangat, 2023). In this context, VMD showed superior performance in removing various types of noise in EMG compared to EMD and ensemble EMD (Ma et al., 2021). In a recent study, entropy-based metrics derived from VMD were shown to be more accurate than existing methods in detecting neuromuscular disorders (Nagineni et al., 2024). In addition to clinical applications, VMD has also been used to study the activation patterns of fine motor skills (Zhai et al., 2023) and to improve the precision of hand gesture recognition (Lin et al., 2022; Yang et al., 2023).

3.1.3 Electroencephalography

An electroencephalogram (EEG) is a non-invasive method of synchronized recording of electrical potentials generated at the cortical surface of the brain by electrodes attached to the scalp. EEG is a widely used diagnostic tool in both clinical and research settings to detect neurological disorders, monitor brain health, and study cognitive processes. In the field of EEG research, the use of EMD covers a variety of topics, with a particular focus on detecting depression (Shen et al., 2023), identifying the onset of epileptic seizures (Li et al., 2021a), differentiating the ictal period (Li et al., 2013), diagnosing early-stage Alzheimer's disease (AlSharabi et al., 2023), and evaluating schizophrenia (Siuly et al., 2020). In addition to clinical studies, EMD has been widely used in the analysis of healthy human EEG data, particularly regarding the detection of emotional states (Zheng & Li, 2024; Salankar et al., 2021). In the context of EEG artifact removal, the application of VMD in conjunction with machine and deep learning mechanisms has been shown to be an effective technique for improving signal quality (Sajed et al., 2021; Liu et al., 2022; Lin et al., 2023; Hao et al., 2022). The efficacy of neurophysiological studies depends on the implementation of effective noise reduction techniques that provide a more comprehensive understanding of brain functionality as reflected in EEG time series. The application of VMD in clinical trials (Sun et al., 2022), advanced signal processing, and cognitive neuroscience (Zhang et al., 2021) has contributed significantly to extracting information from EEG.

3.1.4 Functional Magnetic Resonance Imaging

Functional magnetic resonance imaging (fMRI) time series are generated from changes in cerebral blood flow and oxygen levels in the brain. They provide a dynamic representation of activity that can be used to study brain function, connectivity, and dysfunction in both healthy and diseased states. The use of EMD in fMRI has been explored for denoising (Xu et al., 2021; Zheng et al., 2020), resting-state decomposition, and functional connectivity mapping (Zhou et al., 2022; Yu et al., 2021), and biomarker extraction related to Alzheimer's disease (Li et al., 2021b). To improve the quality of functional connectivity analysis, Wang et al. (2017) used VMD as a denoising tool for fMRI time series. In addition, VMD was integrated with deep learning to enable improved feature extraction for functional connectivity analysis (Zuo et al., 2020).

3.2 SELECTION OF TIME SERIES DECOMPOSITION TECHNIQUES

Decomposing non-stationary biological time series into simpler subcomponents, without relying on a predefined model structure, helps to uncover and interpret hidden features (Eriksen and Rehman, 2023). When the data are assumed to consist of additive components, decomposition methods help to identify the underlying patterns. An important goal is to break down the magnitude spectrum for better analysis. Given the non-negative nature of spectral content, non-negative matrix factorization is often used for feature extraction (Liu et al., 2004). This technique primarily relies on the non-negativity constraint of the input data (Long et al., 2024). Its extensions have been applied in various biomedical contexts, such as quantifying sleep signals using time-frequency analysis (Shokrollahi & Krishnan, 2013), extracting EEG electrode weights for mental image recognition (Gurve et al., 2020), and analyzing noise and sampling variations in ECG biometrics (Huang et al., 2022).

It is reasonable to assume that physiological signals are composed of a mixture of different subcomponents. For example, signals recorded from scalp electrodes may contain contributions from eye blinks, facial muscle activity, and electrical brain activity. These overlapping components contribute to the complexity of the measured signal. Assuming that the components are independent, independent component analysis is an appropriate method to isolate brain activity from other sources. This approach has been extended to a family of techniques for mapping resting-state networks from fMRI signals (To et al., 2022). While independent component analysis produces time series that are statistically independent, principal component analysis can be used to obtain orthogonal components.

When processing signals with components whose frequency content varies or overlaps over short periods of time, the short-time Fourier transform (STFT) and the WT are powerful tools for breaking them down into simpler patterns. These patterns usually consist of single or narrow frequency bands within certain time intervals. The STFT is a simple and effective tool for analyzing stationary signals, such as 1- or 2-second segments of brain electrical activity, which are often considered stationary in such short time intervals (Kaplan et al., 2005; Duru & Assem, 2018). This makes STFT a practical choice for extracting meaningful information about mental states (Al-Jumaili et al., 2023). On the other hand, WT offers better time-frequency localization and provides more detailed insights, especially when the signal-matched wavelet functions are selected. While STFT is excellent for stationary signals, WT excels at capturing the nuances of non-stationary or rapidly changing signals, making it a versatile tool for biomedical signal analysis (Mallat, 1999; Duru et al., 2009).

Biomedical time series, such as EEG signals, are inherently non-stationary, i.e., their statistical properties and frequency content change over time. This temporal variability poses significant challenges to conventional signal processing techniques such as the STFT, which assumes stationarity within the analysis window. To overcome this limitation, these signals need to be decomposed into stationary and oscillating components or analyzed in short, quasi-stationary epochs. These approaches allow for a more accurate interpretation and extraction of meaningful physiological information.

The following section examines EMD, a data-driven and adaptive method developed to deal with the complexity of non-stationary signals. EMD decomposes a time series into a series of IMFs that are locally stationary and oscillatory. These IMFs capture the multiscale characteristics of the signal, which can make EMD particularly effective for analyzing biomedical data. However, if the underlying signal consists of compact components in the frequency domain, VMD offers a more robust alternative. In contrast to EMD, which is heuristic and data-adaptive, VMD works within a variational optimization framework. Its primary objective is to minimize the bandwidth of each derived component while ensuring that their sum accurately reconstructs the original signal. Thanks to this mathematical rigor, VMD is very effective in decomposing signals with well-defined frequency features, which increases its utility in biomedical signal analysis.

3.3 EMD AND VMD TECHNIQUES

In order to interpolate the envelopes of a time series, the EMD method uses local maxima and thus extracts the so-called intrinsic mode functions (IMFs) via the average envelopes. EMD is an inherently adaptive method that proceeds without a priori assumptions about the structure of the data to be analyzed (Huang et al., 1998). In 2004, Wu and Huang demonstrated the effectiveness of the EMD method in accurately characterizing the properties of white noise and distinguishing it from other signal types. In the first step of EMD, all local minima and maxima of the signal ($x(t)$) are determined. Then two separate interpolation polynomials are formed for the upper ($u(t)$) and lower ($l(t)$) local maximum values. The average time series of these two polynomials, the so-called mean envelope ($m(t)$), is then subtracted from the original signal. In Equation 3.1, t stands for the time index.

$$r(t) = x(t) - m(t) \tag{3.1}$$

After subtraction, the $r(t)$ function is evaluated to determine whether it meets the two criteria for classification as an intrinsic mode function. The two criteria are defined as follows: (1) the number of zero crossings and the number of extreme values must not differ by more than one; (2) the average values of the envelopes, which are calculated using the local maxima and minima, should be equal to zero at all times.

If $r(t)$ does not meet these criteria, the above steps are repeated for $r(t)$ until it meets the specified criteria and can be labeled as IMF. This process is called "sifting". Once an IMF has been calculated, it is subtracted from the original signal, whereupon the procedure is applied again to extract another IMF. The process is finalized when the remaining input function has only one extremum or becomes monotonic. Finally, the summation of n parts of the IMF yields the measured original time series data, as shown in Equation 3.2.

$$x(t) = \sum_{j=1}^{n} \text{IMF}(t) + r(t) \tag{3.2}$$

Although EMD is an effective technique for analyzing non-linear and non-stationary signals, its effectiveness depends on the specific nature of the input time series under consideration. Time series that consist of random data or white noise, as well as those that exhibit significant trends or drifts, hinder the proper application of EMD techniques (Zhonghua and Huang, 2004). The absence of oscillating components or the presence of signals with frequencies close to each other makes the identification of IMFs difficult. Due to its iterative nature, EMD can be computationally intensive, especially for large data sets.

On the other hand, VMD uses an optimization-based approach that decomposes a signal into a certain number of modes that oscillate around specific center frequencies (Dragomiretskiy & Zosso, 2014). VMD provides precise control over the number of modes, which can enable more accurate separation results in a given application. In principle, VMD requires fewer iterations than EMD, making it computationally more efficient for large data sets. VMD decomposes the time series into a certain number of mode components. Each mode oscillates around a certain center frequency that is either unknown or known a priori. The initial guess about the center frequencies of the modes can be defined as a random vector, as peak frequencies, or as predefined values with respect to the nature of the data. The VMD method requires the minimization of the cost function defined in Equation 3.3. In Equation 3.3, $f(t)$ stands for the input time series with the time variable t, w_k denotes the center frequency of the kth component, and $u_k(t)$ is the component time series. The below equation is optimized in the spectral domain for K modes, and the summation of all $u_k(t)$ should yield $f(t)$.

$$\min_{\{u_k\},\{w_k\}} \left\{ \sum_{k=1}^{K} \left\| \partial_t \left[\left(\delta(t) + j\frac{1}{\pi t} \right) * u_k(t) \right] e^{-jw_k t} \right\|^2 \right\} \tag{3.3}$$

The use of Lagrange multipliers can solve the above-constrained equation. The augmented Lagrange equation is generally solved based on an iterative approach such as alternating direction method of multipliers, while gradient descent, conjugate gradient method, proximal gradient methods, Newton–Raphson method, and sequential quadratic programming can be used as alternative techniques for the solution.

The parameter "number of modes" of the VMD plays a decisive role in the exact decomposition of the time series. Selecting too many modes can lead to several noisy components being generated. On the other hand, selecting too few modes may result in some important components being overlooked. Therefore, it is crucial to find the optimal number of modes that facilitate the decomposition procedure for further analysis. This parameter can be set to match the expectations of the nature of physiological variability measured or based on an objective measurement of the information contained in the data. In the first approach, the number of mode parameters proposed in the previous literature can be used, or it can be selected by evaluating the decomposed time series after the implementation of VMD for different mode values. In the second approach, an optimal number of modes can be proposed based on objective metrics regarding the physical expectations of the components. By using multiple mathematical variables, physical or physiological indicators hidden in the measured signal can be expressed. For example, each component should have an appropriate amount of energy compared to the others, which can be indicated by the variance of the component. Another approach is to check whether the spectral content of each component contains different frequency bands. Entropy parameters can also provide a basis for indicating the complexity of the time series. For example, parameters with higher entropy indicate more complex time series, especially those with more noise contamination.

We have created several simulated time series to verify the performance of the decomposition algorithms. These time series consist of the summation of pure sine waveforms and additive Gaussian noise. Since the parameters of the basic components of the simulations are known in advance, the performance of the decomposition methods can be evaluated. The simulations and decomposition methods are performed with MATLAB 2023.

3.3.1 Simulation Case 1

To demonstrate the application of the EMD technique, two example time series are generated by adding two oscillations with and without random noise. The fundamental frequency of these oscillations is set to 4 and 13 Hz, for 4 s with a sampling frequency of 1 kHz. The two original time series are shown in the first row of Figure 3.1. After implementing the EMD, several IMFs are obtained, and the resulting waveforms are shown in Figure 3.1 for both input time series. The noise-free signal was decomposed into three waveforms, while the number of decomposed waveforms for the noisy signal was five. The original oscillations used to create the simulation data sets were accurately identified for both input signals. In the input simulation signal, if the center frequency values of the individual oscillations of each

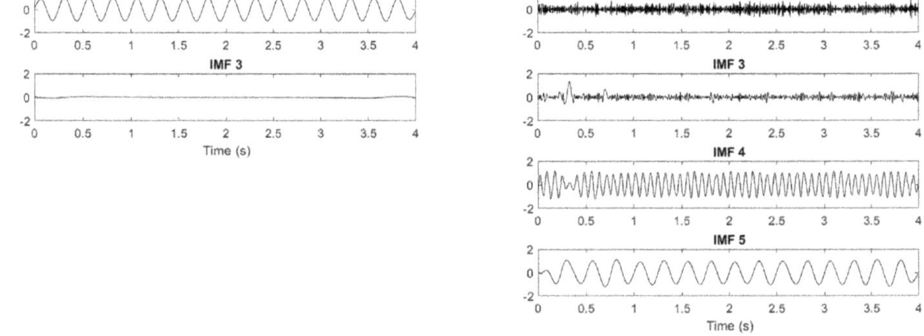

FIGURE 3.1 The simulated sinusoidal signal is plotted in the first line of the left panel, while the noisy simulated data is shown in the first line of the right panel. The left panel contains the three output components of the EMD, while the five components extracted from the noisy signal are shown on the right.

component in the spectrum can be decoupled from each other (at least twice), perfect decomposition is possible, which is directly related to the number of output IMFs.

3.3.2 Simulation Case 2

A simulated data set and the results derived by using the VMD algorithm are shown in Figure 3.2. Three sinusoidal waveforms with individual frequencies of 2, 14, and 38 Hz are generated and summed to the simulated data set. The parameters of this simulation can be observed in running EEG signals. A white Gaussian noise vector is also added to these sinusoidal waveforms. Three fundamental frequency values are shown in the first row of the second column in Figure 3.2. The number of components for the decomposition method for VMD is set to three based on the number of peak frequencies displayed in the spectrum. Using the VMD method, the input time series was successfully decomposed into the simulation components.

3.3.3 Simulation Case 3

Due to the nature of physiological measurements, the presence of components with a single fundamental frequency is often not observed. Most of the measured time series are contaminated with unexpected noise components and unknown physiological components originating from the cross-interaction of frequencies. Therefore, we developed a third scenario to illustrate this fact using four different frequency bands and additive Gaussian noise. The four discrete components exhibited temporal variations in frequency content that were modulated as a band. These components are summed in the time domain, and the time-frequency transform of this waveform

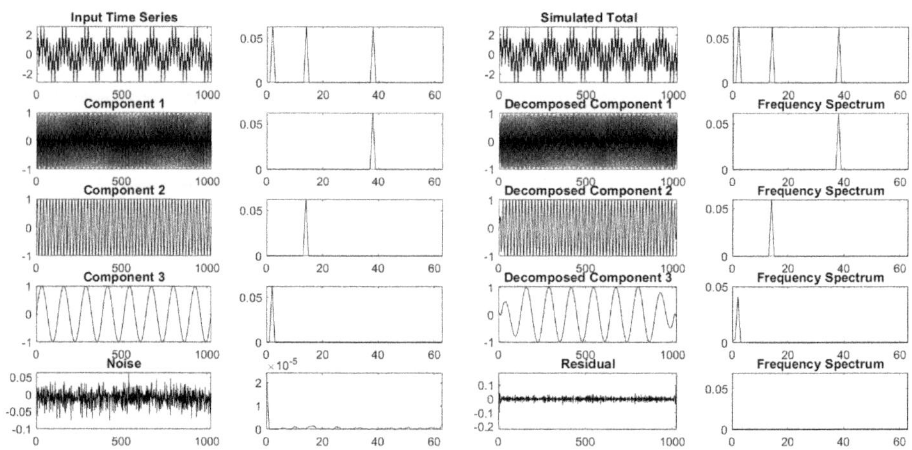

FIGURE 3.2 The leftmost column shows the simulated input time series resulting from the summation of components and noise. The second column shows the frequency spectrum of the corresponding simulated component. The results of the VMD decomposition are shown in the third column, while the summation of the extracted components is shown in the first row of this column. Finally, the last column shows the frequency spectrum calculated from the summation of the decomposed components and the frequency spectrum of the individual components.

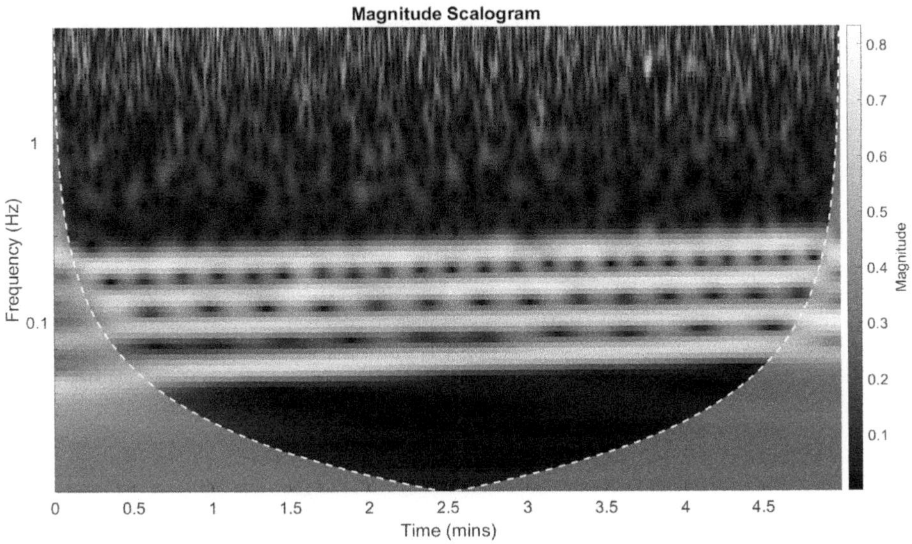

FIGURE 3.3 Time-frequency transformation of the simulated data set was obtained by using wavelet decomposition.

is shown in Figure 3.3. The center frequency values of the components were chosen to be close to 0.05, 0.1, 0.15, and 0.2 Hz, respectively. The sampling rate was set to 10 Hz, and the total duration of the signal was set to 5 min. These parameters can be observed during a PPG measurement.

Based on the results obtained by wavelet decomposition, different frequency components of the input time series can be analyzed. Since there are four individual components, these components can serve as indicators of unique physiological activity patterns. Therefore, the decomposition method should consider these components as a unique output series, and a certain number of components should be added to model the noise. For this scenario, we chose a mode number of six, and VMD was implemented. As a result, the noise-contaminated components of the original time series were successfully decomposed, as shown in Figure 3.4. The frequency content of each component matched the original waveform. Furthermore, the two additional components recovered the noisy part of the input signal and formed a residual time series with a very low variance compared to the other four decomposed components.

3.3.4 Real ERP Data

In the third phase of the implementation, the event-related potentials (ERPs) triggered by the presentation of the Posner Cue Attention Task are subjected to a VMD analysis. Time-locked averages are formed from the EEG measurements, and the ERP of a parietal electrode is given by its time-frequency decomposition (Figure 3.5). In ERP studies, the negative peak of the waveform observed after about 100 ms of the presented stimuli is referred to as N1, while the positive peak after 300 ms is referred to as P300. Similarly, N400 and P600 are examined to analyze the differences caused

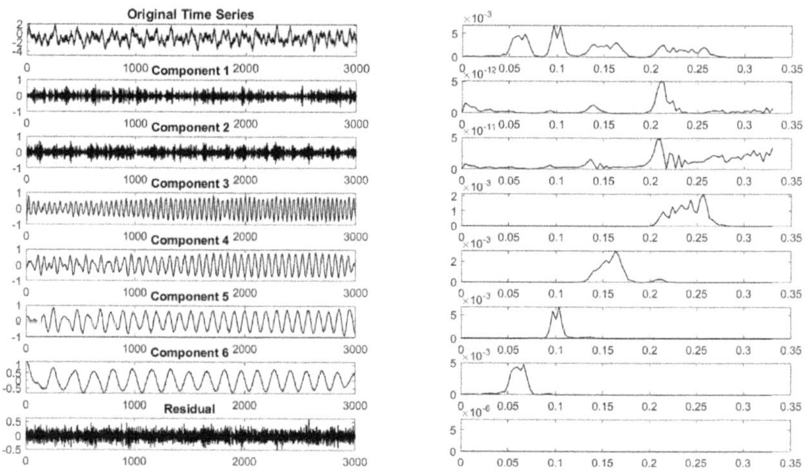

FIGURE 3.4 The time series are plotted in the left-hand column and the corresponding frequency spectra in the right-hand column. The simulated time series are shown in the first row. The remaining rows show the components derived using the VMD method.

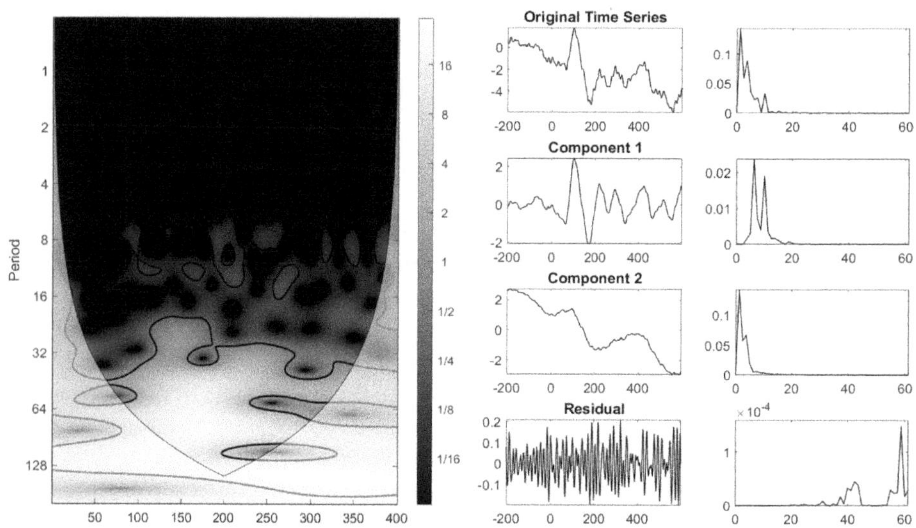

FIGURE 3.5 Time-frequency transformation of the ERP data set obtained by using the wavelet decomposition (left, first column). The original ERP is shown in the first row of the second column. The results of the VMD are plotted in the second column. The values of the frequency spectrum of the time series are plotted in the third column.

by the different types of stimuli. In our test, the ERP time series is recorded from a parietal electrode with a sampling rate of 500 Hz and a duration of 800 ms. Looking at the corresponding time-frequency decomposition, two main components stand out. Therefore, three components are selected as mode numbers in the VMD method in order to extract the components uniquely from the ERP data. The first component that the VMD outputs is a denoised time series, which is also a detrended version of the input signal. The frequency spectrum of the first component confirms that the low-frequency part of the original signal has been extracted, and this spectrum is also present in the second component. The denoising and decomposition approaches facilitate our knowledge of the neurodynamics of responses to the stimuli.

3.3.5 Real EMG Data

In this section, we address the analysis of real electromyographic data collected during an isometric contraction task and use VMD to investigate its spectral properties. The task lasted approximately 312 s, with data acquired at a sampling rate of 2 kHz. First, the spectral content of the EMG time series is visualized using the WT, as shown in Figure 3.6. This initial analysis reveals a prominent frequency component centered around 50 Hz in the early phase of the experiment, providing insight into the underlying neuromuscular dynamics.

In addition, the EMG signal is decomposed into six different components using VMD to examine the underlying sub-series. The WT is then calculated for each of these decomposed time series, revealing their spectral content. Remarkably, the summation of these decomposed components and the residual signal reconstructs the original input signal with high accuracy. A closer look at each decomposed time series reveals components with distinct frequency characteristics. For example, the fourth component has a lower frequency content compared to the original waveform, which can be interpreted as an expression of the power drop associated with the task failure. This finding becomes particularly clear when the decomposed time series are analyzed in isolation, allowing for a deeper understanding of neuromuscular adaptations and fatigue dynamics during the task (Figure 3.7).

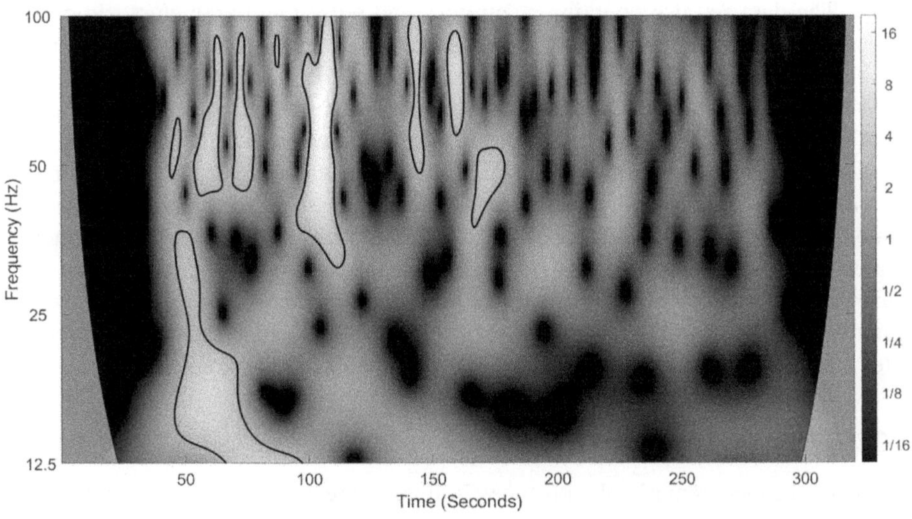

FIGURE 3.6 The time-frequency spectrum of the EMG time series.

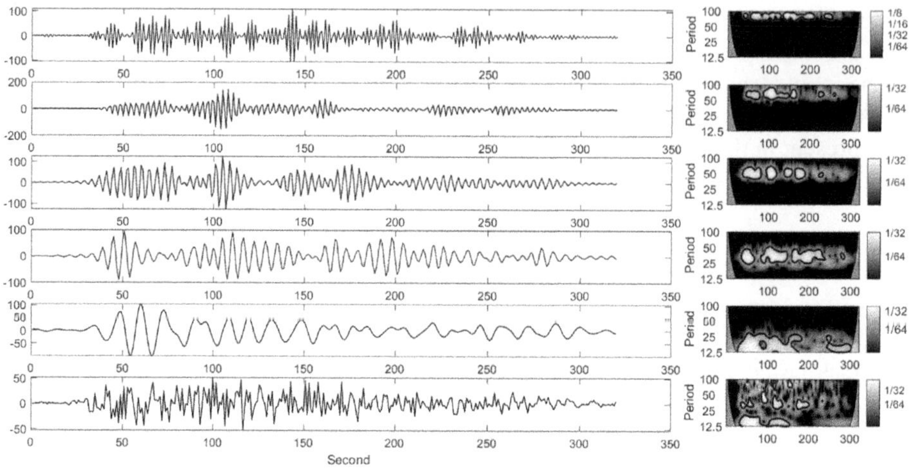

FIGURE 3.7 Six decomposed EMG time series and the residual time series (left column). The time-frequency content of the corresponding components (right column).

3.3.6 Real PPG Data

PPG has proven to be an invaluable, non-invasive, and cost-effective tool for monitoring the activity of the autonomic nervous system. It provides valuable insights into physiological parameters such as oxygen saturation, blood pressure, and pulse rate. In this section, we present an analysis of PPG data from a data set collected by Pimentel et al. (2016) at Beth Israel Deaconess Medical Center in Boston, Massachusetts. The dataset includes physiologic signals from critically ill patients, including PPG recordings sampled at 125 Hz. A 10-second segment of PPG data from a single patient was selected for this demonstration (Goldberger et al., 2000; Pimentel et al., 2016). The data set also contains respiratory and pulse rate parameters sampled at 1 Hz. For the selected time series, the respiratory and pulse rates were recorded as 25 breaths min^{-1} and 93 beats min^{-1}, respectively.

When the PPG signal is decomposed with VMD, the third component (in the fourth row of Figure 3.8) shows a clear concentration of power at about 1.6 Hz. This component corresponds to the pulse rate, as the estimated value can be approximated by multiplying 1.6 Hz by 60 s, leading to a result that matches the recorded pulse rate. In contrast, the residual component of the decomposition has a low-band frequency content, with a local peak at 0.4 Hz, which is consistent with the respiratory rate. This PPG analysis clearly demonstrates how the decomposed components can be directly linked to specific physiological indices such as pulse rate and respiratory rate, providing a nuanced understanding of the underlying physiological processes.

3.3.7 Real fMRI Signal

In recent decades, remarkable advances in imaging technology have revolutionized our ability to image hemodynamic responses with unprecedented spatial resolution, particularly through MRI. In this study section, we focus on the resting-state

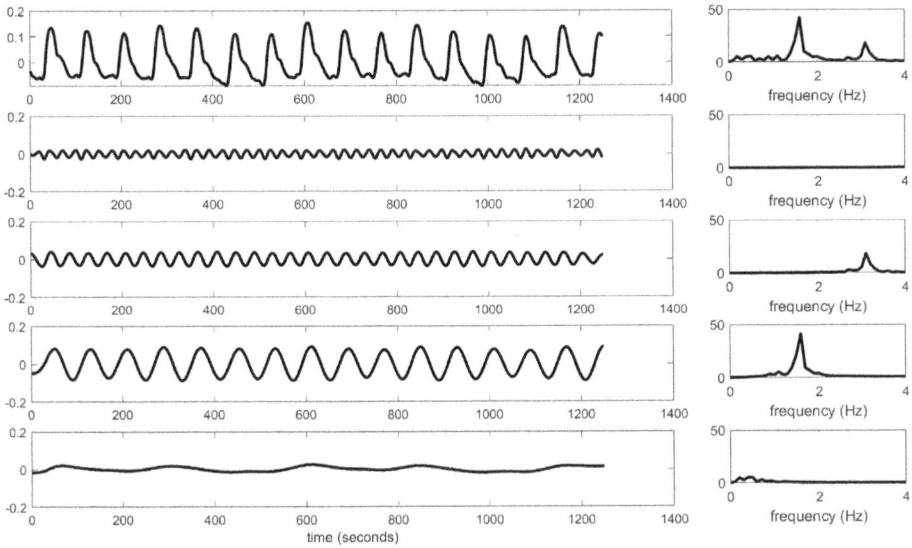

FIGURE 3.8 The raw PPG signal is shown in the first row and the frequency spectrum in the right-hand column. The results of the VMD are shown in the second, third, and fourth rows for the three components, while the last row represents the residual component.

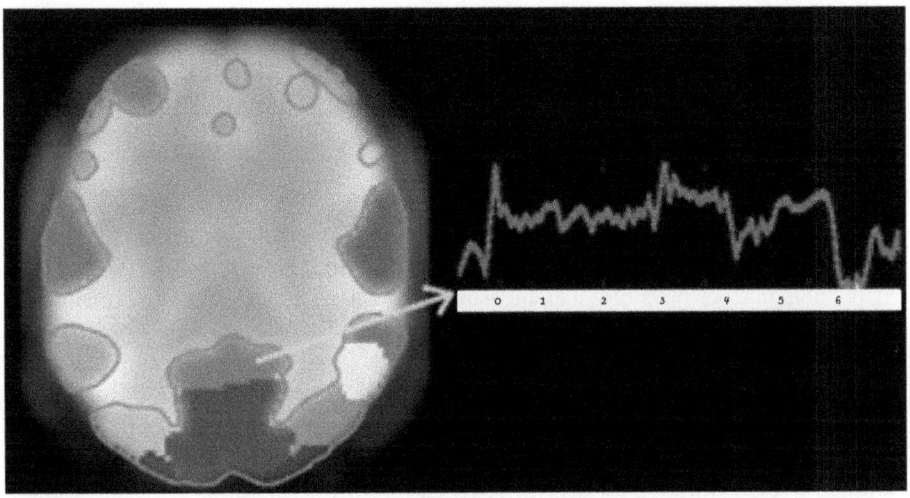

FIGURE 3.9 Time series of the BOLD signal recorded during the resting state.

posterior cingulate cortex time series that serve as input for VMD (Figure 3.9). The data shown in this part were collected from a sedentary participant who had undergone a resting-state fMRI test (Duru & Balcioglu, 2018). The spectral characteristics of the BOLD signal (blood-oxygen-level-dependent) in the resting state show a pronounced dominance of low-frequency oscillations. After the application of VMD, the resulting components exhibit non-overlapping frequency spectra, each

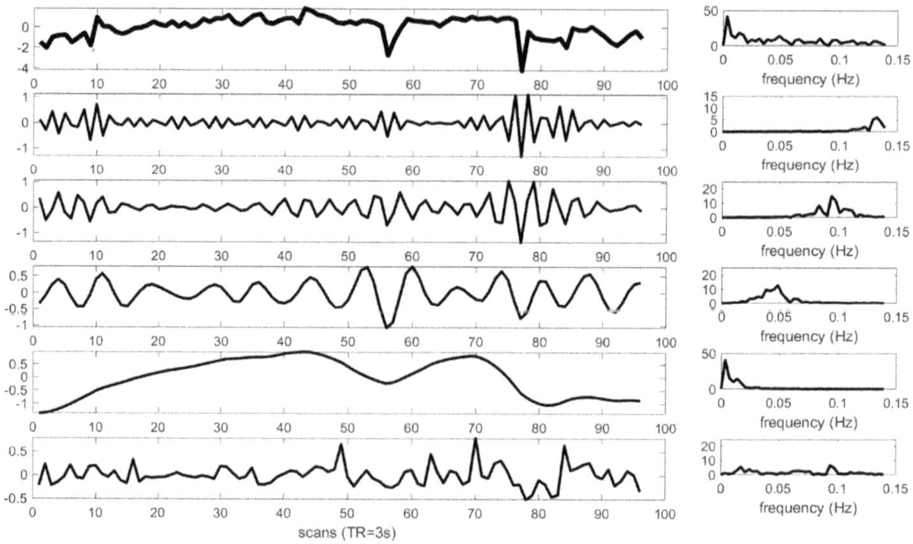

FIGURE 3.10 BOLD time series of the posterior cingulate cortex and the spectral content are shown in the first row. The VMD time components and their frequency content are shown in the second to fifth rows.

capturing different aspects of the signal. The initial complexity of the waveform likely reflects the interplay of the activity of different neural networks in the brain (Leech et al., 2012). These networks are associated with low-frequency spontaneous fluctuations in the BOLD signal, which are thought to underlie crucial aspects of brain function and connectivity (Niazy et al., 2011) (Figure 3.10).

3.4 CONCLUSION

Multiresolution analysis is essential for the investigation of specific physiological phenomena, as physiological measurements often reflect the superposition of multiple components with different time-frequency characteristics. For example, transient events or time-varying phenomena such as sudden changes in heart rate, muscle fatigue, or the components of ERPs require a detailed and adaptive analysis approach. In PPG, an IMF can represent the respiratory rate, while in EEG, the IMFs can correspond to a time series of different power bands (e.g., alpha, beta, or gamma waves). Similarly, an IMF in EMG can show the activation patterns of motor units. In addition, research has shown that functional connectivity networks exhibit pronounced non-stationarity and non-linear properties at rest (Guan et al., 2020).

In general, the process of time series decomposition aims to extract relevant information that elucidates the physiological state and plays a crucial role in noise reduction, improving diagnostic accuracy, and better understanding the underlying physiological mechanisms. In contrast to conventional methods such as STFT or WT, VMD decomposes physiological time series into subcomponents, preserving the local features of the data without relying on predefined basis functions.

This makes VMD a superior approach for analyzing transient events and time-varying phenomena, as described above. Considering the inherent complexity of physiological data, often characterized by non-stationarity and non-linearity, VMD provides an interpretable and robust framework for decomposing such signals and enables deeper insights into the dynamics of physiological systems.

Decomposing a time series facilitates a more comprehensive understanding and interpretation of the data, as it reduces the complexity of the signal by breaking it down into more straightforward and comprehensible components. Physiological time series decomposition is an important analysis technique for gaining meaningful insights from complex and often noisy biological data sets. It enables more optimal healthcare outcomes, improved treatment efficacy, and sophisticated medical research. This chapter discussed the use of EMD and VMD algorithms, with illustrations of their implementations using simulated datasets and a real data sample. The discussion focused on the decomposition of the components and examined their characteristics. The most important attribute of the decomposed signals was their distinct frequency content. The physiological and physical aspects of the data to be analyzed should be subjected to rigorous scrutiny.

REFERENCES

Al-Jumaili S, Duru AD, Ibrahim AA, Uçan ON. (2023). Investigation of epileptic seizure signatures classification in EEG using supervised machine learning algorithms. *Traitement du Signal*, 40(1): 43–54.

AlSharabi K, Salamah YB, Aljalal M, Abdurraqeeb AM, Alturki FA. (2023). EEG-based clinical decision support system for Alzheimer's disorders diagnosis using EMD and deep learning techniques. *Frontiers in Human Neuroscience*, 17: 1190203. https://doi.org/10.3389/fnhum.2023.1190203. PMID: 37719771; PMCID: PMC10501399.

Centeno-Bautista MA, Rangel-Rodriguez AH, Perez-Sanchez AV, Amezquita-Sanchez JP, Granados-Lieberman D, Valtierra-Rodriguez M. (2023). Electrocardiogram analysis by means of empirical mode decomposition-based methods and convolutional neural networks for sudden cardiac death detection. *Applied Sciences*, 13(6): 3569. https://doi.org/10.3390/app13063569

Dragomiretskiy K, Zosso D. (2014). Variational mode decomposition. *IEEE Transactions on Signal Processing*, 62(3): 531–544. https://doi.org/10.1109/TSP.2013.2288675

Dubey R, Kumar M, Upadhyay A, Pachori RB. (2022). Automated diagnosis of muscle diseases from EMG signals using empirical mode decomposition based method. *Biomedical Signal Processing and Control*, 71(Part A): 103098.

Duru AD, Ademoglu A, Demiralp T. (2009). Analysis of brain electrical topography by spatio-temporal wavelet decomposition. *Mathematical and Computer Modelling*, 49(11–12): 2224–2235. https://doi.org/10.1016/j.mcm.2008.07.017

Duru AD, Assem M. (2018). Investigating neural efficiency of elite karate athletes during a mental arithmetic task using EEG. *Cognitive Neurodynamics*, 12: 95–102. https://doi.org/10.1007/s11571-017-9464-y

Duru AD, Balcioglu TH. (2018). Functional and structural plasticity of brain in elite karate athletes. *Journal of Healthcare Engineering*, 2018: 8310975.

Eriksen T, Rehman NU. (2023). Data-driven nonstationary signal decomposition approaches: A comparative analysis. *Scientific Reports*, 13(1971): 1798. https://doi.org/10.1038/s41598-023-28390-w

Goldberger A, Amaral L, Glass L, et al. (2000). PhysioBank, PhysioToolkit, and PhysioNet: Components of a new research resource for complex physiologic signals. *Circulation*, 101(23): e215–e220.

Guan S, Jiang R, Bian H, Yuan J, Xu P, Meng C, Biswal B. (2020). The profiles of non-stationarity and non-linearity in the time series of resting-state brain networks. *Frontiers in Neuroscience*, 14, 493. https://doi.org/10.3389/fnins.2020.00493. PMID: 32595440; PMCID: PMC7300942.

Gurve D, Delisle-Rodriguez D, Romero-Laiseca M, Cardoso V, Loterio F, Bastos T, Krishnan S. (2020). Subject-specific EEG channel selection using non-negative matrix factorization for lower-limb motor imagery recognition. *Journal of Neural Engineering*, 17(2): 026029. https://doi.org/10.1088/1741-2552/ab4dba. PMID: 31614343.

Hao Y, Zhang X, Tang M, et al. (2022). EEG artifact removal using variational mode decomposition and deep learning. *Neurocomputing*, 481: 1–12.

Huang NE, Shen Z, Long SR, et al. (1998). The empirical mode decomposition and the Hilbert spectrum for nonlinear and non-stationary time series analysis. *Proceedings of the Royal Society of London. Series A: Mathematical, Physical and Engineering Sciences*, 454(1971): 903–995.

Huang Y, Yang G, Wang K, Liu H, Yin Y. (2022). Robust multi-feature collective non-negative matrix factorization for ECG biometrics. *Pattern Recognition*, 123: 108376. https://doi.org/10.1016/j.patcog.2021.108376

Kaplan AY, Fingelkurts AA, Fingelkurts AA, Borisov SV. (2005). Nonstationary nature of the brain activity as revealed by EEG/MEG: Methodological, practical and conceptual challenges. *Signal Processing*, 85(11): 2190–2212.

Karlen W, Garde A, Myers D, Scheffer C, Ansermino JM, Dumont GA. (2015). Estimation of respiratory rate from photoplethysmographic imaging videos compared to pulse oximetry. *IEEE Journal of Biomedical and Health Informatics*, 19(4): 1331–1338. https://doi.org/10.1109/JBHI.2015.2429746

Koppolu PK, Chemmangat K. (2023). Automatic selection of IMFs to denoise the sEMG signals using EMD. *Journal of Electromyography and Kinesiology*, 73: 102834.

Leech R, Braga R, Sharp DJ. (2012). Echoes of the brain within the posterior cingulate cortex. *Journal of Neuroscience*, 32(1): 215–222. https://doi.org/10.1523/JNEUROSCI.3689-11.2012

Li C, Wu Y, Lin H, Li J, Zhang F, Yang Y. (2022). ECG denoising method based on an improved VMD algorithm. *IEEE Sensors Journal*, 22(23): 22725–22733. https://doi.org/10.1109/JSEN.2022.3214239

Li C, Zhou W, Liu G, et al. (2021a). Seizure onset detection using empirical mode decomposition and common spatial pattern. *IEEE Transactions on Neural Systems and Rehabilitation Engineering*, 29: 458–467. https://doi.org/10.1109/TNSRE.2021.3055276

Li J, Liu C, Xu C, et al. (2021b). Empirical mode decomposition based functional MRI connectivity analysis for Alzheimer's disease identification. *Neurocomputing*, 438: 229–240.

Li S, Zhou W, Yuan Q, Geng S, Cai D. (2013). Feature extraction and recognition of ictal EEG using EMD and SVM. *Computers in Biology and Medicine*, 43(7): 807–816.

Li Y, Luo J-H, Dai Q-Y, Eshraghian JK, Ling BW-K, Zheng C-Y, Wang X-L. (2023). A deep learning approach to cardiovascular disease classification using empirical mode decomposition for ECG feature extraction. *Biomedical Signal Processing and Control*, 79: 2.

Lin Z, Li S, Zhang X, et al. (2022). Surface electromyography feature extraction based on variational mode decomposition and convolutional neural network for hand gesture recognition. *IEEE Access*, 10: 7976–7987.

Lin Z, Xie Y, Xiong P, et al. (2023). EEG denoising based on multiscale variational mode decomposition and convolutional neural network. *Biomedical Signal Processing and Control*, 70: 103004.

Liu W, Zheng N, Li X. (2004). Nonnegative matrix factorization for EEG signal classification. In: Yin FL, Wang J, Guo C. (eds) *Advances in Neural Networks*. Springer, Berlin, Heidelberg. https://doi.org/10.1007/978-3-540-28648-6_75

Liu X, Ma X, Wang Y, et al. (2022). EEG denoising method based on adaptive variational mode decomposition and feature fusion. *IEEE Access*, 10: 6510–6522.

Long Z, Xu Y, Zou W, et al. (2024). Extended nonnegative matrix factorization for dynamic functional connectivity analysis of fMRI data. *Cognitive Neurodynamics*, 18: 1651–1669. https://doi.org/10.1007/s11571-023-10039-z

Lu L, Liu Y, Wang Z, et al. (2021). A robust ECG denoising algorithm based on empirical mode decomposition and dual-tree complex wavelet transform. *IEEE Access*, 9: 18615–18625.

Ma S, Lv B, Lin C, Sheng X, Zhu X. (2021). EMG signal filtering based on variational mode decomposition and sub-band thresholding. *IEEE Journal of Biomedical and Health Informatics*, 25(1): 47–58. https://doi.org/10.1109/JBHI.2020.2987528

Mallat S. (1999). *A Wavelet Tour of Signal Processing* (2nd ed.). Academic Press, New York.

Motin MA, Karmakar CK, Palaniswami M. (2019). Selection of empirical mode decomposition techniques for extracting breathing rate from PPG. *IEEE Signal Processing Letters*, 26(4): 592–596. https://doi.org/10.1109/LSP.2019.2900923

Nagineni S, Taran S, Polat K. (2024). Variational mode decomposition based entropy features for classification of myopathy, neuropathy, and normal EMG signals. *Data Analytics for Intelligent Systems*, 2024: 2053–2563.

Nazari M, Sakhaei SM. (2018). Variational mode extraction: A new efficient method to derive respiratory signals from ECG. *IEEE Journal of Biomedical and Health Informatics*, 22(4): 1059–1067. https://doi.org/10.1109/JBHI.2017.2734074

Niazy RK, Xie J, Miller K, Beckmann CF, Smith SM. (2011). Spectral characteristics of resting state networks. *Progress in Brain Research*, 193: 259–276.

Oh S, Choi Y-S. (2024). Detection of ventricular fibrillation using ensemble empirical mode decomposition of ECG signals. *Electronics*, 13(4): 695. https://doi.org/10.3390/electronics13040695

Pachori D, Tripathy RK, Jain TK. (2024). Detection of atrial fibrillation from PPG sensor data using variational mode decomposition. *IEEE Sensors Letters*, 8(3): 1–4. https://doi.org/10.1109/LSENS.2024.3358589

Pimentel MAF, Johnson AEW, Charlton PH, et al. (2016). Towards a robust estimation of respiratory rate from pulse oximeters. *IEEE Transactions on Biomedical Engineering*, 64(8): 1914–1923. https://doi.org/10.1109/TBME.2016.2613124

Pollreisz D, TaheriNejad N. (2022). Detection and removal of motion artifacts in PPG signals. *Mobile Networks and Applications*, 27: 728–738. https://doi.org/10.1007/s11036-019-01323-6

Qi P, Xu H, Zhang H, Tong J, Xia S. (2023). Residual neural networks based on empirical mode decomposition for mitral regurgitation prediction. *Biomedical Signal Processing and Control*, 86(Part B): 105265.

Sajed S, Movahedi A, Tafreshi R. (2021). EEG signal denoising using variational mode decomposition: A comparative study. *Biomedical Signal Processing and Control*, 69: 102922.

Salankar N, Mishra P, Garg L. (2021). Emotion recognition from EEG signals using empirical mode decomposition and second-order difference plot. *Biomedical Signal Processing and Control*, 65: 102389.

Sharma H, Sharma KK. (2020). Sleep apnea detection from ECG using variational mode decomposition. *Biomedical Physics & Engineering Express*, 6(1): 015026.

Sharma H. (2019). Heart rate extraction from PPG signals using variational mode decomposition. *Biocybernetics and Biomedical Engineering*, 39(1): 75–86.

Shen J, Zhang Y, Liang H, et al. (2023). Exploring the intrinsic features of EEG signals via empirical mode decomposition for depression recognition. *IEEE Transactions on Neural Systems and Rehabilitation Engineering*, 31: 356–365.

Shokrollahi M, Krishnan S. (2013). Non-negative matrix factorization and sparse representation for sleep signal classification. *Annual International Conference of the IEEE Engineering in Medicine and Biology Society*, 2013: 4318–4321. https://doi.org/10.1109/EMBC.2013.6610501. PMID: 24110688.

Siuly S, Khare SK, Bajaj V, Wang H, Zhang Y. (2020). A computerized method for automatic detection of schizophrenia using EEG signals. *IEEE Transactions on Neural Systems and Rehabilitation Engineering*, 28(11): 2390–2400. https://doi.org/10.1109/TNSRE.2020.3022715

Sun Y, Liu L, Zhang Y, et al. (2022). EEG seizure detection and brain state analysis using multi-objective variational mode decomposition. *IEEE Transactions on Biomedical Engineering*, 69: 720–729.

Thayer JF, Ahs F, Fredrikson M, et al. (2012). A meta-analysis of heart rate variability and neuroimaging studies: Implications for heart rate variability as a marker of stress and health. *Neuroscience & Biobehavioral Reviews*, 36(2): 747–756.

To XV, Vegh V, Nasrallah FA. (2022). Towards data-driven group inferences of resting-state fMRI data in rodents: Comparison of group ICA, GIG-ICA, and IVA-GL. *Journal of Neuroscience Methods*, 366: 109411.

Wang Y, Yang C, Tian Y, et al. (2017). Variational mode decomposition for denoising fMRI data. *NeuroImage*, 147: 964–975.

Xu W, Gao S, Jin Y, et al. (2021). EMD-based fMRI denoising with adaptive integration of multiple time scales. *NeuroImage*, 237: 118079.

Yang X, Guo J, Li Z, et al. (2023). Multiscale variational mode decomposition and recurrent neural network for electromyography-based gesture recognition. *IEEE Transactions on Neural Systems and Rehabilitation Engineering*, 31: 361–371.

Yu Y, Guo X, Wang P, et al. (2021). Graph theoretical analysis of resting-state fMRI using EMD-based network approach. *Neuroinformatics*, 19: 479–493.

Zhai X, Wu T, Su Y, et al. (2023). A novel approach to extract muscle activation patterns from high-density surface electromyography signals using variational mode decomposition. *Journal of Electromyography and Kinesiology*, 64: 102595.

Zhang L, Wei D, Lu G, et al. (2022). A novel ECG signal denoising algorithm based on empirical mode decomposition and deep learning. *Computational and Mathematical Methods in Medicine*, 2022: 1–10.

Zhang X, Cheng J, Chen Z, et al. (2021). Variational mode decomposition of EEG signals for different brain states in cognitive tasks. *Neuroscience Letters*, 761: 136078.

Zheng H, Li X. (2024). An EEG-based framework of EMD and CNN for arousal and valence recognition. *Research on Biomedical Engineering*, 40: 387–395. https://doi.org/10.1007/s42600-024-00351-w

Zheng W, Chai C, Tian Y, et al. (2020). A comparative study on fMRI denoising methods: EMD versus wavelet transform. *Frontiers in Neuroscience*, 14: 145.

Zhonghua W, Huang NE. (2004). A study of the characteristics of white noise using the empirical mode decomposition method. *Proceedings of the Royal Society of London. Series A: Mathematical, Physical and Engineering Sciences*, 460(2046): 1597–1611.

Zhou F, Wu S, Lin X, et al. (2022). Multiscale empirical mode decomposition for resting-state fMRI data decomposition and functional connectivity mapping. *Neuroinformatics*, 20: 61–74.

Zuo C, Jiang Y, Luo H, et al. (2020). Variational mode decomposition and deep learning for brain functional connectivity analysis from fMRI data. *IEEE Transactions on Neural Networks and Learning Systems*, 31(4): 1298–1309.

Section III

Applications in Neurological Signal Processing

4 Denoising of Single-Channel EEG Signals Using Wavelet Transform with the Krawtchouk Functions

Natalia Filimonova, Mykola Makarchuk, and Elfriede Friedmann

4.1 INTRODUCTION

The EEG captures brain activity during various functional tests and is an essential objective tool for evaluating and treating physiological, psychological, and pathological conditions in humans. In an ideal scenario, EEG signals reflect changes in the synchrony of elementary cortical potentials, whether originating from postsynaptic activity or other sources. However, experimental EEG is often distorted by various types of artifacts, which are generated by non-neural sources such as muscle movements, heartbeats, eye movements, and respiration. They can be classified according to their source of generation [1–3]. Epochs contaminated by artifacts can be manually removed, but this involves considerable manual effort and results in substantial data loss. Hence, methods have been developed to automate this process and remove various types of artifacts and noise. For example, multiple techniques have been proposed, including the Fourier transform, time-frequency analysis, wavelet transform (WT), neural networks (NNs), regression, adaptive filtering, principal component analysis (PCA), independent component analysis (ICA), and others [4–10]. It is essential to note that, by definition, PCA inherently assumes the orthogonality of the data components—a condition that is often not met in practice. The algebraic relationship between the source vectors is determined by the spatial location and orientation of each source, as well as, to some extent, the conductivity properties of the head. The orientation of certain ocular generators (e.g., blinks) can closely align with the orientation of frontal EEG generators [11]. Additionally, PCA involves the decomposition of a covariance or correlation matrix into its eigenvectors. The calculation of this matrix assumes that each of the analyzed signals has a Gaussian distribution, which does not coincide with a real EEG recording. The algorithm also uses signal averaging, which implies the assumption of the ergodic property. Ergodicity means that the time average of the rhythm is equal to the space (or ensemble) average,

DOI: 10.1201/9781003611448-7

which means that one long-lasting signal contains the same statistical information about the rhythm as several short-lasting signals. Ensuring ergodicity in EEG signals remains a significant challenge.

ICA is usually considered an extension of PCA. It is known that there are over 100 different algorithms for ICA, with the main assumption being that the source components must be approximately independent and follow non-Gaussian distributions. The accuracy of the method diminishes when the sources of EEG activity are not statistically independent. For example, the electrooculography (EOG) signal for eye blinks and movements of the eyeballs originates from both cerebral and ocular sources. Furthermore, the requirement for the space-fixed position of EEG activity sources practically doesn't take the time structure of EEG into account. Additionally, most ICA algorithms are not automated, as they require visual inspection to select their independent components for correction. Moreover, ICA correction has been shown to distort signal power within the frequency range of 5–20 Hz [12]. These algorithms may overestimate the spectrum of artifacts and thus cause spectral distortion in EEG signals [12]. Moreover, such methods require multi-channel data and long data epochs to produce reliable results [13]. ICA does not consistently identify the same components across different experiments, making it difficult to determine which components are relevant, as the method extracts them blindly and in a different manner each time.

The quality of algorithms based on blind source separation, such as ICA, is significantly dependent on the number of electrodes, and they become more effective at eliminating artifacts as the number of EEG channels increases. If there is only one channel, ICA cannot be applied to remove artifacts.

Multi-channel EEG data is highly effective in clinical diagnostics, prosthetics, and wheelchair control, among other applications. However, in recent years, single-channel EEG devices have gained popularity due to their ease of measurement and portability. Currently, devices are being actively developed for monitoring sleep quality and diagnosing related sleep disorders, recognizing emotional and mental states, monitoring anesthesia depth, detecting driver drowsiness, providing home medicine, and implementing brain–computer interface applications, among others, which typically utilize only one EEG electrode. Therefore, there is a high demand for EEG artifact removal algorithms that can operate effectively using only a single EEG channel [14].

In recent years, several methods for analyzing single-channel EEG recordings have been proposed. For example, methods based on a hidden Markov model [15], a multi-branch convolutional NN [16], and a hybrid attention NN [17] have been used for single-channel sleep EEG analysis. For studying working memory load, Maimon et al. [18] proposed the convolution–decimation operators with a simple Haar wavelet, while Jiao et al. [19] used a vector machine for the identification of multiple types of mental stress based on single-channel EEG.

Conducting a valid analysis and classification of human states based on single-channel EEG requires prior cleaning from artifacts. Several methods have been proposed to remove specific types of artifacts from single-channel EEG [20–24].

The effectiveness of the WT in analyzing time-frequency patterns of signals across multiple scales has made it a widely used technique for noise removal, particularly in the analysis of physiological signals. WT decomposes a signal into several sets of coefficients representing the time-frequency patterns of the signal. Signal denoising is usually achieved by setting thresholds for wavelet coefficients, eliminating those corresponding to noise. Due to their flexibility, efficiency, and robustness, wavelet methods are widely used to remove EEG artifacts. Compared to blind source separation, the wavelet method has a low computational cost, can be applied online, and does not require multiple channels [25]. In previous studies [25–30], proposed methods for removing specific types of artifacts from single-channel EEG are based on wavelet analysis. However, selecting an appropriate mother wavelet for different types of artifacts remains an open challenge, particularly when artifacts overlap or exhibit sudden amplitude changes, making it difficult to minimize information loss [25].

Thus, the task is to develop a method that accounts for the non-stationary nature of EEG signals and is universal for different types of artifacts and their possible overlap when cleaning a single-channel EEG.

4.2 METHOD

The single-channel EEG artifact removal method proposed in this paper is based on identifying the time-frequency features of impulse components in EEG signals [31]. We developed a method that integrates a mathematical model of pattern recognition invariance by the visual system into wavelet analysis. This approach enables the simultaneous and accurate determination of both the time coordinate of local impulses in the EEG signal and their frequency characteristics. We refer to our approach as wavelet analysis-based filtering for localization.

As a model of the response of the receptive fields of the visual system and, accordingly, of the mother wavelet, we used the Krawtchouk functions [32–42] (Figure 4.1).

We performed wavelet transformations following a commonly accepted scheme, as illustrated in Figure 4.2. Notably, unlike the traditional approach, our method begins the WT with the largest window.

A distinct feature of our method is that, within each WT window, we do not compute a single convolution of the input signal with the corresponding mother wavelet; instead, we convolve the signal with a set of mother wavelets parameterized by both the shift within the given window and an asymmetry parameter derived from the

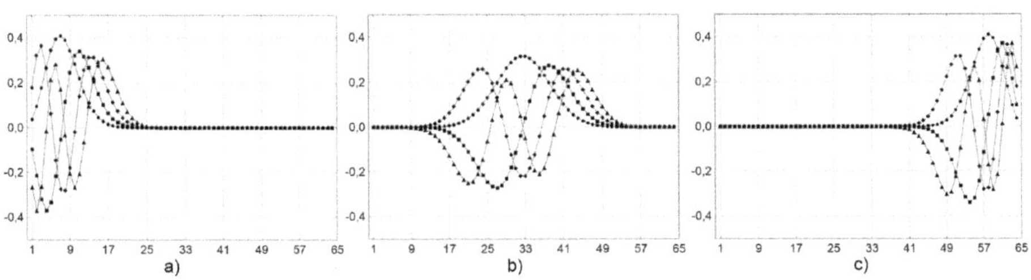

FIGURE 4.1 The first four Krawtchouk functions for values (a) $p = 0.1$, (b) $p = 0.5$, (c) $p = 0.9$; $i = 1, 64$.

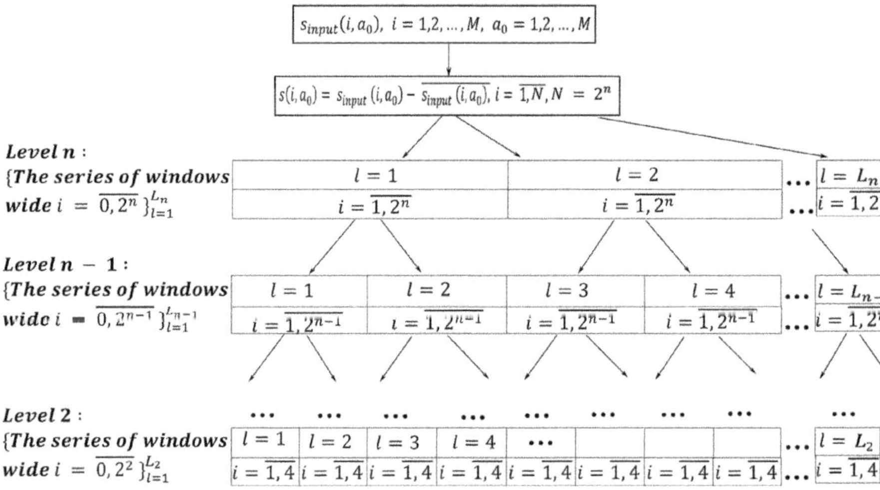

FIGURE 4.2 Signal wavelet transformation scheme.

definition of the Krawtchouk functions. We then identify the maximum decomposition coefficient as a function of these shift and asymmetry coefficients.

The maximum values of the decomposition coefficients define the invariant, essential features of the corresponding impulse components. At the same time, we separate the information about the shape of the impulse components from their temporal localization.

We analyzed EEG data recorded at a discrete frequency of 500 Hz with filters for high and low frequencies, respectively, of 0.5 and 200 Hz. We analyze EEG data recorded at a sampling rate of 500 Hz, using high-pass and low-pass filters set at 0.5 and 200 Hz, respectively.

The input signal may contain an impulse component of an artifact whose time of appearance is unknown to us—this is a hidden parameter $a_0\left(a_0 \in \{1,2,\ldots,M\}\right)$ that we will find later. So as an input signal, we have $s_{\text{input}}\left(i, a_0\right), i = 1,2,\ldots,M, \quad a_0 = 1,2,\ldots,M$, M being the size of the signal.

The proposed method can be divided into the following steps, as outlined in the signal processing algorithm presented in Figure 4.4 [31].

Step 1.

We define the scale of the WT with window widths of N points, $N = 2^n, \quad n = 10,9,\ldots,2$. According to the parameters of the EEG recording, we begin from window widths of 1,024 points.

Step 2.

We create a new zero level for the input signal in the largest window by shifting the signal to its average. The resulting signal is: $s(i,a_0) = s_{\text{input}}(i,a_0) - \overline{s_{\text{input}}(i,a_0)}, \quad i = \overline{1,N}$, where $\overline{s_{\text{input}}(i,a_0)} = \sum s_{\text{input}}(i,a_0)/N$ is the average of the signal for the biggest window. We start from the $N = 2^{10} = 1{,}024$ points.

Step 3.

We calculate the first four Krawtchouk functions for the windows $N = 2^n, n = 10, 9, \ldots, 2$ and for $p = 0.1, 0.2, \ldots, 0.9$, since we use 9 asymmetry coefficients of the Krawtchouk functions for a better approximation of non-symmetrical impulse components. Because the Krawtchouk functions $F_n^{(p)}(i,N)$ are symmetric, we only need to calculate the Krawtchouk functions by their recursive rule for p = 0.1, 0.2, 0.3, 0.4 and obtain all the others according to this symmetry property

$$F_n^{(p)}(i,N) = (-1)^n F_n^{(p-1)}(N-i,N).$$

At p = 0.5 the Krawtchouk functions are symmetric with respect to the middle of the interval; therefore, they are calculated only for $i = 0, 1, \ldots, N/2$. This reduces the accumulation of error in the recurrent calculation of subsequent function values. Fast algorithms for the computation of the Krawtchouk functions can be found in previous studies [32–42]. In addition, we can calculate the function values in advance and save them in the corresponding file. In each window, we use the first four Krawtchouk functions, as numerical experiments on various signals have shown that this is the optimal number. The use of a smaller number of functions does not allow for a sufficient analysis of the shape of complex impulse components, and a larger number of functions captures the frequencies of the next level. Also, the size of the last smallest window is four points. Thus, the four Krawtchouk functions in the final window form a complete set of basis functions, enabling an exhaustive signal analysis and minimizing information loss.

Step 4.

We compute the generalized spectral coefficients with respect to the set of shifted Krawtchouk basis functions:

$$c_k^{(p)}(a,a_0) = s(i,a_0) * F_k^{(p)}((i-a)\bmod N), \quad a = \overline{1,N};\ k = 1,2,3,4; \quad p = 0.1, 0.2, \ldots, 0.9 \tag{4.1}$$

by applying the convolution theorem [43],

$$c_k^{(p)}(a,a_0) = s(i,a_0) * F_k^{(p)}(i-a) = F^{-1}[F(s(i,a_0)) \cdot F(F_k^{(p)}(i-a))], \tag{4.2}$$

where F is the Fourier transformation, $i, a = \overline{1,N}$; $k = 1,2,3,4$; $p = 0.1, 0.2, \ldots, 0.9$. We account for the fact that the boundaries of the observation window have been joined to form a continuous loop.

Step 5.

We calculate the energy functional:

$$W^{(p)}(a,a_0) = \sum_{k=1,2,3,4} c_k^{(p)}(a,a_0)^2. \tag{4.3}$$

Step 6.

We find the maximum of the energy functional:

$$\max_{p=0.1,0.2,\ldots,0.9;a=\overline{1,N}} W^{(p)}(a,a_0) = W_{\max}^{(p_0)}(a_0,a_0). \tag{4.4}$$

In this step, we find the temporal localization of the impulse component in the signal, along with the asymmetry coefficients of the Krawtchouk basis functions that best match the impulse's asymmetry. Convolutional neural networks (CNNs) are widely and effectively used for EEG analysis and classification. Unlike traditional CNNs, our approach convolves the input signal with a set of basis functions parameterized by shift and asymmetry, rather than a single fixed basis. At this step, we identify the unique basis function that produces the impulse response with the highest energy.

Step 7.

Now we determine whether the extremum found in the previous step has to be interpreted as an artifact. If $W_{\max}^{(p_0)}(a_0,a_0) \geq T_n$, where T_n is the threshold value of the energy functional for the window of level n, then we define this part of the signal as an artifact and proceed to the next step. The threshold values T_n are found in advance using a training sequence. This training algorithm for classifiers finds the minimum from the set of energy functionals of the artifacts identified by experts:

$$\min_{d=\overline{1,D}} W_{\max}^{(p_0)}(a_0,a_0) = T_n,\ d = \overline{1,D}. \tag{4.5}$$

In the case $W_{\max}^{(p_0)}(a_0,a_0) < T_n$, we proceed with signal processing within the next-level windows and set the artifact value to 0: $s_{ar,l}^{n1}(i) = 0, i = \overline{1,2^{n1}}$.

Step 8.

We determine the time of the artifact within window number l at level number n as follows: $a_l^{n_11} = a_0, n_1 = \overline{n,2}; l = \overline{1,L_n}$, where $L_n = N/2^n$ is the number of windows at level l. With this, we find the amplitudes and frequencies of the artifact in the corresponding window:

$$\left\{c_k^{(p_0)}(a_0,a_0)\right\}_{k=1}^{4}, n_1 = \overline{n,2}. \tag{4.6}$$

Step 9.

If $W_{\max}^{(p_0)}(a_0,a_0) \geq T_n$, then we restore the artifact of the EEG $s_{\text{ar},l}^{n_1}$:

$$s_{\text{ar},l}^{n_1}(i) = \sum_{k=1,2,3,4} c_k^{(p_0)}(a_0,a_0)\cdot F_k^{(p_0)}((i-a_0)\bmod N),\quad i = \overline{1,2^{n1}}. \tag{4.7}$$

Step 10.

We calculate the residual signal $\left\{s_{r,l}^{n_1}(i),\quad i = \overline{1,2^{n_1}}\right\}$ as the difference between the input signal at the previous level and the detected artifact $\left\{s_{\text{ar},l}^{n_1}(i),\quad i = \overline{1,2^{n_1}}\right\}$:

$$s_{r,l}^{n_1}(i) = s_{r,l}^{n_1+1}(i) - s_{\text{ar},l}^{n_1}(i),\quad i = \overline{1,2^{n_1}}. \tag{4.8}$$

Note that for the first largest window, $s_{r,l}^{n_1+1}(i) = s(i)$.

Thus, in each window of the WT, we decompose the signal into two components $s_{\text{ar},l}^{n_1}$ and $s_{r,l}^{n_1}$ (Figure 4.3).

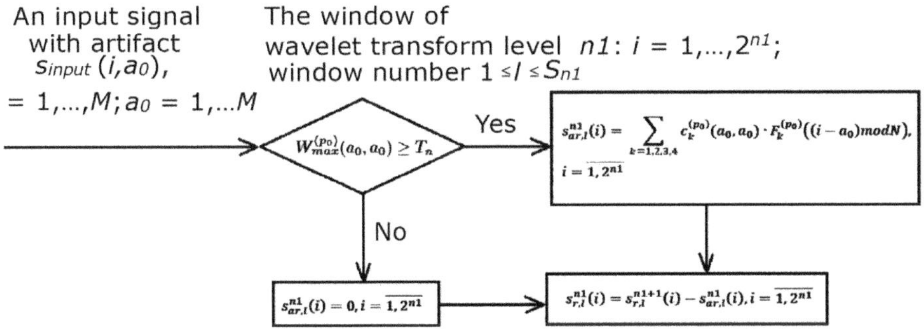

FIGURE 4.3 Decomposition of the EEG in each window of the wavelet transform.

Thus, we consider oscillations at low frequencies as drift of the isoline for the oscillations of high frequencies. In addition, this eliminates the repeated finding of the frequency characteristics of the same impulse component in subsequent windows of the wavelet analysis. This approach also represents a key distinguishing feature of our algorithm compared to conventional WT methods.

Step 11.

We proceed to the next window $l = l + 1$ at level n of the wavelet analysis. If the remaining length of the signal is less than the window width, we pad the signal with zeros: if $M < (L_n - 1) \cdot 2^n$, then $s(i) = 0, i = \overline{M, L_n \cdot 2^n}$. In case we have reached the end of the sequence of windows of a given level $l > L_n$, we go to the next level of windows $n = n - 1$. In the case $n < 2$, the algorithm completes successfully.

Step 12.

As a result we obtain the decomposition of the initial EEG signal $s(i), i = \overline{1, M}$ into artifacts $s_{ar}(i), i = \overline{1, M}$ and a clean signal $s_{cl}(i), i = \overline{1, M}$ (Figure 4.4):

$$s_r(i), i = \overline{1, M}$$

$$s^n_{ar} = s^n_{ar,1}\left(i = 1, 2, \ldots, 2^n\right); \tag{4.9}$$

$$s^{n-1}_{ar} = \left\{s^{n-1}_{ar,1}\left(i = 1, 2, \ldots, 2^{n-1}\right)\right\}\left\{s^{n-1}_{ar,2}\left(i = 2^{n-1} + 1, 2^{n-1} + 2, \ldots, 2^n\right)\right\}; \tag{4.10}$$

$$s^{n-2}_{ar} = \left\{s^{n-2}_{ar,1}\left(i = 1, 2, \ldots, 2^{n-2}\right)\right\}\left\{s^{n-2}_{ar,2}\left(i = 2^{n-2} + 1, 2^{n-2} + 2, \ldots, 2 * 2^{n-2}\right)\right\}$$

$$\left\{s^{n-2}_{ar,3}\left(i = 2 * 2^{n-2} + 1, 2 * 2^{n-1} + 2, \ldots, 3 * 2^{n-2}\right)\right\}$$

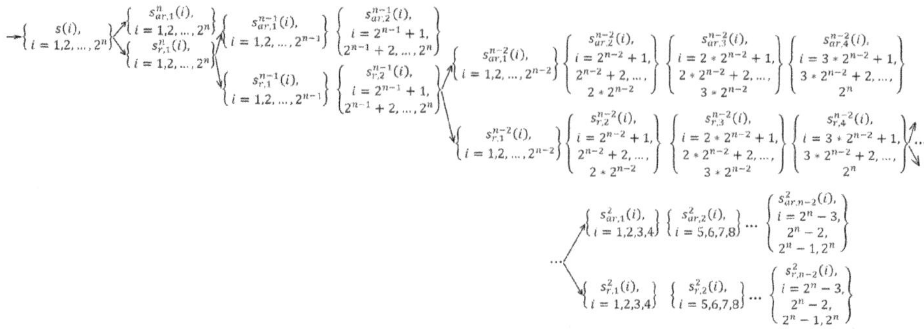

FIGURE 4.4 Decomposition of the EEG into artifacts $s_{\text{ar}}(i)$, $i=\overline{1,M}$ and residual signal.

$$\left\{s_{\text{ar},4}^{n-2}\left(i=3*2^{n-2}+1,3*2^{n-2}+2,\ldots,2^{n}\right)\right\}; \tag{4.11}$$

...

$$s_{\text{ar}}^{2}=\left\{s_{\text{ar},1}^{2}(i=1,2,3,4)\right\}\ \left\{s_{ar,2}^{2}(i=5,6,7,8)\right\}\ldots$$

$$\left\{s_{\text{ar},n-2}^{2}\left(i=2^{n}-3,2^{n}-2,2^{n}-1,2^{n}\right)\right\}; \tag{4.12}$$

$$s_{\text{cl}}=\left\{s_{r,1}^{2}(i=1,2,3,4)\right\}\ \left\{s_{r,2}^{2}(i=5,6,7,8)\right\}\ldots$$

$$\left\{s_{r,n-2}^{2}\left(i=2^{n}-3,2^{n}-2,2^{n}-1,2^{k}\right)\right\}. \tag{4.13}$$

4.3 EXAMPLE

We demonstrate the effectiveness of our wavelet localization-based filtering algorithm using an example of an EEG recording in lead Fp2, which contains a mixture of muscle and blink artifacts. The EEG recording consists of 1,024 data points, corresponding to a duration of 2.048 s, as the sampling rate was 500 Hz. To provide a more detailed view of the analysis results, the data were divided into two segments, which are shown in Figures 4.5 and 4.6.

In Figure 4.5a, we present the original EEG recording, $s(i), i=1,2,\ldots,1{,}024$, along with the first artifact component $s_{\text{ar}}^{10}=s_{\text{ar},1}^{10}(i=1,2,\ldots,1{,}024)$ identified in the largest window $N=2^{10}=1{,}024$ at the initial level of wavelet analysis $l=10$. Figure 4.5b shows the artifact $s_{\text{ar}}^{10}(i), i=\overline{1,1024}$ at level 10, which we calculated according to equation (4.7). Figure 4.5f shows the residual signal at this level $s_{r,10}^{10}(i)=s(i)-s_{\text{ar}}^{10}(i)$, $i=\overline{1,1024}$ according to equation (4.8). In Figure 4.5c, we display the combined artifact components $s_{\text{ar},1}^{9}$ and $s_{\text{ar},2}^{9}$, identified at the next level

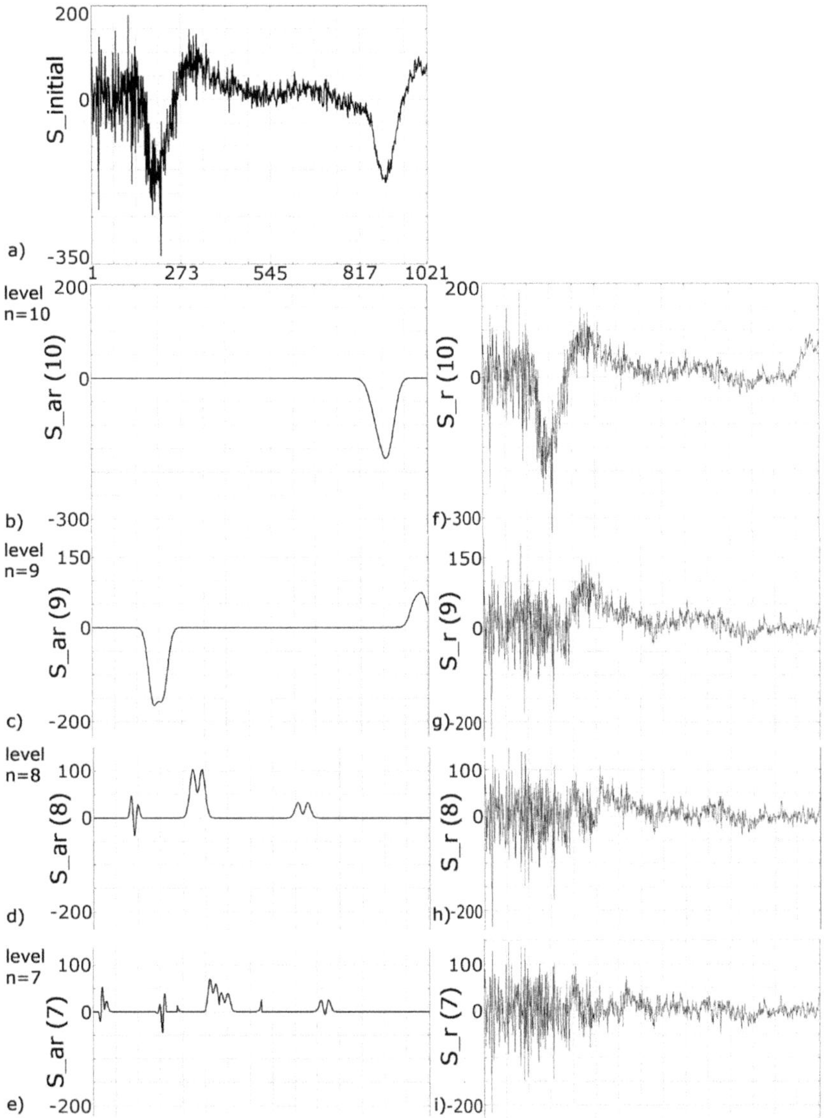

FIGURE 4.5 Consecutive decomposition of the EEG signal from the Fp2 lead, containing mixed muscle artifacts and blinks, at scales 10–7.

$l = 9$, $s_{\text{ar}}^{9}, (i = 1, 2, \dots, 1{,}024)$. Figure 4.5g presents the corresponding residual signal $s_{r,9}^{9}(i) = s_{r,10}^{10}(i) - s_{\text{ar}}^{9}(i), \quad i = \overline{1,1024}$. Figures 4.5d,e and 4.6a–e illustrate the artifact components at their respective wavelet levels, while Figures 4.5h,i and 4.6f–j display the associated residual EEG signals. At the final level, we obtain the clean EEG signal, which is equal to the residual signal $s_{\text{cl}}(i) = s_{r,2}^{2}(i), \quad i = \overline{1,1024}$ of that level $l = 2$ according to equation 4.13. Figure 4.6k shows the total artifact signal,

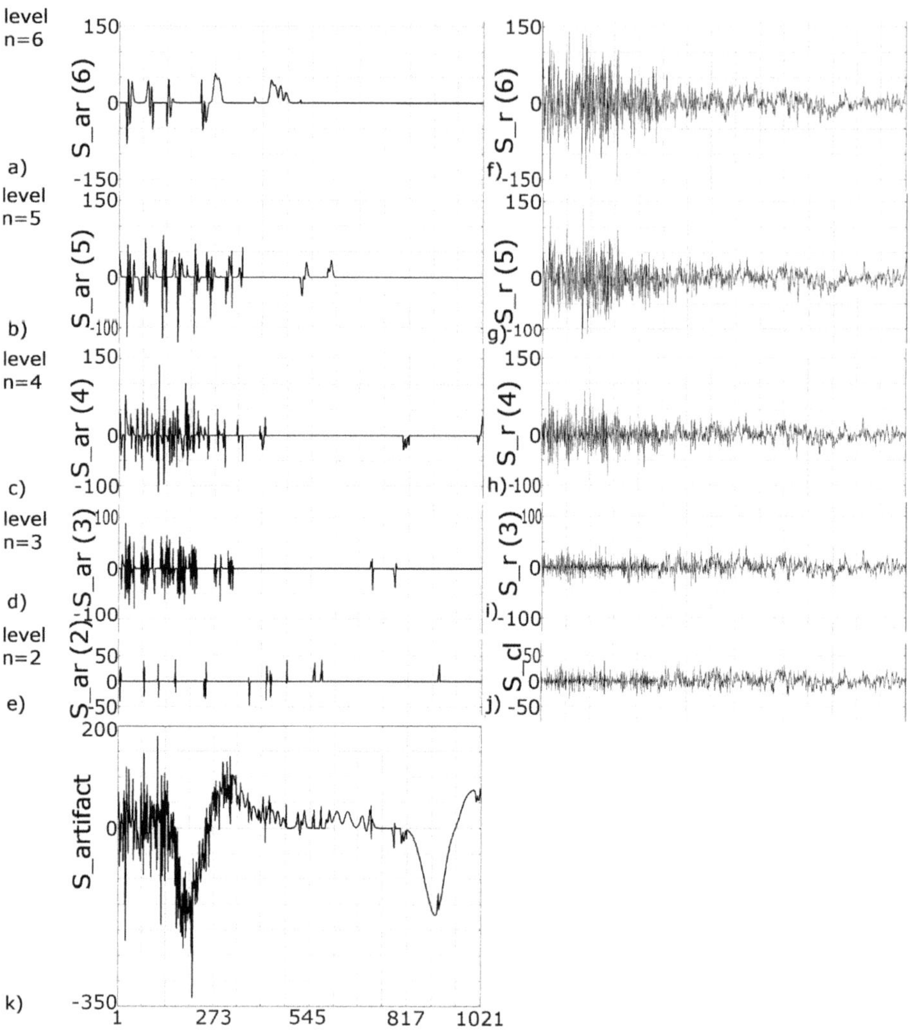

FIGURE 4.6 Consecutive decomposition of the EEG signal from the Fp2 lead, containing mixed muscle artifacts and blinks, at scales 6–2.

calculated as the arithmetic sum of all artifact components across the levels. Thus, we have decomposed the original EEG signal (Figure 4.5a) into two parts: the artifact signal (Figure 4.6k) and the clean EEG signal (Figure 4.6j). The original signal is reconstructed as the sum of these two components.

We used the following thresholding system to identify artifacts: $T_{10} = 60{,}000.0$; $T_9 = 50{,}000.0$; $T_8 = 40{,}000.0$; $T_7 = 20{,}000.0$; $T_6 = 10{,}000.0$; $T_5 = 8{,}000.0$; $T_4 = 7{,}000.0$; $T_3 = 5{,}000.0$; $T_2 = 3{,}000.0$. These threshold values are universal for all artifact types, which we will present next. The only exception concerns artifacts related to the cardiac component, which will be described in detail in Section 4.4.7: Cardiac.

4.4 RESULTS

The proposed method was evaluated using real EEG data collected from 23 healthy volunteers (10 women), aged 19–22 years, with no reported health issues, who were students at Taras Shevchenko National University of Kyiv, Ukraine.

4.4.1 Blinking

One of the most significant artifacts affecting EEG recordings is ocular artifacts, which cause substantial signal distortions. Ocular artifacts, caused by eye movements and blinks, can propagate across the scalp and be detected in EEG recordings. Ocular artifacts can notably diminish the effectiveness of methods such as event-related potential method, among others. For example, the analysis of neurocognitive processes in face recognition training based on single-channel recording of the N250 event-related potential is sensitive to ocular artifacts [44]. Such eye signals can be recorded using an EOG. Hence, bidirectional interference will introduce a removal error when we remove EOG artifacts from EEG [1,3,9–12,44–47]. This bidirectional interference is most clearly manifested in the regression method, since the relationship between EEG and one or more EEG channels is determined by calculating propagation coefficients or transmission coefficients. By estimating the proportion of EOG to EEG, we can remove EOG artifacts; however, this method has a major drawback [47]. Subtracting a linear combination of the recorded EOG from the EEG may not only remove ocular artifacts but also interesting cerebral activity and influence the accuracy of diagnostics [12]. Special methods have been developed to remove ocular artifacts from single-channel EEG [48–50]. We demonstrate the effective removal of ocular artifacts from single-channel EEG using our wavelet-based localization filtering method, without the need for additional tuning (Figures 4.7–4.9). We used the same thresholding system as we described in the example.

4.4.2 Lateral Eye Movements Artifact

Eye movement artifacts arise from changes in the orientation of the retinal and corneal dipole, and blink artifacts are caused by ocular conduction due to changes in corneal contact with the eyelid [51]. In addition, due to the effect of bulk conduction, both ocular artifacts and EEG activity propagate to the surface of the head and are recorded by electrodes [4]. We demonstrate effective removal of eye movement artifacts from single-channel EEG using our wavelet-based filtering method that enables precise localization without requiring additional tuning (Figures 4.10–4.12). We employed the same thresholding system as described in the example.

4.4.3 Muscle Contractions

Contamination of EEG data by muscle activity is a well-recognized challenging problem because it occurs due to different types of muscle groups [52]. These artifacts can be caused by any muscles that are close to the signal recording sites. The degree of muscle contraction and stretching determines the amplitude and shape of

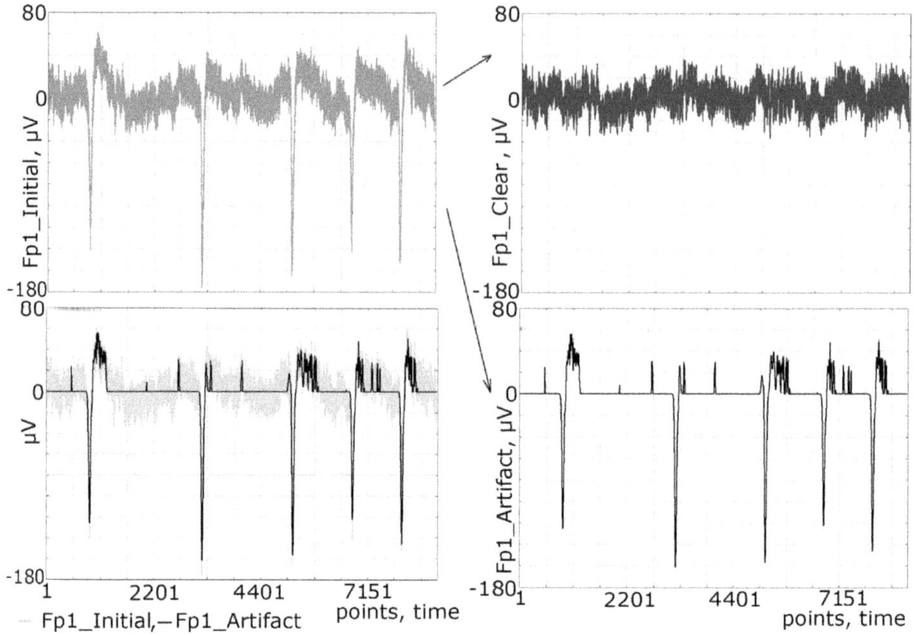

FIGURE 4.7 Removing ocular artifacts from Fp1 with the presented algorithm: original signal (light gray), artifacts (black), clean (residual) signal (dark gray).

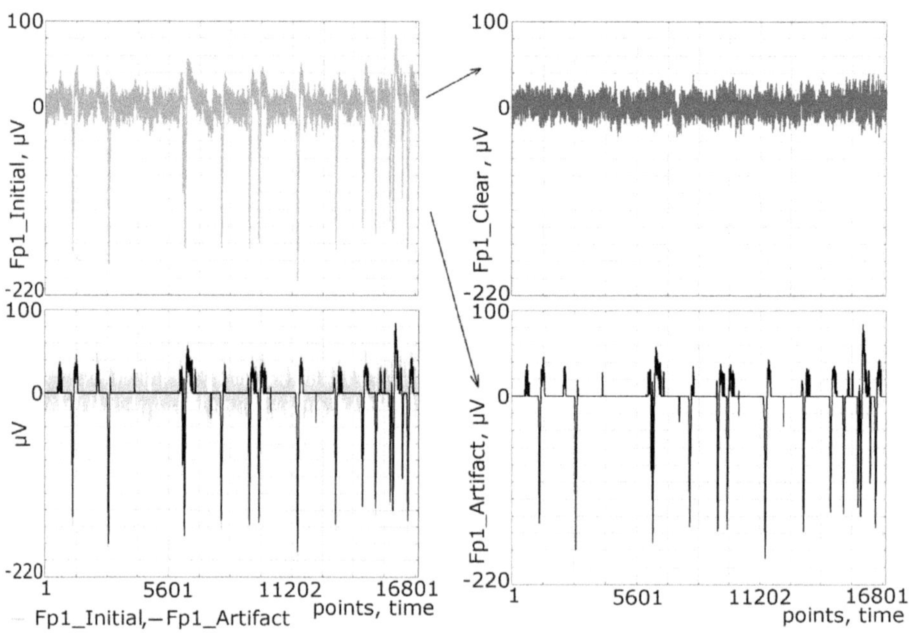

FIGURE 4.8 Removing ocular artifacts from Fp1 with the presented algorithm: original signal (light gray), artifacts (black), clean (residual) signal (dark gray).

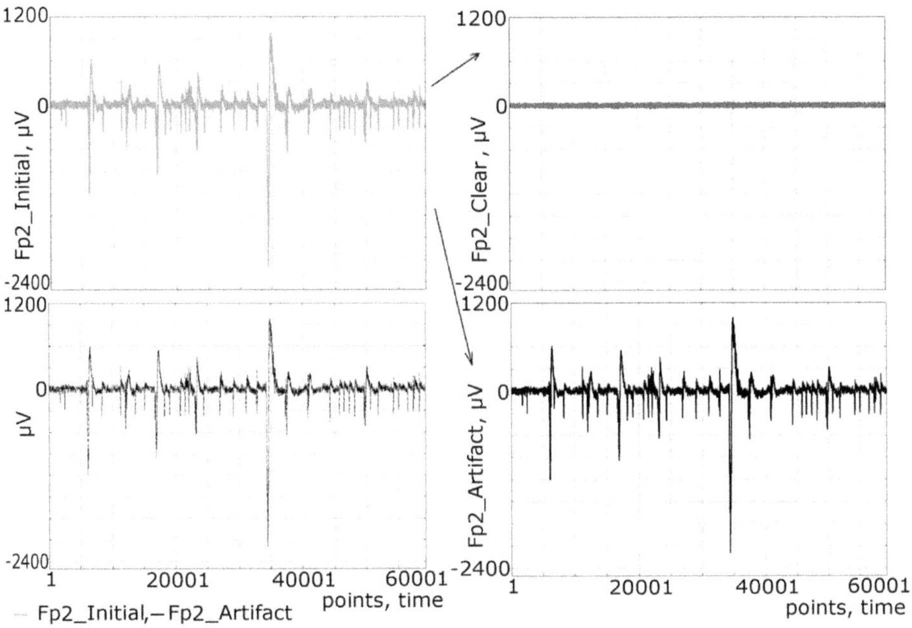

FIGURE 4.9 Removing ocular artifacts from Fp2 with the presented algorithm: original signal (light gray), artifacts (black), clean (residual) signal (dark gray).

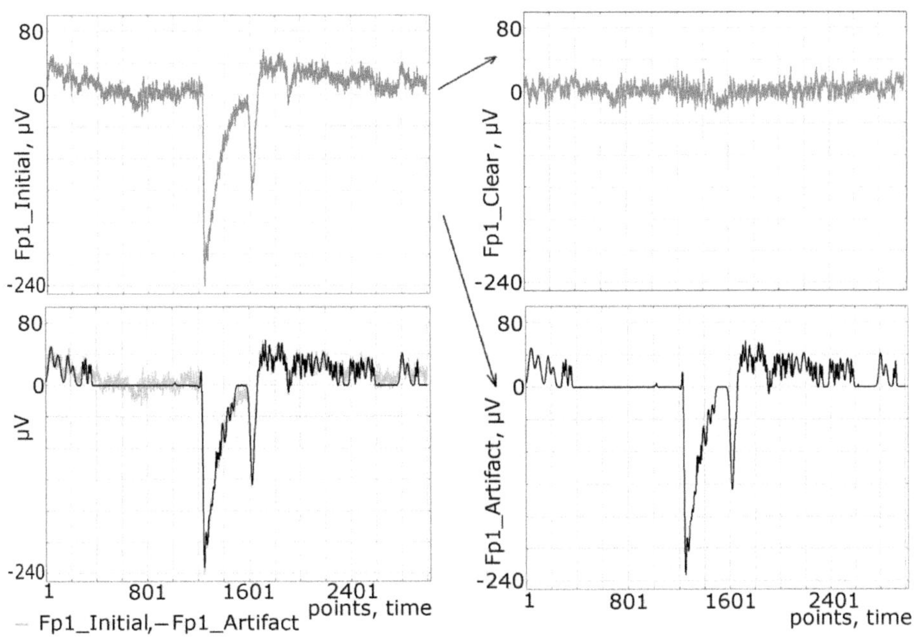

FIGURE 4.10 Removing eye movement artifacts from Fp1: original signal (light gray), artifacts (black), clean (residual) signal (dark gray).

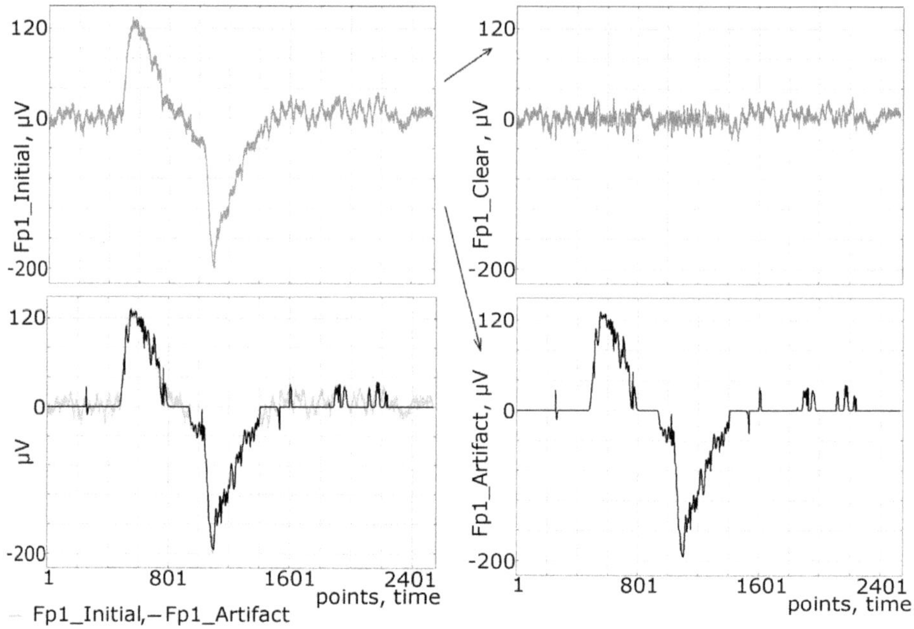

FIGURE 4.11 Filtering eye movement-related noise from the Fp1 signal: original signal (light gray), artifacts (black), clean (residual) signal (dark gray).

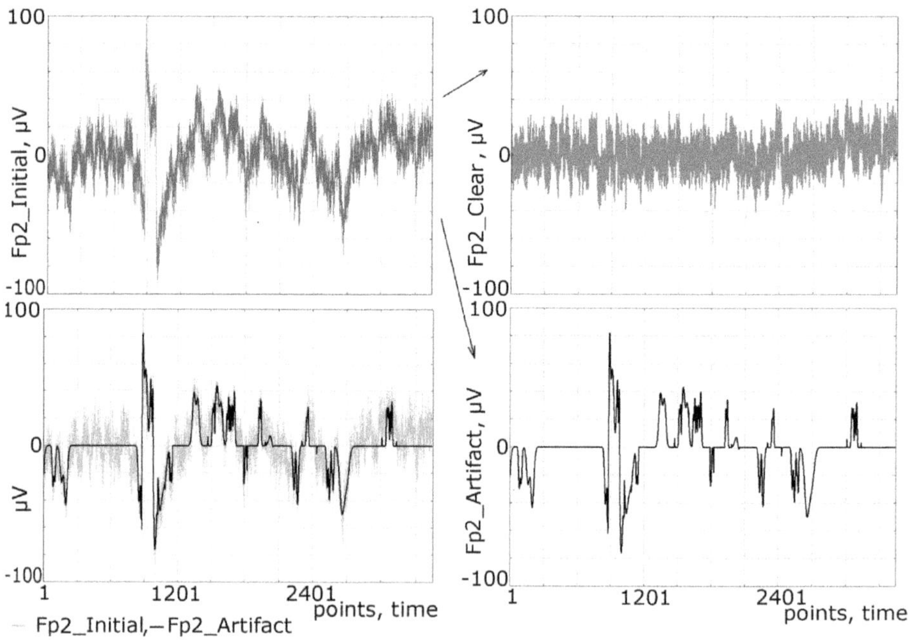

FIGURE 4.12 Filtering eye movement-related noise from the Fp2 signal: original signal (light gray), artifacts (black), clean (residual) signal (dark gray).

the artifacts. It is challenging to obtain brain activity from a single-channel EEG with muscle artifacts compared to EOG and eye movements. Muscle artifacts are critical but difficult to eliminate [4]. Muscle EEG artifacts can be removed using well-known deep learning models, including autoencoders, CNNs, and recurrent NNs. To remove muscle artifacts from contaminated EEG signals, Hossain et al. [21], applied the Feature Pyramid Network, U-Net, U-Net Multilevel Contextual Gating, and LinkNet. The large variability in the shape of muscle artifacts makes it difficult to extract specific parameters that would help in identifying and removing artifacts. Therefore, various methods for removing muscle artifacts have been proposed; however, they require preliminary training and tuning for specific types of artifacts [53–57]. We demonstrate effective removal of muscle artifacts from single-channel EEG with our method that filters based on wavelet analysis for localization without additional tuning (Figures 4.13–4.15). We used the same thresholding system as we described in the example.

4.4.4 Chewing

EEG records can also contain artifacts caused by tongue movements and swallowing [58]. Chewing and tongue (hypoglossal) artifact are muscle artifacts from the temporalis muscle and are marked by a sudden onset and intermittent bursts of generalized very fast activity. These artifacts exhibit high-frequency oscillations localized in time. In Figures 4.16–4.18, we demonstrate the effective removal of

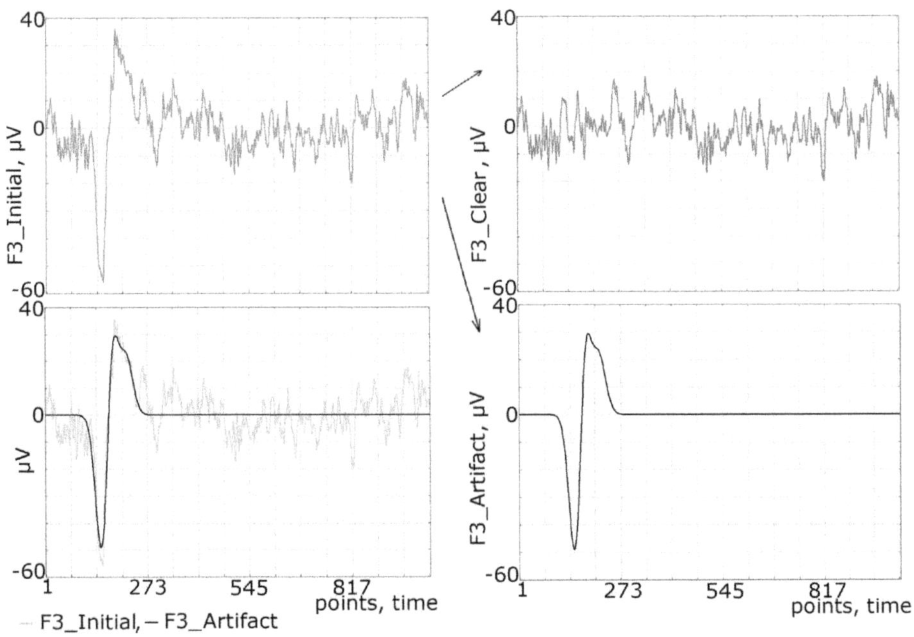

FIGURE 4.13 Filtering out muscle activity artifacts from F3: original signal (light gray), artifacts (black), clean (residual) signal (dark gray).

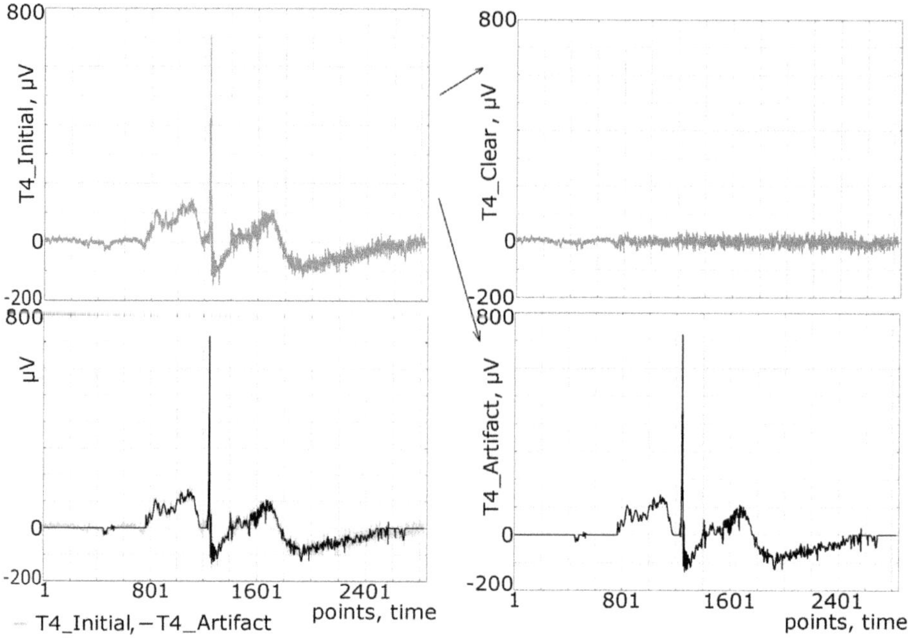

FIGURE 4.14 Removal of muscle artifacts from T4: original signal (light gray), artifacts (black), clean (residual) signal (dark gray).

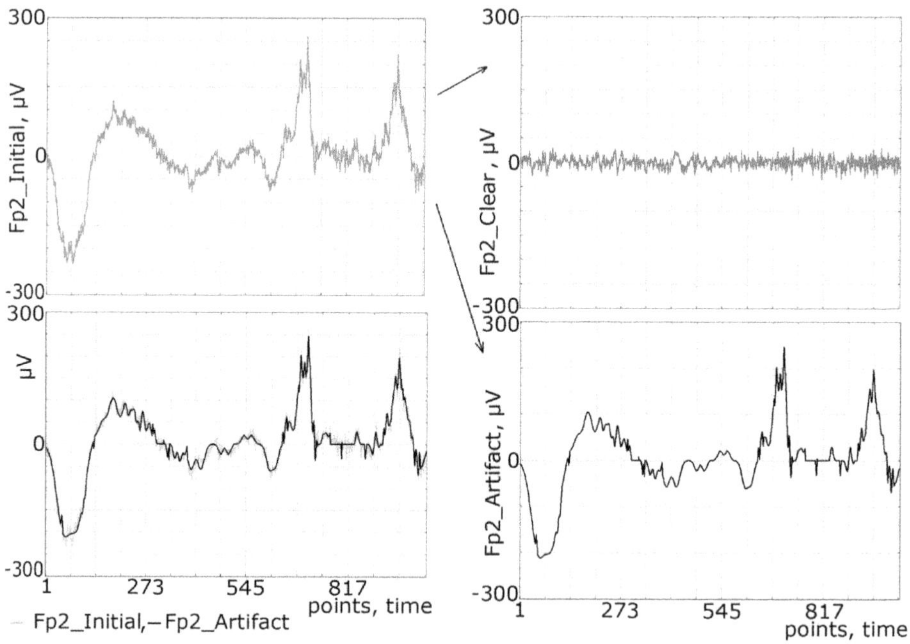

FIGURE 4.15 Filtering out muscle activity artifacts from the Fp2 electrode: original signal (light gray), artifacts (black), clean (residual) signal (dark gray).

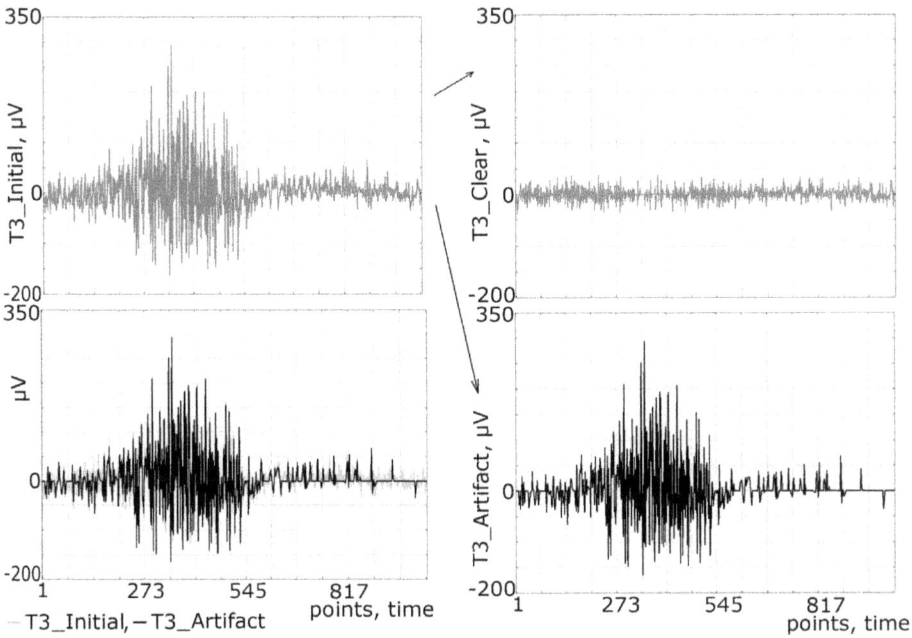

FIGURE 4.16 Filtering chewing-related artifacts from the T3 signal: original signal (light gray), artifacts (black), clean (residual) signal (dark gray).

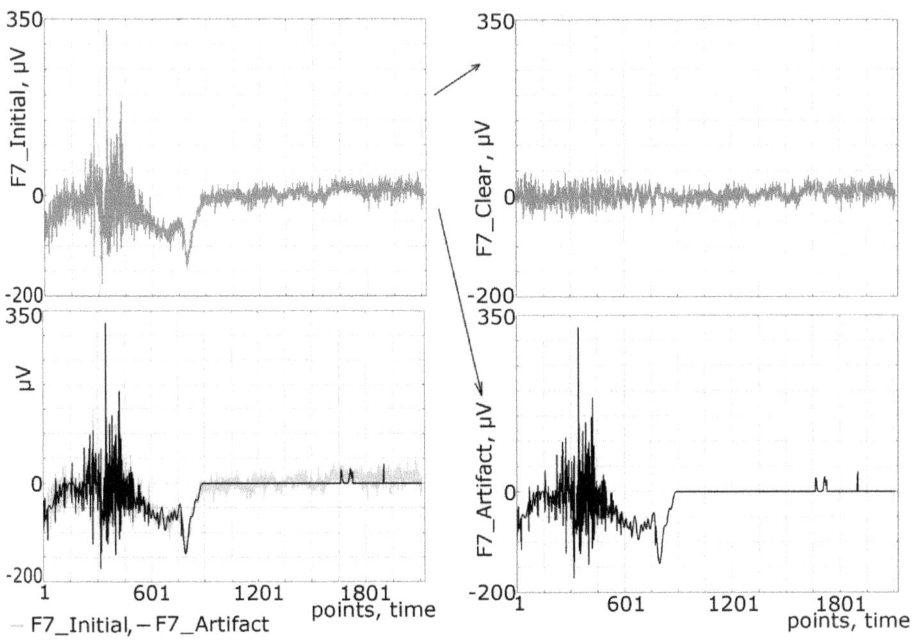

FIGURE 4.17 Filtering chewing-related artifacts from the F7 signal: original signal (light gray), artifacts (black), clean (residual) signal (dark gray).

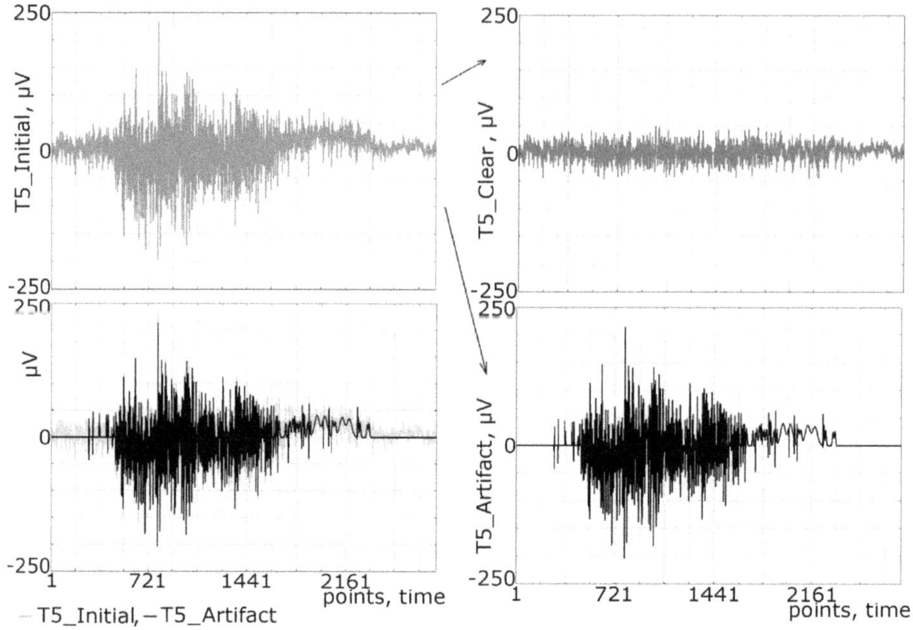

FIGURE 4.18 Filtering chewing-related artifacts from the T5 signal: original signal (light gray), artifacts (black), clean (residual) signal (dark gray).

chewing artifacts from single-channel EEG using our method, which filters based on wavelet analysis for localization without requiring additional tuning. We would like to note that we used the same thresholding system as we described in the example.

4.4.5 Electrode Pop

The intrinsic physiological artifacts mentioned above, as well as external sources of artifacts, also have a damaging effect on EEG measurements. Instrument artifacts, as a type of extrinsic artifact, originate from electrode misplacement and cable movements. A common finding is electrode pop, which is most often caused by a loose electrode [1,4,58,59]. The electrode pop artifact removal efficiency from single-channel EEG is demonstrated with our method, which utilizes filters based on wavelet analysis for localization, as shown in Figures 4.19–4.21. Note that we used the same thresholding system as described in the example.

4.4.6 Drift

Non-physiological signals such as impedance variations at the electrode–scalp interface are typically undesired and classified as artifacts. The impedance variations at the electrode–scalp interface manifest as pops and drifts in the recorded EEG signals

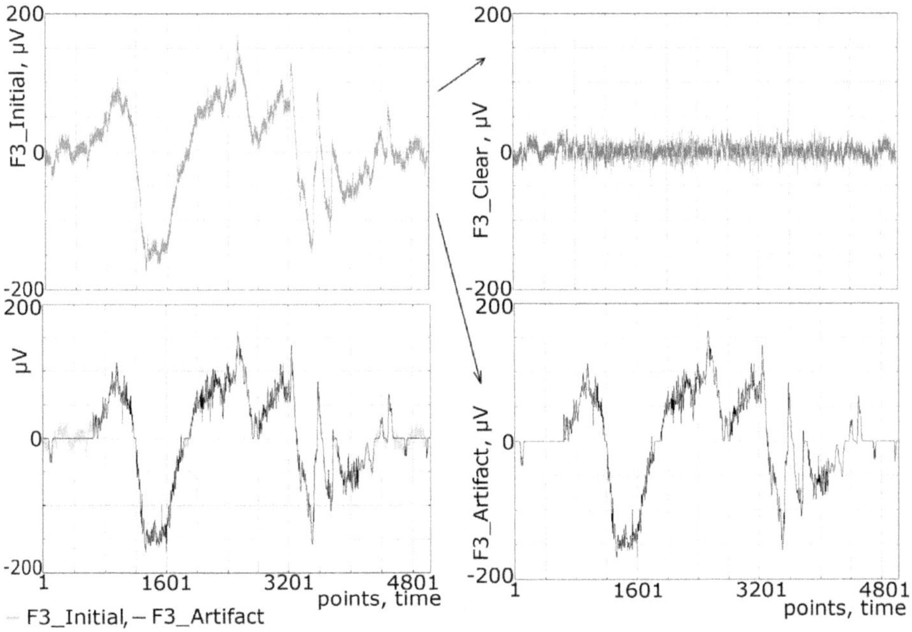

FIGURE 4.19 Removal of electrode pop artifacts from F3 using a wavelet-based filtering method for enhanced localization.

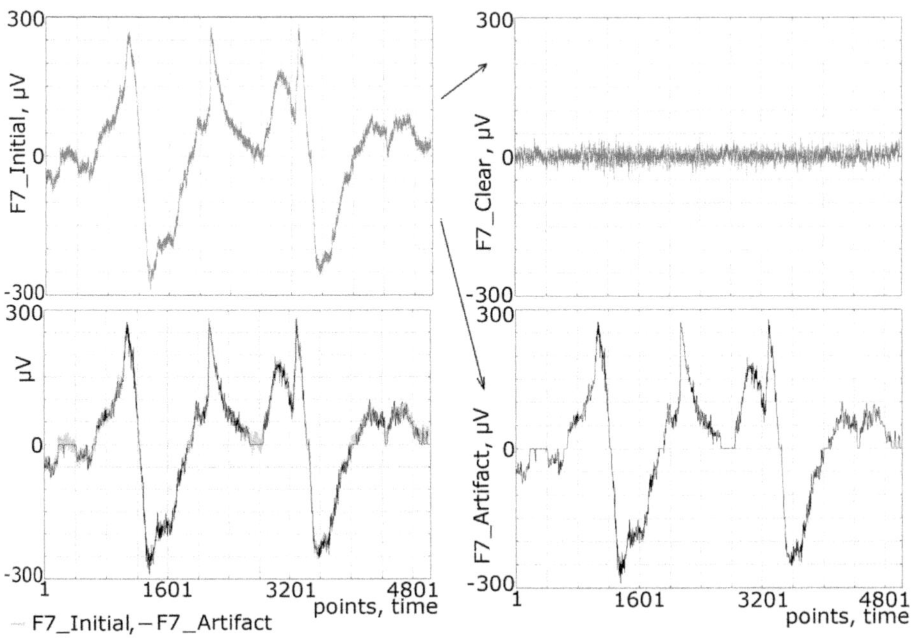

FIGURE 4.20 Filtering out electrode pop artifacts from F7: original signal (light gray), artifacts (black), clean (residual) signal (dark gray).

[59]. In Figures 4.22–4.24, we demonstrate the efficiency of removing drift artifacts from single-channel EEG using the same thresholding system.

4.4.7 Cardiac

Because scalp electrodes also capture the electric field generated by the activity of the cardiovascular system, the cardiac artifacts arise as a heavy contaminant of the EEG data [60]. Cardiac artifacts are internal artifacts that originate from the body's own activity [4]. However, we consider them separately because they have specific frequencies and require special adjustments to our algorithm. We successfully identified and removed cardiac artifacts from the signals T_4 and T_6 (Figures 4.25 and 4.26). We used the following special threshold system: $T_{10}=7{,}000.0$; $T_9=7{,}000.0$; $T_8=7{,}000.0$; $T_7=6{,}000.0$; $T_6=5{,}000.0$; $T_5=3{,}000.0$; $T_4=2{,}000.0$; $T_3=1{,}000.0$; $T_2=500.0$. Unfortunately, this system turned out not to be universal. In cases where the amplitudes of cardiac artifacts differ significantly from those of brain activity, they can be isolated and removed. However, suppose the amplitudes of the EEG signal are close enough to the amplitudes of cardiac artifacts. In that case, they are very difficult to identify them (Figure 4.27). For the signal Fz (Figure 4.27), we used the following special threshold system: $T_{10}=7{,}000.0$; $T_9=6{,}000.0$; $T_8=5{,}000.0$; $T_7=3{,}000.0$; $T_6=1{,}000.0$; $T_5=800.0$; $T_4=700.0$; $T_3=600.0$; $T_2=300.0$. However, along with cardiac artifacts, we also removed a part of the EEG signal. However, the signal was of such high quality, with additional artifacts, that we did not fully

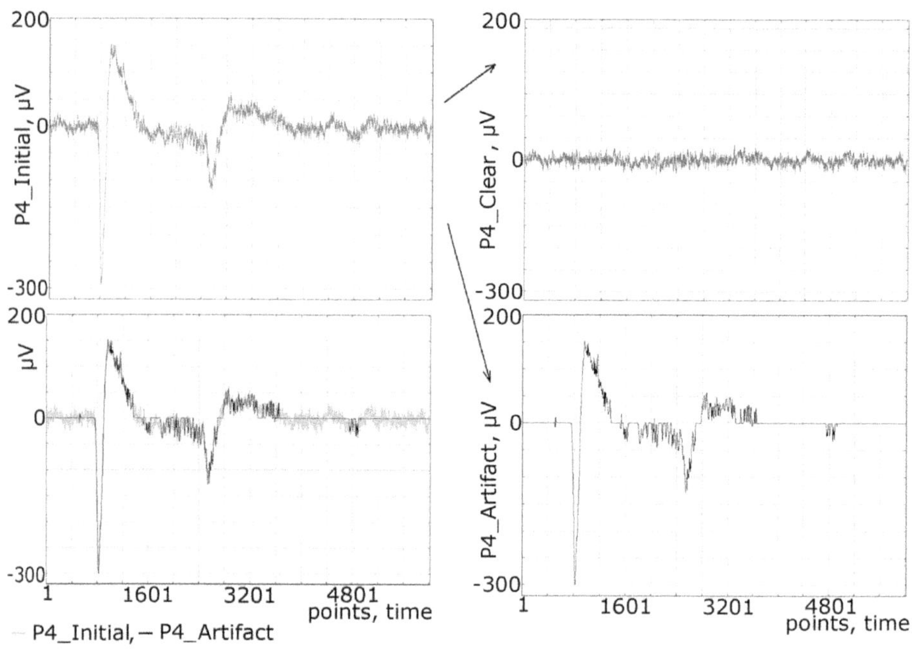

FIGURE 4.21 Electrode pop artifact removal at P4: original signal (light gray), artifacts (black), clean (residual) signal (dark gray).

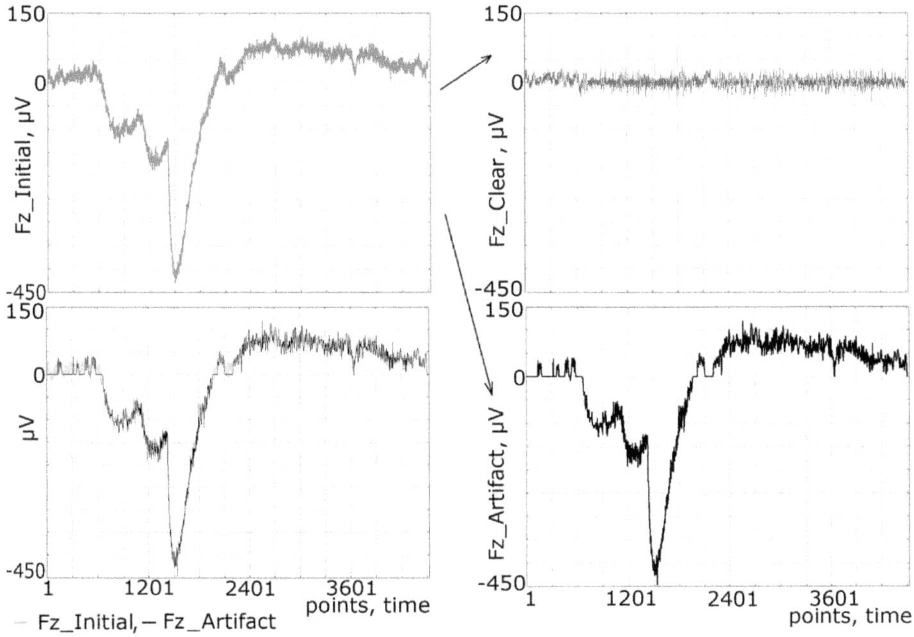

FIGURE 4.22 Drift artifact removal at Fz: original signal (light gray), artifacts (black), clean (residual) signal (dark gray).

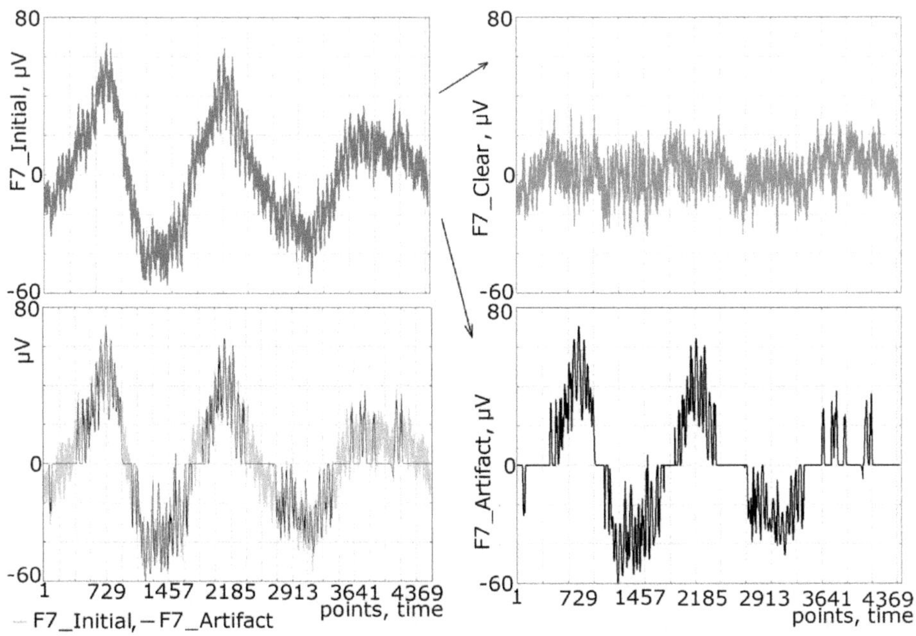

FIGURE 4.23 A wavelet analysis-driven filter was used to eliminate drift artifacts in F7 and improve localization accuracy.

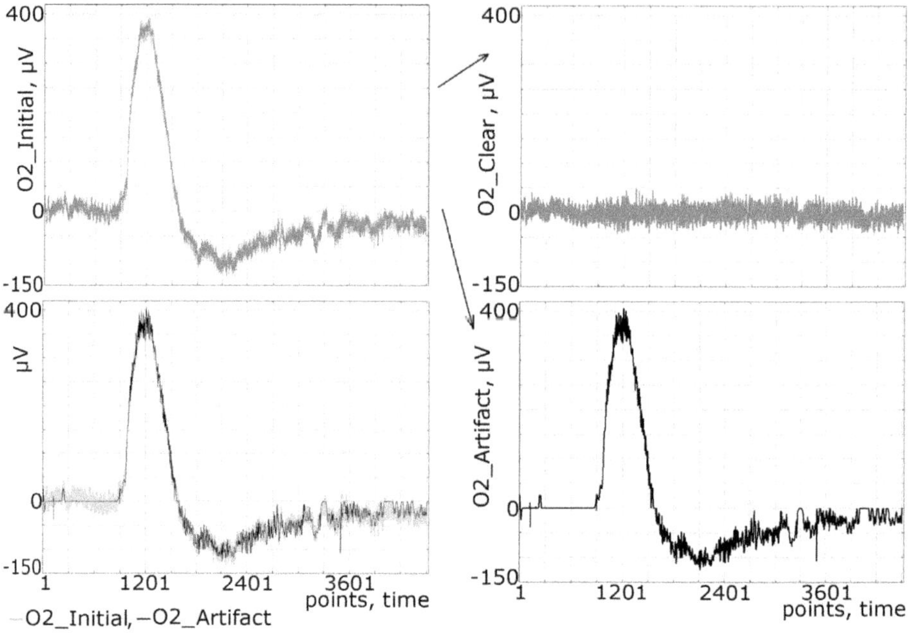

FIGURE 4.24 Elimination of drift artifacts from O_2: original signal (light gray), artifacts (black), clean (residual) signal (dark gray).

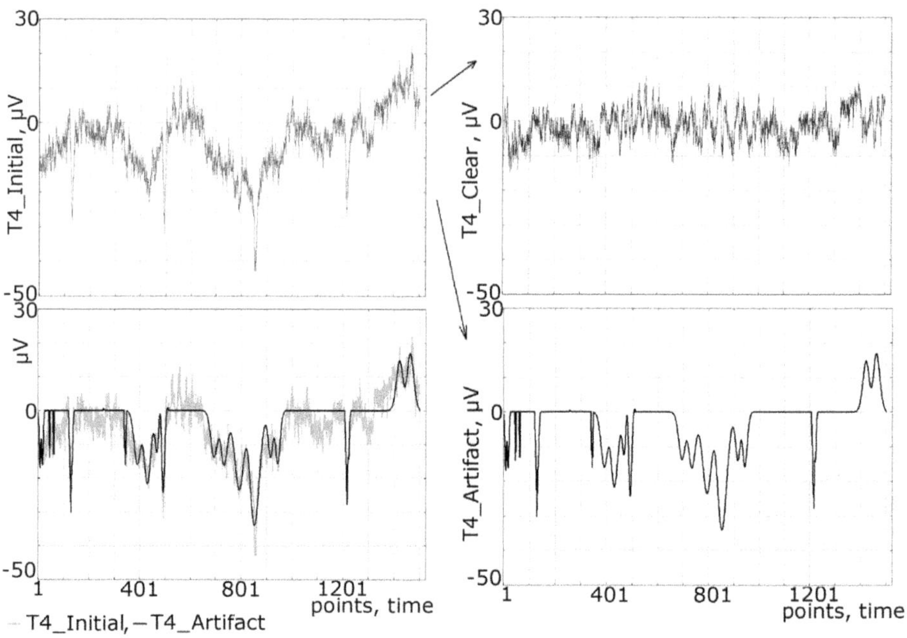

FIGURE 4.25 Eliminating cardiac artifacts from T_4: original signal (light gray), artifacts (black), clean (residual) signal (dark gray).

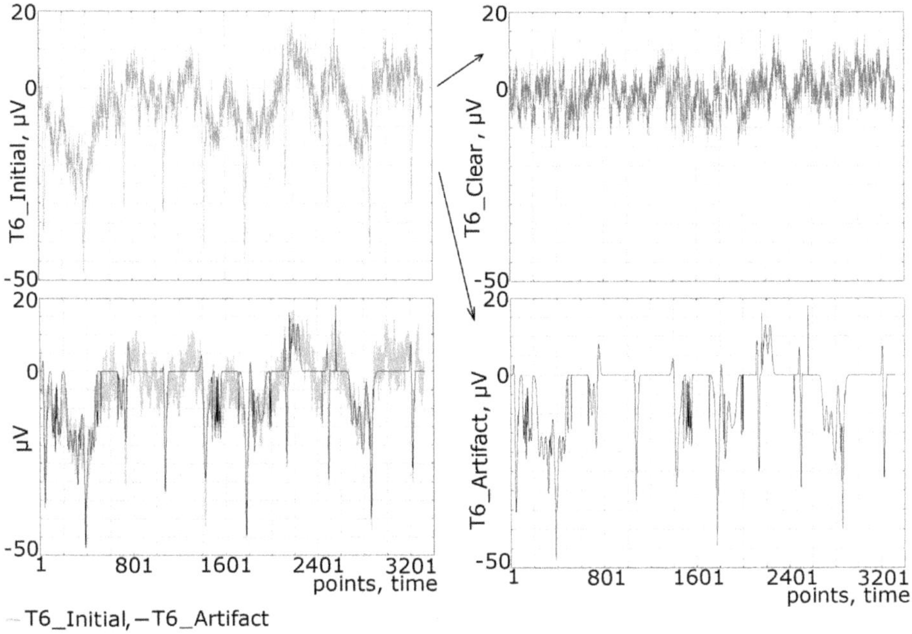

FIGURE 4.26 Removing cardio artifacts from T_6: original signal (light gray), artifacts (black), clean (residual) signal (dark gray).

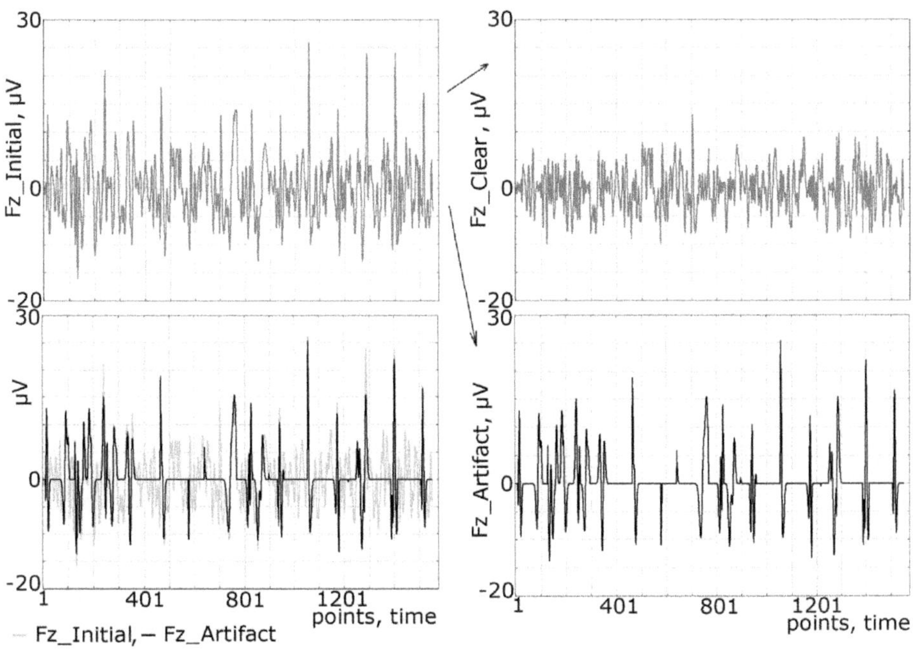

FIGURE 4.27 Removing cardiac artifacts from Fz: original signal (light gray), artifacts (black), clean (residual) signal (dark gray).

understand what a clean EEG should look like. For our method, cardiac artifacts are the most difficult to remove.

4.5 DISCUSSION

The results of testing the method on EEG artifacts of different types showed its universality and efficiency. Only cardiac artifacts required additional special adjustments.

Unlike other artifact removal methods, we recognize the impulse component of the artifact and its localization. This allows us to remove the artifact locally without changing the initial EEG in the parts that are not contaminated by artifacts. We consider the artifacts as an EEG baseline drift and remove this drift, preserving all higher frequencies of brain activity.

The methods most closely related to the proposed approach, which uses wavelet-based filtering for localization, are wavelet-based artifact removal techniques for single-channel EEG and CNN-based methods [9,10,21,24,26–30,47,54]. Unlike approaches that rely on standard wavelet decompositions with fixed basis functions, our method employs a set of Krawtchouk basis functions parameterized by time shift and asymmetry coefficient [31]. This formulation enables a separation between the shape characteristics of an impulse and its temporal localization.

Deep CNNs have demonstrated outstanding performance in handling high-dimensional data, delivering state-of-the-art results in both classification and regression tasks across diverse fields [61]. To optimize CNN architectures, evolutionary algorithms are frequently employed due to their strengths in global convergence, ergodicity, robustness, and adaptability [62]. However, most NN approaches do not explicitly assess the ergodicity of the underlying data—a critical property with implications for data integration and interpretation [63]. Ergodicity implies that the statistical properties of a system remain consistent across individuals and time. More formally, it means the time average of a process (rhythm) equals its ensemble (or space) average; that is, the statistical characteristics derived from a single long-term signal are equivalent to those from many short-term signals. If data are ergodic, one can generalize findings across individuals and time—observing one person multiple times or many people once yields comparable insights [64]. For deep learning models, this assumption is vital: the training data must adequately represent the unseen data the model will encounter [64]. Ergodic theory also contributes to stabilizing training by reducing chaotic dynamics, resulting in more predictable learning behavior [65]. Nonetheless, the assumption of ergodicity is often incompatible with the actual properties of EEG data, which tend to be highly individual-specific and temporally variable.

Moreover, because many algorithms extract artifact-related features within fixed temporal windows during training, shifts in the timing or structure of artifacts can alter these features—potentially requiring an impractically large or even infinite training set to achieve robust generalization.

In contrast, our method does not rely on assumptions of stationarity, Gaussianity, or ergodicity. It is grounded in the invariance model of pattern recognition inspired by the visual system, enabling the extraction of artifact features independently of their temporal occurrence.

4.6 COMPARISON WITH ICA

One of the most commonly used techniques for artifact removal is the advanced source analysis method. However, it can only be applied to a limited subset of signals. To illustrate this, we selected an eight-channel EEG segment recorded from leads F3, F4, C3, C4, P3, P4, T3, and T4 (Figure 4.28a). The artifact was present exclusively in lead F3. Figure 4.28b shows the result of artifact removal using ICA. While the artifact in F3 was effectively suppressed, all other channels were also altered. Notably, ICA caused a significant reduction in the amplitude of the originally clean signals.

Figure 4.28c displays the difference between the original and the processed signals, highlighting that a substantial portion of the original signal was lost during artifact removal. ICA treats the EEG as a linear mixture of statistically independent components and attempts to recover the original sources [66]. However, artifacts—such as eye blinks—may not be statistically independent from brain activity. Furthermore, the goal of artifact removal differs from that of source separation or minimizing mutual information [67]. For instance, channels unaffected by artifacts should ideally remain unaltered, yet ICA transforms all inputs indiscriminately.

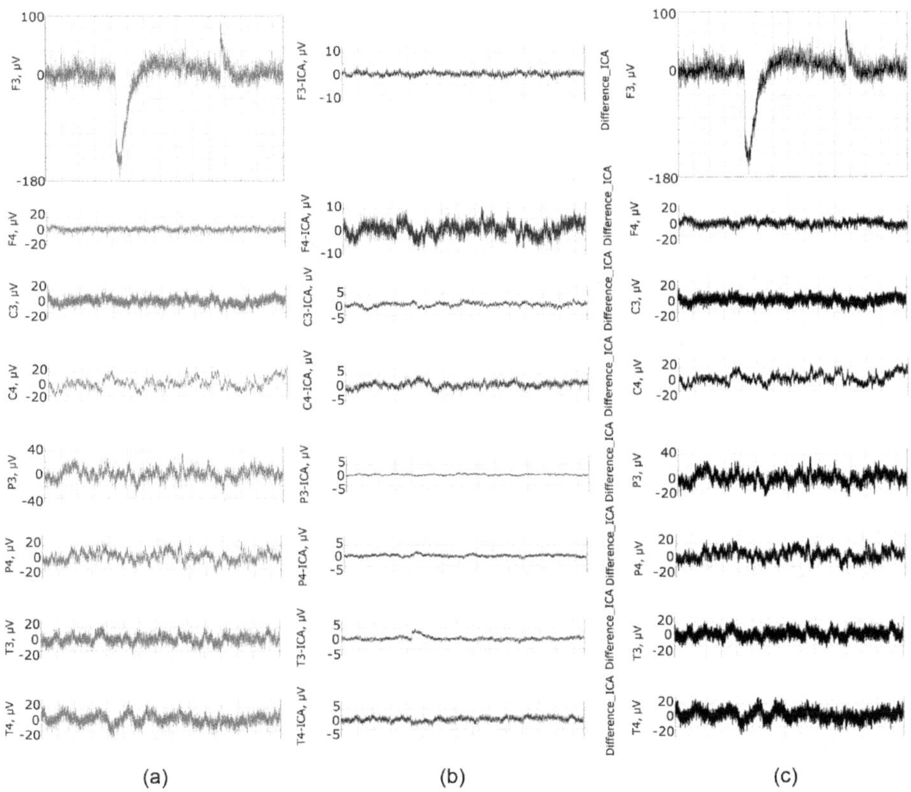

FIGURE 4.28 (a) Initial EEG recorded in the leads F3, F4, C3, C4, P3, P4, T3, and T4 with artifacts; (b) cleaned EEG by the ICA method; and (c) the difference between the initial EEG recordings and those cleaned by the ICA method.

Figure 4.29b shows the result of applying the wavelet localization-based filtering algorithm. Our method was consistently applied across an eight-channel EEG recording, with each lead analyzed independently. Artifact-free segments in each channel were preserved and remained unaltered. Figure 4.29c displays the difference between the original and the cleaned signals. The presence of zero-valued segments in Figure 4.29c confirms that portions of the signal without artifacts were retained for further analysis.

This example highlights the effectiveness of the wavelet localization-based filtering algorithm in removing artifacts from both single-channel and multi-channel EEG recordings while preserving clean signal segments.

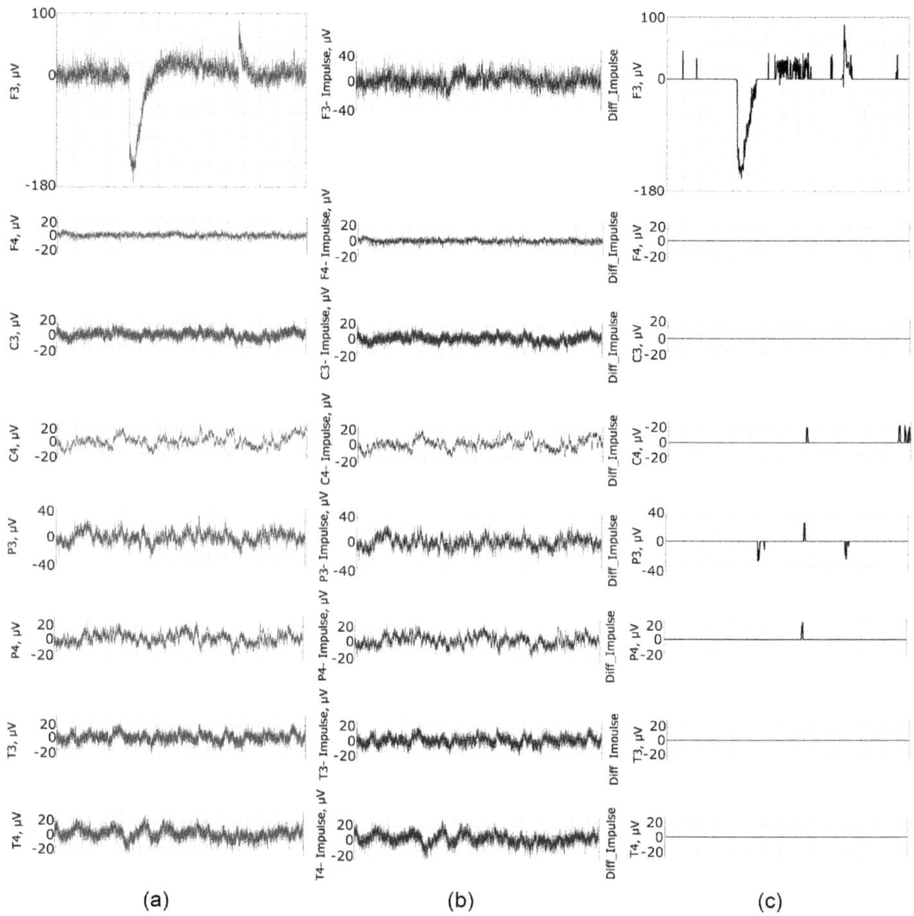

FIGURE 4.29 (a) Initial EEG recorded in the leads F3, F4, C3, C4, P3, P4, T3, and T4 with artifacts; (b) cleaned EEG by using a method that filters based on wavelet analysis for localization; and (c) the difference between the initial EEG recordings and the cleaned EEG obtained by the method that filters based on wavelet analysis for localization.

4.7 CONCLUSIONS

We presented a universal method for artifact removal in EEG signals, capable of addressing a wide range of artifact types without requiring assumptions such as stationarity, Gaussianity, or ergodicity. The algorithm is applicable to both single-channel and multi-channel EEG recordings and operates independently across channels, ensuring that artifact-free segments remain intact.

A key strength of the proposed approach lies in its generality and adaptability: it does not rely on prior knowledge of the artifact's nature or fixed temporal windows. Instead, it uses wavelet-based localization to detect and remove artifacts with high temporal precision, preserving the integrity of the underlying neural signal.

In future work, we aim to integrate this filtering method with machine learning techniques to automatically detect, classify, and adaptively respond to different artifact types. This hybrid system would enable the real-time identification and removal of artifacts during EEG acquisition, paving the way for robust online EEG analysis in clinical and research settings.

ACKNOWLEDGMENTS

This work is part of BiTWerK, an interdisciplinary collaborative research initiative of the University of Kassel, and of the DFG Research Training Group 2749 Multiscale Clocks. We thank Dorothee Knees and Caroline Disser for their support.

REFERENCES

1. Tandle, A., Jog, N., D'cunha, P., & Chheta, M. (2016). Classification of artefacts in EEG signal recordings and EOG artefact removal using EOG subtraction. *Communications on Applied Electronics*, *4*(1), 12–19. https://doi.org/10.5120/cae2016651997
2. Walters-Williams, J., & Li, Y. (2011). A new approach to denoising EEG signals-merger of translation invariant wavelet and ICA. *International Journal of Biometrics and Bioinformatics (IJBB)*, *5*(2), 130–148. https://www.cscjournals.org/manuscript/Journals/IJBB/volume5/Issue2/IJBB-65.pdf
3. Croft, R. J., & Barry, R.J. (2000). Removal of ocular artifact from the EEG: A review. *Clinical Neurophysiology*, *30*, 5–19. https://doi.org/10.1016/S0987-7053(00)00055-1
4. Jiang, X., Bian, G. B., & Tian, Z. (2019). Removal of artifacts from EEG signals: A review. *Sensors*, *19*(5), 987. https://doi.org/10.3390/s19050987
5. Kotte, S., & Dabbakuti, J. K. (2020). Methods for removal of artifacts from EEG signal: A review. *Journal of Physics: Conference Series*, *1706*(1), 012093. https://doi.org/10.1088/1742-6596/1706/1/012093
6. Mumtaz, W., Rasheed, S., & Irfan, A. (2021). Review of challenges associated with the EEG artifact removal methods. *Biomedical Signal Processing and Control*, *68*, 102741. https://doi.org/10.1016/j.bspc.2021.102741
7. Coelli, S., Calcagno, A., Cassani, C. M., Temporiti, F., Reali, P., Gatti, R., Galli, M., & Bianchi, A. M. (2024). Selecting methods for a modular EEG pre-processing pipeline: An objective comparison. *Biomedical Signal Processing and Control*, *90*, 105830. https://doi.org/10.1016/j.bspc.2023.105830
8. Sadiya, S., Alhanai, T., & Ghassemi, M. M. (2021). Artifact detection and correction in EEG data: A review. *10th International IEEE/EMBS Conference on Neural Engineering (NER)*, Italy, pp. 495–498. https://doi.org/10.1109/NER49283.2021.9441341

9. Kalpakam, N. V., & Venkataramanan, S. (2004). Haar wavelet decomposition of EEG signal for ocular artifact de-noising: a mathematical analysis. *The 2nd Annual IEEE Northeast Workshop on Circuits and Systems, 2004. NEWCAS 2004*, New York, pp. 141–144. https://doi.org/10.1109/NEWCAS.2004.1359042
10. Vázquez, R. R., Velez-Perez, H., Ranta, R., Dorr, V. L., Maquin, D., & Maillard, L. (2012). Blind source separation, wavelet denoising and discriminant analysis for EEG artefacts and noise cancelling. *Biomedical Signal Processing and Control*, *7*(4), 389–400. https://doi.org/10.1016/j.bspc.2011.06.005
11. Joyce, C. A., Gorodnitsky, I. F., & Kutas, M. (2004). Automatic removal of eye movement and blink artifacts from EEG data using blind component separation. *Psychophysiology*, *41*(2), 313–325. https://doi.org/10.1111/j.1469-8986.2003.00141.x
12. Wallstrom, G. L., Kass, R. E., Miller, A., Cohn, J. F., & Fox, N. A. (2004). Automatic correction of ocular artifacts in the EEG: A comparison of regression-based and component-based methods. *International Journal of Psychophysiology*, *53*(2), 105–119. https://doi.org/10.1016/j.ijpsycho.2004.03.007
13. Delorme, A., & Makeig, S. (2004). EEGLAB: An open source toolbox for analysis of single-trial EEG dynamics including independent component analysis. *Journal of Neuroscience Methods*, *134*(1), 9–21. https://doi.org/10.1016/j.jneumeth.2003.10.009
14. James, C. J., Gibson, O., & Davies, M. (2006). On the analysis of single versus multiple channels of electromagnetic brain signals. *Artificial Intelligence in Medicine*, *37*(2), 131–143. https://doi.org/10.1016/j.artmed.2006.03.003
15. Chen, W., Cai, Y., Li, A., Su, Y., & Jiang, K. (2024). Single-channel sleep EEG classification method based on LSTM and hidden markov model. *Brain Sciences*, *14*(11), 1087. https://doi.org/10.3390/brainsci14111087
16. Siddiqa, H. A., Tang, Z., Xu, Y., Wang, L., Irfan, M., & Abbasi, S. F. (2024). Single-channel EEG data analysis using a multi-branch CNN for neonatal sleep staging. *IEEE Access*, *12*, 29910–29925. https://doi.org/10.1109/ACCESS.2024.3365570
17. Zhang, X., He, G., Shang, T., & Fan, F. (2024). Comparative analysis of single-channel and multi-channel classification of sleep stages across four different data sets. *Brain Sciences*, 14(12), 1201. https://doi.org/10.3390/brainsci14121201
18. Maimon, N. B., Molcho, L., Intrator, N., & Lamy, D. (2021). Novel single-channel EEG features correlate with working memory load. *2021 IEEE 3rd Global Conference on Life Sciences and Technologies (LifeTech)*, New York, pp. 471–472. https://doi.org/10.1109/LifeTech52111.2021.9391963
19. Jiao, Y., DU, G., Wang, X., Liu, C., & Liu, Y. (2024). Single-channel EEG for the identification of multiple types of mental stress. *Journal of Mechanics in Medicine and Biology*, *24*, 2440031. https://doi.org/10.1142/s0219519424400311
20. Higashi, H. (2025). Single-channel electroencephalography decomposition by detector-atom network and its pre-trained model. *Journal of Neuroscience Methods*, *414*, 110323. https://doi.org/10.1016/j.jneumeth.2024.110323
21. Hossain, M. S., Mahmud, S., & Khandakar, A. (2022). Deep learning technique to denoise electromyogram artifacts from single-channel electroencephalogram signals. *8th International Conference on Signal, Image Processing and Embedded Systems (SIGEM 2022)*, Selangor, Malaysia, pp. 85–98. https://doi.org/10.5121/csit.2022.122006
22. Chen, X., He, C., & Peng, H. (2014). Removal of muscle artifacts from single-channel EEG based on ensemble empirical mode decomposition and multiset canonical correlation analysis. *Journal of Applied Mathematics*, *261347*, 10. https://doi.org/10.1155/2014/261347
23. Dora, C., Patro, R. N., Rout, S. K., Biswal, P. K., & Biswal, B. (2021). Adaptive SSA based muscle artifact removal from single channel EEG using neural network regressor. *IRBM*, *42*(5), 324–333. https://doi.org/10.1016/j.irbm.2020.08.002

24. Paissan, F., Kumaravel, V., & Farella, E. (2022). Interpretable CNN for single-channel artifacts detection in raw EEG signals. *2022 IEEE Sensors Applications Symposium (SAS)*, Sundsvall, Sweden, pp. 1–6. https://doi.org/10.1109/SAS54819.2022.9881381
25. Dora, M., & Holcman, D. (2022). Adaptive single-channel EEG artifact removal with applications to clinical monitoring. *IEEE Transactions on Neural Systems and Rehabilitation Engineering, 30*, 286–295. https://doi.org/10.1109/TNSRE.2022.3147072
26. Dora, M., Jaffard, S., & Holcman, D. (2022). The WQN algorithm to adaptively correct artifacts in the EEG signal. *Applied and Computational Harmonic Analysis, 61*, 347–356. https://doi.org/10.1016/j.acha.2022.07.007
27. Hu, Q., Li, M., & Li, Y. (2022). Single-channel EEG signal extraction based on DWT, CEEMDAN, and ICA method. *Frontiers in Human Neuroscience, 16*, 1010760. https://doi.org/10.3389/fnhum.2022.1010760
28. Khatun, S., Mahajan, R., & Morshed, B. I. (2015). Comparative analysis of wavelet based approaches for reliable removal of ocular artifacts from single channel EEG. *2015 IEEE International Conference on Electro/Information Technology (EIT)*, New York, pp. 335–340. https://doi.org/10.1109/EIT.2015.7293364
29. Bajaj, N., Bajaj, N., Requena Carrion, J., Bellotti, F., Berta, R., & De Gloria, A. (2020). Automatic and tunable algorithm for EEG artifact removal using wavelet decomposition with applications in predictive modeling during auditory tasks. *Biomedical Signal Processing and Control, 55*, 101624. https://doi.org/10.1016/J.BSPC.2019.101624
30. Phadikar, S., Sinha, N., Ghosh, R., & Ghaderpour, E. (2022). Automatic muscle artifacts identification and removal from single-channel EEG using wavelet transform with meta-heuristically optimized non-local means filter. *Sensors, 22*(8), 2948. https://doi.org/10.3390/s22082948
31. Filimonova, N., Specovius-Neugebauer, M., & Friedmann, E. (2025). Determination of the time-frequency features for impulse components in EEG signals. *Neuroinform, 23*, 17. https://doi.org/10.1007/s12021-024-09698-y
32. Krawtchouk, M. (1929). Sur une generalisation des polynomes d'Hermite. *Comptes rendus de l'Académie des Sciences, 189*(17), 620–622.
33. Vainerman, L., & Filimonova, N. (1994). Hyperspectral imagery with the application of Krawtchouk polynomials. *Journal of Electrical and Computer Engineering, 2231*, 148–155. https://doi.org/10.1117/12.179775
34. Al-Utaibi, K. A., Abdulhussain, S. H., Mahmmod, B. M., Naser, M. A., Alsabah, M., & Sait, S. M. (2021). Reliable recurrence algorithm for high-order Krawtchouk polynomials. *Entropy, 23*(9), 1162. https://doi.org/10.3390/e23091162
35. Jahid, T., Karmouni, H., Hmimid, A., Sayyouri, M., & Qjidaa, H. (2017). Image moments and reconstruction by Krawtchouk via Clenshaw's reccurence formula. *2017 International Conference on Electrical and Information Technologies (ICEIT)*, Washington, DC, pp. 1–7. https://doi.org/10.1109/EITech.2017.8255265
36. Karmouni, H., Jahid, T., Lakhili, Z., Hmimid, A., Sayyouri, M., Qjidaa, H., & Rezzouk, A. (2017). Image reconstruction by Krawtchouk moments via digital filter. *Intelligent Systems and Computer Vision (ISCV)*, New York, pp. 1–7. https://doi.org/10.1109/ISACV.2017.8054958
37. Venkataramana, A., & Raj, P. A. (2011). Recursive computation of forward Krawtchouk moment transform using Clenshaw's recurrence formula. *2011 Third National Conference on Computer Vision, Pattern Recognition, Image Processing and Graphics*, Washington, DC, pp. 200–203. https://doi.org/10.1109/NCVPRIPG.2011.66
38. Idan, Z. N., Abdulhussain, S. H., & Al-Haddad, S. A. R. (2020). A new separable moments based on Tchebichef-Krawtchouk polynomials. *IEEE Access*, 8, 41013–41025. https://doi.org/10.1109/ACCESS.2020.2977305

39. Yap, P. T., Paramesran, R., & Ong, S. H. (2003). Image analysis by Krawtchouk moments. *IEEE Transactions on Image Processing*, 12(11), 1367–1377. https://doi.org/10.1109/TIP.2003.818019
40. Zhang, G., Luo, Z., Fu, B., Li, B., Liao, J., Fan, X., & Xi, Z. (2010). A symmetry and bi-recursive algorithm of accurately computing Krawtchouk moments. *Pattern Recognition Letters*, *31*(7), 548–554. https://doi.org/10.1016/j.patrec.2009.12.007
41. Abdulhussain, S. H., Ramli, A. R., Al-Haddad, S. A. R., Mahmmod, B. M., & Jassim W. A. (2018). Fast recursive computation of Krawtchouk polynomials. *Journal of Mathematical Imaging and Vision*, *60*(3), 285–303. https://doi.org/10.1007/s10851-017-0758-9
42. Asli, B. H. S., & Flusser, J. (2014). Fast computation of Krawtchouk moments. *Information Sciences*, *288*, 73–86. https://doi.org/10.1016/j.ins.2014.07.046
43. Cooley, J. W., & Tukey, J. W. (1965). An algorithm for the machine calculation of complex Fourier series. *Mathematics of Computation*, *19*(90), 297–301. https://api.semanticscholar.org/CorpusID:121744946
44. Kotowski, K., Ochab, J. K., Stapor, K., & Sommer, W. (2023). The importance of ocular artifact removal in single-trial ERP analysis: The case of the N250 in face learning. *Biomedical Signal Processing and Control*, *79*, 104115. https://doi.org/10.1016/j.bspc.2022.104115
45. Chang, W. D., Lim, J. H., & Im, C. H. (2016). An unsupervised eye blink artifact detection method for real-time electroencephalogram processing. *Physiological Measurement*, *37*(3), 401–417. https://doi.org/10.1088/0967-3334/37/3/401
46. Paissan, F., Kumaravel, V., & Farella, E. (2022). Interpretable CNN for single-channel artifacts detection in raw EEG signals. *2022 IEEE Sensors Applications Symposium (SAS)*, Sundsvall, Sweden, pp. 1–6. https://doi.org/10.1109/SAS54819.2022.9881381
47. Gorji, H. T., Koohpayezadeh, A., & Haddadnia, J. (2013). Ocular artifact detection and removing from EEG by wavelet families: A comparative study. *Journal of Information Engineering and Applications*, *3*(13), 39–47.
48. Madhavi, K. S., Deeksha, M., Gayathri, S., Venkat, N. U. N., & Goru, H. K. (2022). Denoising of ocular artifacts from single-channel EEG signals: A review. *2022 8th International Conference on Advanced Computing and Communication Systems (ICACCS)*, New York, Vol. 1, pp. 1002–1007. https://doi.org/10.1109/ICACCS54159.2022.9785129
49. Chang, W. D., Cha, H. S., Kim, K., & Im, C. H. (2016). Detection of eye blink artifacts from single prefrontal channel electroencephalogram. *Computer Methods and Programs in Biomedicine*, *124*, 19–30. https://doi.org/10.1016/j.cmpb.2015.10.011
50. Noorbasha, S.K., & Sudha, G.F. (2020). Removal of EOG artifacts from single channel EEG – An efficient model combining overlap segmented ASSA and ANC. *Biomedical Signal Processing and Control*, *60*, 987. https://doi.org/10.1016/j.bspc.2020.101987
51. Schlögl, A., Keinrath, C., Zimmermann, D., Scherer, R., Leeb, R., & Pfurtscheller, G. (2007). A fully automated correction method of EOG artifacts in EEG recordings. *Clinical Neurophysiology*, *118*(1), 98–104. https://doi.org/10.1016/j.clinph.2006.09.003
52. Hamal, A. Q., & bin Abdul Rehman, A. W. (2013). Artifact processing of epileptic EEG signals: An overview of different types of artifacts. *2013 International Conference on Advanced Computer Science Applications and Technologies*, New York, pp. 358–361. https://doi.org/10.1109/ACSAT.2013.77
53. Gorjan, D., Gramann, K., De Pauw, K., & Marusic, U. (2022). Removal of movement-induced EEG artifacts: Current state of the art and guidelines. *Journal of Neural Engineering*, *19*(1), 542. https://doi.org/10.1088/1741-2552/ac542c
54. Grobbelaar, M., Phadikar, S., Ghaderpour, E., Struck, A. F., Sinha, N., Ghosh, R., & Ahmed, M. Z. I. (2022). A survey on denoising techniques of electroencephalogram signals using wavelet transform. *Signals*, *3*(3), 577–586. https://doi.org/10.3390/signals3030035

55. Chen, J., Pi, D., Jiang, X., Xu, Y., Chen, Y., & Wang, X. (2023). Denosieformer: A transformer-based approach for single-channel EEG artifact removal. *IEEE Transactions on Instrumentation and Measurement, 2023*, 1–1. https://doi.org/10.1109/TIM.2023.3341114
56. Mourad, N. (2018). Automatic correction of short-duration artefacts in single-channel EEG recording: A group-sparse signal denoising algorithm. *IET Signal Process, 12*, 549–557. https://doi.org/10.1049/iet-spr.2017.0344
57. Chen, X., He, C., & Peng, H. (2014). Removal of muscle artifacts from single-channel EEG based on ensemble empirical mode decomposition and multiset canonical correlation analysis. *Journal of Applied Mathematics, 261347*, 10. https://doi.org/10.1155/2014/261347
58. Britton J.W., Frey L.C., Hopp J.L., Korb P., Koubeissi M.Z., Lievens W.E., Pestana-Knight E.M., St. Louis E.K. (2016). *Electroencephalography (EEG): An Introductory Text and Atlas of Normal and Abnormal Findings in Adults, Children, and Infants* [Internet]. St. Louis EK, Frey LC, editors. Chicago: American Epilepsy Society. https://doi.org/10.5698/978-0-9979756-0-4
59. Kobler, R. J., Sburlea, A. I., Mondini, V., & Muller-Putz, G. R. (2019). HEAR to remove pops and drifts: The high-variance electrode artifact removal (HEAR) algorithm. *Annual International Conference of the IEEE Engineering in Medicine and Biology Society. IEEE Engineering in Medicine and Biology Society. Annual International Conference*, New York, pp. 5150–5155. https://doi.org/10.1109/EMBC.2019.8857742
60. Arnau, S., Sharifian, F., Wascher, E., & Larra, M. F. (2023). Removing the cardiac field artifact from the EEG using neural network regression. *Psychophysiology, 60*(10), e14323. https://doi.org/10.1111/psyp.14323
61. Mallat, S. (2016). Understanding deep convolutional networks. *Philosophical Transactions. Series A, Mathematical, Physical, and Engineering Sciences, 374*(2065), 20150203. https://doi.org/10.1098/rsta.2015.0203
62. Zhang, Y., Wang, L., Zhao, J., Han, X., Wu, H., Li, M., & Deveci, M. (2024). A convolutional neural network based on an evolutionary algorithm and its application. *Information Sciences, 670*, 120644. https://doi.org/10.1016/j.ins.2024.120644
63. Medaglia, J. D., Ramanathan, D. M., Venkatesan, U. M., & Hillary, F. G. (2011). The challenge of non-ergodicity in network neuroscience. *Network, 22*(1–4), 148–153. https://doi.org/10.3109/09638237.2011.639604
64. Sakkalis, V., Oikonomou, T., Pachou, E., Tollis, I., Micheloyannis, S., & Zervakis, M. (2006). Time-significant wavelet coherence for the evaluation of schizophrenic brain activity using a graph theory approach. *Conference proceedings: Annual International Conference of the IEEE Engineering in Medicine and Biology Society. IEEE Engineering in Medicine and Biology Society. Annual Conference*, New York, pp. 4265–4268. https://doi.org/10.1109/IEMBS.2006.260680
65. Alvarado, N., & Burgos, S. (2004). Convergence properties of hyperbolic neural networks on riemannian manifolds. *NeurIPS 2024 Workshop on Mathematics of Modern Machine Learning*, Deutschland.
66. Stone, J. V. (2004). Independent component analysis: A tutorial introduction. *Technometrics, 49*(3), 357–359. https://doi.org/10.7551/mitpress/3717.001.0001
67. Hyvärinen, A., & Oja, E. (2000). Independent component analysis: Algorithms and applications. *Neural Networks, 13*(4–5), 411–430. https://doi.org/10.1016/S0893-6080(00)00026-5

5 Optimized Feature Selection and Neural Network-Based Classification of Motor Imagery Using EEG Signals

A Time-Frequency Approach

Muhammad Sudipto Siam Dip, Mohammod Abdul Motin, Md. Anik Hasan, and Sumaiya Kabir

5.1 INTRODUCTION

Motor imagery is a process that captures brain activity and analyzes it to interpret user actions. It is a spontaneous potential generated by the brain without external stimulus, with potential applications in brain–computer interfaces (BCIs), medical rehabilitation, brain-controlled devices, and prosthetic hands (Chen et al., 2018; Graimann et al., 2010; Sereshkeh et al., 2017). Since the electroencephalogram (EEG) is non-invasive and used in several types of brain applications, it is continuously gaining attention in developing different applications of BCIs, such as motor imagery classification (Zhang et al., 2023). However, the non-stationary nature of the brain signal picked up by the EEG makes it hard to recognize patterns in the signal and classify them as motor imagery (Sanei, 2013).

The automated methods for EEG-based motor imagery classification are broadly divided into two categories: those using hand-crafted features combined with traditional machine learning classifiers such as support vector machines (SVMs), linear discriminant analysis (LDA), and k-nearest neighbors, and those employing neural network-based automated feature extraction and classifiers (Bishop, 1995). The conventional features of EEG for motor imagery classification include spatial features (Brunner et al., 2007; Thomas et al., 2009; Wang et al., 2006), frequency-spatial features (Ang et al., 2008; Chin et al., 2009; Kumar et al., 2017), and temporal-frequency features (Hsu & Sun, 2009; Xu & Song, 2008). Among spatial feature

 DOI: 10.1201/9781003611448-8

extraction techniques, common spatial patterns and independent component analysis aim to enhance motor imagery-related features by extracting spatial patterns and separating EEG sources (Blankertz et al., 2007). Moreover, different statistical and entropy-based features have been previously reported (Hu et al., 2009; Khorshidtalab et al., 2013; Uktveris & Jusas, 2017).

Apart from the hand-crafted features with traditional classifiers, different researchers have also explored advanced deep learning-based frameworks (Al-Saegh et al., 2021). However, the direct input of raw EEG signals often suffers from several issues, including low signal-to-noise ratio, inherent non-stationarities, and baseline problems (Sun & Mou, 2023), and has limited interpretability. To overcome these problems, different signal processing techniques were applied to reduce or partially remove those noises before feature extraction and feed the features to deep learning models as classifiers. Among the different preprocessing techniques, bandpass filtering, independent component analysis, and artifact subspace reconstruction (ASR), were applied to improve signal quality (Al-Saegh et al., 2021; Chang et al., 2019).

In Corralejo et al. (2011), a genetic algorithm (GA) was incorporated with different types of features such as continuous and discrete wavelet transforms, spectral features, and other statistical techniques such as an autoregressive model and μ-based rhythm matched filter. However, due to GA's stochastic (random) nature, it can get stuck in what may be a good set but not necessarily the best/optimal set of features (Tang et al., 1996). Xinyang Yu et al. used principal component analysis (PCA) as their feature reduction technique with spatial filter-based feature extraction methods (Yu et al., 2014). However, PCA focuses on variance to reduce dimensionality and doesn't necessarily select the most discriminative features when classifying different motor imagery tasks. The sequential forward floating selection (SFFS) method has previously been used in channel selection for motor imagery-based BCI (Qiu et al., 2016). Bhatti et al. used sequential backward floating selection, an alternative version of SFFS, which provides effective feature selection with the RBFNN classifier and performs better than recursive feature elimination with the SVM classifier (Bhatti et al., 2019). Liu et al. showed that the firefly algorithm can be utilized for motor imagery classification by selecting the best subset of features (Liu et al., 2017). They implemented spectral regression discriminant analysis as their classifier and obtained an average accuracy of 70.20%. Malan and Sharma (2019) introduced regularized neighborhood component analysis (RNCA) for motor imagery classification with an SVM classifier, where RNCA with SVM outperformed ReliefF, GA, and PCA. However, RNCA required multiple parameters to be optimized to generalize the model. Jing Jiang et al. compared four different feature selection techniques, namely mutual information, least absolute shrinkage and selection operator (LASSO), PCA, and step-wise LDA, which were incorporated with common spatial pattern-based feature extraction. Among these, LASSO with the SVM achieved the best performance with 88.58% average accuracy (Jiang et al., 2020). Furthermore, Yao Guo et al. used the minimal-redundancy-maximal-relevance technique to eliminate features that were either irrelevant or redundant from the overall feature subset. Their method used an LDA classifier to enhance the classification performance (Guo et al., 2020).

Most of these previous studies mainly focused on improving motor imagery classification using existing feature selection techniques. However, few have explored efficient methods that target optimal information for the learning algorithm. As the number of features increases, dimensionality issues arise and computational complexity increases. The hypothesis is that employing a large number of relevant features specific to specific channels can resolve these issues without compromising accuracy. In this work, we extracted hand-crafted features from the statistical time domain, spectral domain, and time-frequency from EEG. We propose a hybrid feature selection technique based on modified SFFS to find out the optimal number of features from each channel to be fed into the neural network for EEG-based motor imagery classification. This paper is organized as follows: Section 5.2 presents our overall approach in a nutshell and details our strategy. Section 5.3 describes the experimental results and evaluation. Section 5.4 presents a brief analysis/outcome of the feature selection approach. Finally, the conclusion is given in Section 5.5.

5.2 DATASET AND PROPOSED FRAMEWORK

5.2.1 DATASET

In this work, we used the publicly available BCI Competition 2008—Graz dataset (Brunner et al., 2008). This dataset contains nine participants in total. The experimental setup for this dataset includes 22 channels of EEG with the left mastoid as a reference. As our analysis aimed at EEG signals, we only used the EEG channel recordings. The dataset consists of recordings from two sessions for each subject, with six runs per session, totaling 12 runs for each participant. Each run contains 48 trials, with 12 trials for each class, resulting in 288 trials in a single session. The sampling frequency for the recorded signal was 250 Hz. A bandpass filter was applied between 0.5 and 100 Hz, with a 50 Hz notch filter used to remove line noise. The amplifier voltage is constantly maintained at 100 μV during the session. The recording paradigm is illustrated in Figure 5.1.

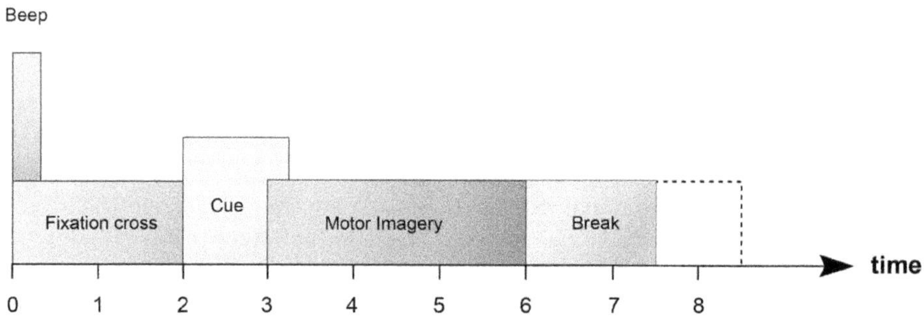

FIGURE 5.1 Experimental paradigm: Participants were instructed to complete the motor imagery task until the fixation cross disappeared from the screen at 6 s.

5.2.2 Proposed Framework

The block diagram of the proposed model is illustrated in Figure 5.2. The proposed model consists of four parts: (1) preprocessing, (2) feature extraction, (3) feature selection, and (4) model architecture and hyperparameter tuning.

5.2.3 Preprocessing

A finite impulse response bandpass filter was applied between 0.5 and 35 Hz. The ASR method was employed to clean up the noisy portion of the data (Chang et al., 2019). This process involved identifying and eliminating segments that were either discontinuous or noisy. Furthermore, the utilization of the independent component analysis technique was employed to identify and classify components that are associated with eye movements and muscle artifacts. EEGLAB (Delorme & Makeig, 2004) was used to preprocess the signal for these stages. Previous studies have demonstrated that the duration of motor imagery does not exceed 0.5 s (Neuper et al., 2009). We extracted epochs associated with the target label. These epochs were selected 0.2 s before the event onset and finished 0.5 s after the event onset for all the events, as shown in Figure 5.3. To capture the pre-stimulus activity, an additional 0.2 s before the event initiation was included in our study. This uniform temporal segmentation allowed us to capture any subtle pre-event neural activity preceding the actual stimulus.

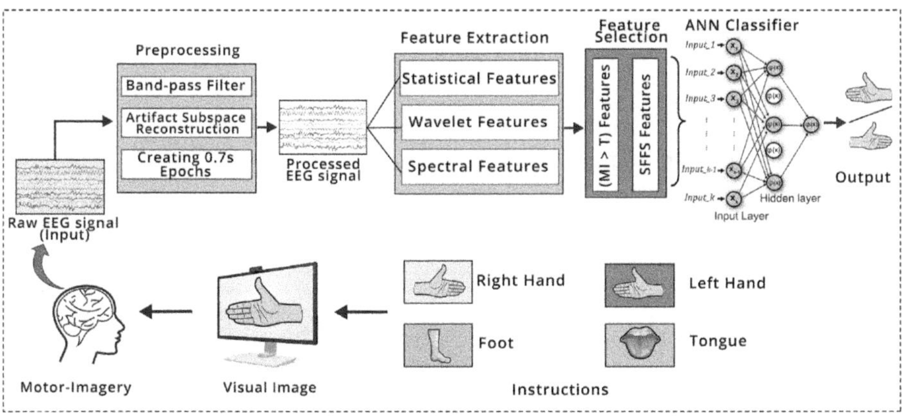

FIGURE 5.2 The overall framework of our approach is divided into two sections. Firstly, the raw dataset preprocessing takes place to improve the signal. This includes unwanted channel removal, filtering the signal to keep it within the low-frequency band region, and ASR to improve signal quality, followed by uniform temporal segmentation to create a 3D dataset. In the second section, the signal is then separated for validation, with one part for training the model and another for evaluation. Feature extraction and standardization are applied to both. Two-step feature selection takes place, and then, with the optimal set, a neural network is constructed, trained, and evaluated on the validation set using the same optimal feature set selected.

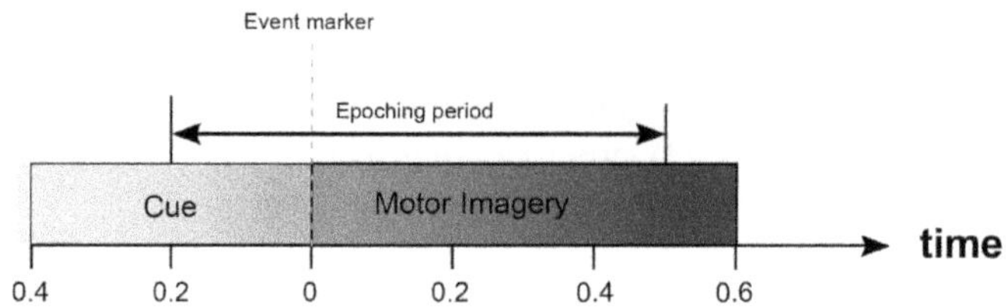

FIGURE 5.3 Uniform temporal segments of EEG data before and after stimuli onset.

5.2.4 Feature Extraction

A set of 418 distinctive features from the statistical time domain, spectral domain, and time-frequency domain were extracted from 22-channel EEG. The feature extraction pipeline encompassed three categories. In the first category, the power spectral density features using the Welch method (Welch, 1967) were explored. The spectral power distribution in delta, theta, alpha, low beta, mid beta, and beta was computed. The frequency range for each band is shown in Table 5.1. The calculation of the power spectral density of EEG epochs occurred in four steps, which include (1) overlapping segmentation, (2) applying a window function, (3) computing the periodogram, and (4) averaging the periodograms.

Let's assume $x[n]$ is an EEG epoch of length N divided into K overlapping segments, where each segment is of length L, and the overlap between segments is D. Then, a window function $w[m]$ is applied to each segment. Here, the kth segment of the signal is

$$x_k[n] = x[n+kD].\ w[n] \tag{5.1}$$

For $n=0, 1 \ldots L-1$ and $k=0, 1, \ldots K-1$.

Then, the periodogram for each of the segments was computed by performing the discrete Fourier transform of each windowed segment as follows:

$$P_k(f) = \frac{1}{L\sum_{n=0}^{L-1} w[n]^2} \left| \sum_{n=0}^{L-1} x_k[n] e^{-\frac{j2\pi fn}{L}} \right|^2 \tag{5.2}$$

where f is considered as the frequency bin. The average periodogram of the K segments was computed to get Welch's power spectral density estimate:

$$\hat{P}(k) = \frac{1}{K} \sum_{k=0}^{K-1} P_k(f) \tag{5.3}$$

In the second category of feature sets, the time-frequency domain features were extracted using the wavelet transform. The Morlet mother wavelet was used to extract wavelet coefficient features. The mother wavelet was utilized by the function:

$$\psi(t) = \pi^{-\frac{1}{4}} e^{i\omega_0 t} e^{-\frac{t^2}{2}} \tag{5.4}$$

The Morlet mother wavelet was used to decompose each channel's signal into a set of wavelet coefficients that represent the signal using the equation below:

$$W(a, b) = \frac{1}{\sqrt{a}} \int_{-\infty}^{+\infty} x(t)\psi^*\left(\frac{t-b}{a}\right) dt \tag{5.5}$$

where *W(a, b)* is the wavelet coefficient at scale *a* and translation *b*, *x(t)* is the input signal, and $\psi(t)$ is the Morlet wavelet function.

To enhance the relevancy of these coefficients, we compressed them using PCA into six distinct components that encapsulated the essence of the original wavelet coefficients. An array of six scales for the continuous wavelet transform (CWT), ranging from 1 to 128 on a logarithmic scale with a base of 2, was computed. The features from the first principal component were calculated.

Eventually, statistical time domain features were computed to capture essential information about limb movement-related signals' temporal dynamics and statistical properties. The features include mean, standard deviation, variance, root mean square, absolute difference signal, skewness, and kurtosis, as depicted in Table 5.1. Data standardization was applied with a standard scalar after constructing the tabular feature set. It centers and scales the data so that the mean is 0 and the standard deviation is 1.

TABLE 5.1
The List of Statistical and Spectral Features Extracted from Each EEG Channel

Feature Names	Frequency Range	Feature Names	Equations
Delta	0.5–4 Hz	Mean	$\mu = \frac{1}{n}\sum_{i=1}^{n} x_i$
Theta	4–8	Standard deviation	$\sigma = \sqrt{\sum_{i=1}^{n}(x_i - \mu)^2}$
Alpha	8–13	Variance	$\sigma^2 = \sum_{i=1}^{n}(x_i - \mu)^2$
Low beta	13–20	Root mean square (RMS)	$\text{RMS} = \sqrt{\frac{1}{n}\sum_{i=1}^{n} x_i^2}$
Mid beta	20–26	Absolute difference signal	$\text{ABS} = \lvert x_i - x_{i-1} \rvert$
High beta	26–35	Skewness	$\text{E}\left(\frac{x-\bar{x}}{\sigma}\right)^3$
–	–	Kurtosis	$\text{E}\left(\frac{x-\bar{x}}{\sigma}\right)^4$

5.2.5 Feature Selection

To determine the optimal number of features, we introduced a hybrid feature selection technique, as illustrated in Figure 5.4. The feature selection process was divided into three steps.

In the first step, mutual information-based feature selection was applied. Mutual information (MI) $I(X; Y)$ measures the mutual dependence between two variables, i.e., X and Y. Here, X represents the independent variable/feature matrix and Y represents the dependent variable or target variable. It is computed for each feature matrix with respect to the target variable.

The MI between those two was calculated as follows:

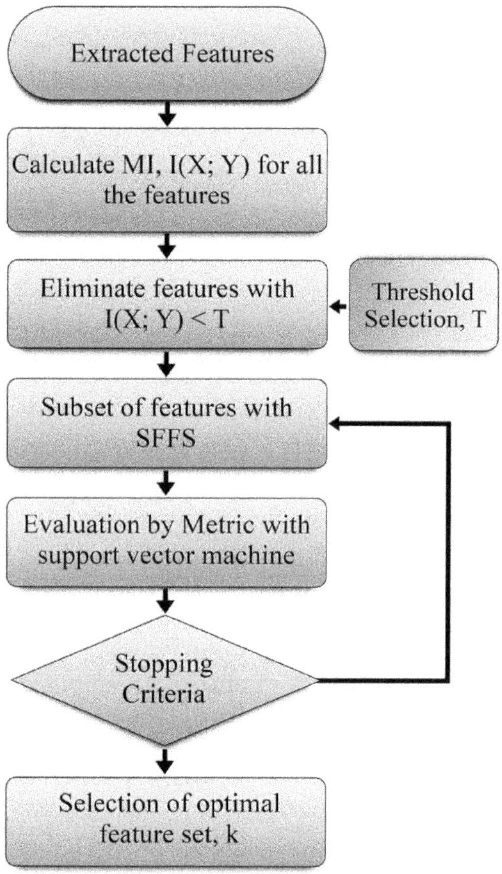

FIGURE 5.4 An overview of our hybrid feature selection procedure. The extracted features are used to calculate individual mutual information. Then, features are dropped based on a predefined threshold, and finally, the selected features are fed into the SFFS algorithm, which finds the optimal set of features with the help of SVM.

$$I(X;Y) = H(X) - H(X \mid Y) \tag{5.6}$$

Here, $H(X)$ represents the entropy of the feature matrix X, and $H(X, Y)$ is the combined/joint entropy of X and Y. The most informative features were selected by defining a threshold (T). Here, T is the minimum value of MI for which a feature is considered to be informative. Any value of mutual information below this is considered less informative, and the features that hold the MI value below this were rejected. Mathematically, for a given feature X_i and a defined threshold T, if $I(X_i;Y) < T$, then X_i was removed. We set the threshold to $T = 0.03$ and kept all the features that were greater than the threshold.

In the second step, we included SFFS-based feature selection. SFFS takes the set of all features, $Y = [y_1, y_2 \ldots y_d]$ of d-dimensions as input taken from the output from the first step, and it returns the final output as a subset of features $X_k = [x_j \mid j = 1, 2, \ldots, k; x_j \in Y]$, where the number of selected feature $k = (1, 2, \ldots, d)$. The size of the returned output from the SFFS algorithm is a subset of all the features and $k < d$. The algorithm initializes at $X_0 = \Phi$ and $k=0$, which denotes an empty set, also known as a "null set." Then, we performed sequential inclusion of each feature. The algorithm selected a feature x^+ from the remaining feature space that maximized the criterion function $J(X_{k+x})$, where x belongs to the set of unselected features $Y - X_k$. The selected feature was then added to the current subset (X_k), and k was incremented by one. Then, the algorithm proceeded to step 2.

In the third step, the conditional exclusion of features was applied. The algorithm evaluates whether removing a feature x^- from the current subset (X_k) would lead to an improvement in performance, as determined by the criterion function ($X_k - k$), where x belongs to the current subset (X_k). If $J((X_k - x) > J((X_k)$, indicating a potential performance gain, the feature x^- is removed from the subset (X_k), and k is decremented by one. The algorithm then returns to step 1. If no improvement can be made or if the subset size k reaches a minimum value of 2 (meaning the algorithm has reduced the feature set to only two features), the algorithm terminates. This implies that no further performance improvement could be achieved. The overall approach is outlined in Algorithm 1.

Step 1: Feature elimination with Mutual Information

For

$$I_i(X_i;Y_i) = H(X_i) - H(X \mid Y_i)$$

If

$$I_i > T \text{ then } S = S + S_i$$

end For

Initialize $Y = \{S_1, S_2, S_3 \ldots S_d\}$, d = no of selected feature

While the stop criterion has not been fulfilled

Step 2 (Inclusion)

$$x^+ = \arg\max J(X_k + x^+), \quad \text{where } x \in Y - X$$

(Continued)

(Continued)

$$K = k + 1$$

$$J_{\max} = \max\left(J\left(X_{k+1}\right), J_{\max}\right)$$

Step 3 (Exclusion)

$$x^{-} = \arg\max J\left(X_k - x^{-}\right), \quad \text{where } x \in X_k$$

If

$$J\left(X_k - x^{-}\right) > J\left(X_k\right):$$

Then

$$X_{k-1} = X_k - x^{-}$$

$$K = k - 1$$

Go to step 3
Else
Go to step 2

Algorithm I Modified Sequential Forward Floating Selection

5.2.6 Model Architecture and Hyperparameter Tuning

We applied a multi-layer perceptron of a fully connected neural network as a classifier. Keras Tuner (Team, 2022) was used to estimate different hyperparameters, as shown in Table 5.2. We tuned every model for every one-*vs*-one task separately. Among these, the best hyperparameters were selected based on randomly constructing neural networks with different combinations of these hyperparameters. The evaluations were done with a holdout validation of 80% training and 20% testing on all data. A random search was conducted in 100 trials for each classification task. The number of units in the hidden layers was set to be selected by random search between 20 and 30 neurons. To prevent over-fitting our model, we utilized dropout regularization. It randomly sets a fraction of the units (neurons) of a specific layer to zero during each training step. We set the dropout for random search to be chosen from the range of 0.1 (10%) to 0.9 (90%). We applied "ReLU" and "LeakyReLU" activation functions for the hidden layers and "sigmoid" for the output layer. The Adam and RMSprop-based optimizers were deployed in the hidden layer to optimize the cost function.

We incorporated binary cross-entropy as the loss function of our deep learning model. The loss function measures the dissimilarity between predicted probabilities and actual binary labels (0 or 1) for each sample. Mathematically, it is defined as

$$L\left(y, \hat{y}\right) = -\left[y \log\left(\hat{y}\right) + \left(1 - y\right) \log\left(1 - \hat{y}\right)\right] \tag{5.10}$$

TABLE 5.2
Estimated Architecture and Their Hyperparameters for Each Task

Task	No. of Hidden Layers	Dropout	Units	Activation Function	Optimizer
I	2	0.1	28	LeakyReLU	Adam
		0.5	27	LeakyReLU	
II	1	0.2	25	ReLU	RMSprop
III	2	0.6	26	LeakyReLU	Adam
		0.9	10	LeakyReLU	
IV	2	0.2	26	ReLU	Adam
		0.9	26	LeakyReLU	
V	1	0.4	23	ReLU	RMSprop
VI	2	0.1	26	ReLU	Adam
		0.6	8		

where y is the actual binary label of the sample and $\hat{y}$ is the predicted probability of the samples that belong to class 1. An $L2$ regularization was employed to control the complexity of the weights. We kept the regularization strength (or lambda) at 0.01. Table 5.2 shows the estimated outputs for each hyperparameter across all tasks.

5.2.7 Performance Evaluation

We implemented a leave-one-subject-out cross-validation technique to obtain the performance of our proposed method. The accuracy of the model was tested with one subject's data, and the rest were used for training. The accuracy was used as the performance metric to measure the model's performance.

5.3 RESULTS

This study focused on using EEG signals to identify four distinct motor imagery tasks: left-hand movement (Task I), right-hand movement (Task II), feet movement (Task III), and tongue movement (Task IV). Furthermore, each of the paired classification tasks is divided into six categories, as shown in Table 5.3.

The average accuracy of motor imagery classification using the proposed feature selection method for Tasks I, II, III, IV, V, and VI is 73.19%, 87.48%, 65.21%, 86.86%, 83.05%, and 82.38%, respectively. The classification accuracy of each subject using an optimal number of features is illustrated in Table 5.4. The average classification accuracy for all tasks without feature selection, with features selected by MI and with the proposed feature selection technique is 69.68 ± 3.21, 76.85 ± 5.64, and 79.69 ± 7.98, respectively. The average classification accuracy for each task without feature selection, using mutual information-based feature selection, and using the proposed feature selection is demonstrated in Figure 5.5. In contrast, Figure 5.6 illustrates the average accuracies of each task for individual subjects for different sets of features.

TABLE 5.3
Each Pairwise Classification Denoted as Tasks

Task No.	Classification
Task I	Left hand vs. right hand
Task II	Left hand vs. foot
Task III	Left hand vs. tongue
Task IV	Right hand vs. foot
Task V	Right hand vs. tongue
Task VI	Foot vs. tongue

TABLE 5.4
Validation Accuracies Obtained from the Optimal Number of Features

Sub	Task I	Task II	Task III	Task IV	Task V	Task VI
1	65.71	82.14	65.95	83.8	65.73	63.63
2	65.03	78.32	65.73	80.55	72.22	72.22
3	72.72	84.5	55.24	79.72	85.41	76.92
4	78.41	89.99	65.46	81.11	78.87	73.42
5	75.17	80.98	70.62	89.36	89.43	88.11
6	64.58	86.8	65.73	85.41	81.11	87.41
7	84.5	96.47	67.13	98.57	94.32	97.16
8	84.5	95.77	66.43	97.85	92.9	97.16
9	68.05	92.36	64.58	85.41	87.5	85.41
AVG	73.19	87.48	65.21	86.86	83.05	82.38
STD	7.98	6.55	4.11	7.08	9.54	11.56

The classification performance incorporated with our proposed hybrid feature selection techniques was compared with the state-of-the-art feature selection techniques-based work in Table 5.5. Our proposed approach outperformed the existing alternatives for all tasks except Task III.

In the initial stage of feature selection with MI, the choice of a suitable threshold is a challenging task. We empirically selected the threshold, which was 0.03. The performance of the model varies with the variation of the threshold, which is shown in Figure 5.7.

The proposed hybrid feature selection method also reduces the high dimensionality of features while working with a large number of feature sets. From the total set of extracted features, the number of features selected in Step 1 and Step 2 is compared in Figure 5.8. According to the figure, the first step achieves a significant reduction in features, around half of the total number of features.

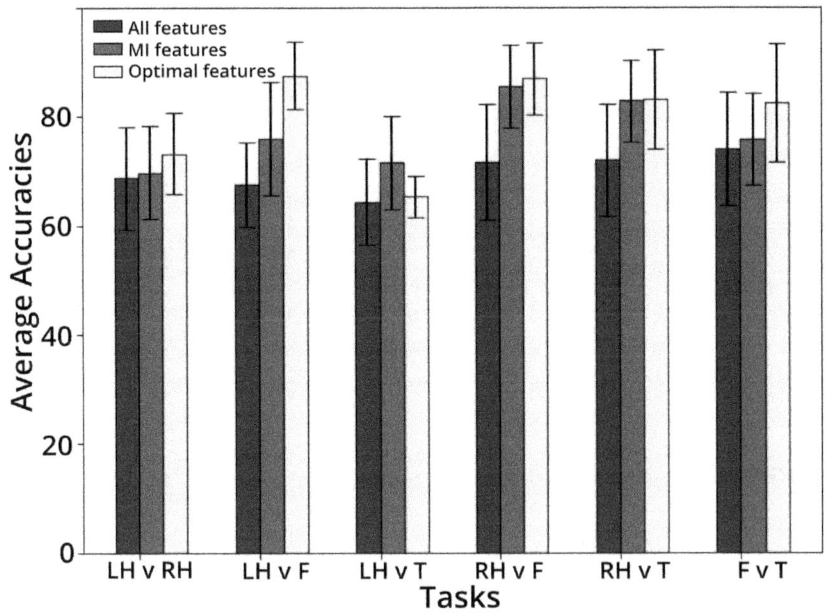

FIGURE 5.5 The average accuracies of each classification task across all subjects are depicted with the help of a comparative bar plot comparing three sets of features. Error bars are also included to show the deviation of accuracies for each.

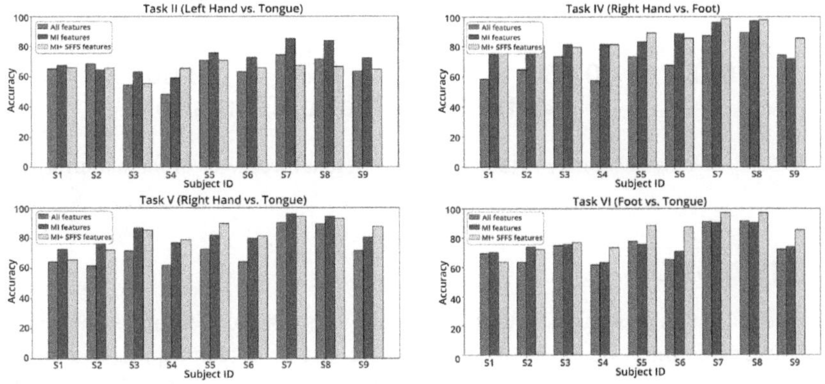

FIGURE 5.6 The best-performing accuracy for each task was compared among all features with those selected manually with higher mutual gain and the optimal features for each subject. The proposed feature selection techniques outperformed all tasks except Task III (LH v T).

TABLE 5.5
Performance Comparison of the Different Feature Selection Algorithms in Classifying Different Motor Imagery Tasks

Feature Selection	Task I	Task II	Task III	Task IV	Task V	Task VI	Avg
Mutual information	69.73	75.89	71.48	85.46	82.82	75.76	76.85
MRMR	71.27	62.26	65.87	64.47	64.66	66.34	65.81
RFE-SVM	70.71	63.82	70.02	62.49	66.05	70.03	67.18
Firefly algorithm	71.87	67.79	62.49	65.89	70.76	72.42	68.53
SFS-SVM	70.63	81.88	65.73	62.79	65.16	75.09	70.21
Proposed	73.19	87.48	65.21	86.86	83.05	82.38	79.70

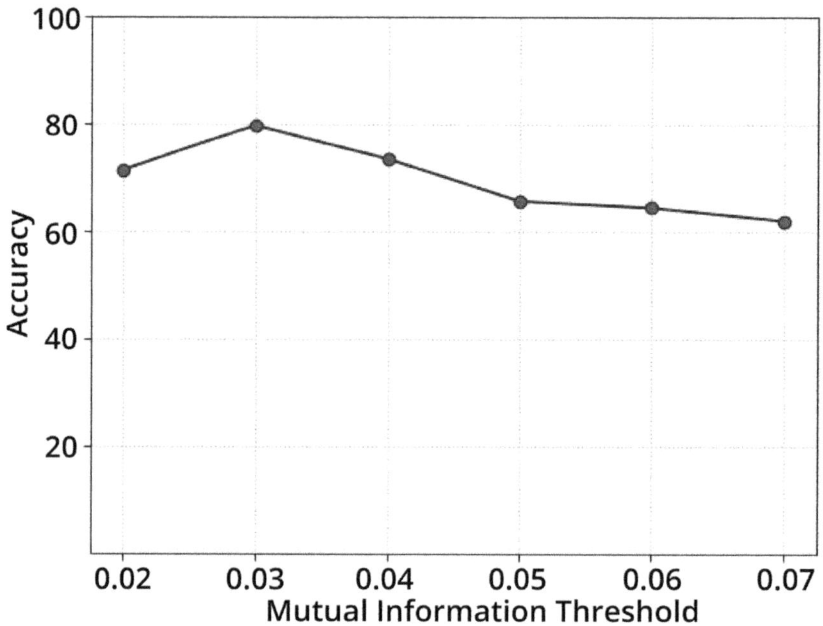

FIGURE 5.7 Average accuracy obtained using different thresholds of mutual information.

5.4 DISCUSSION

We propose a hybrid approach that combines two distinct feature selection techniques, leverages the robustness of those feature selection techniques, and enhances classification accuracy. In the initial stage, feature selection based on MI is employed to decrease additional dimensions and eliminate redundant features. Subsequently, the SFFS method identifies the optimal subset of features from those selected in the first stage. Since redundant features have already been eliminated in the initial

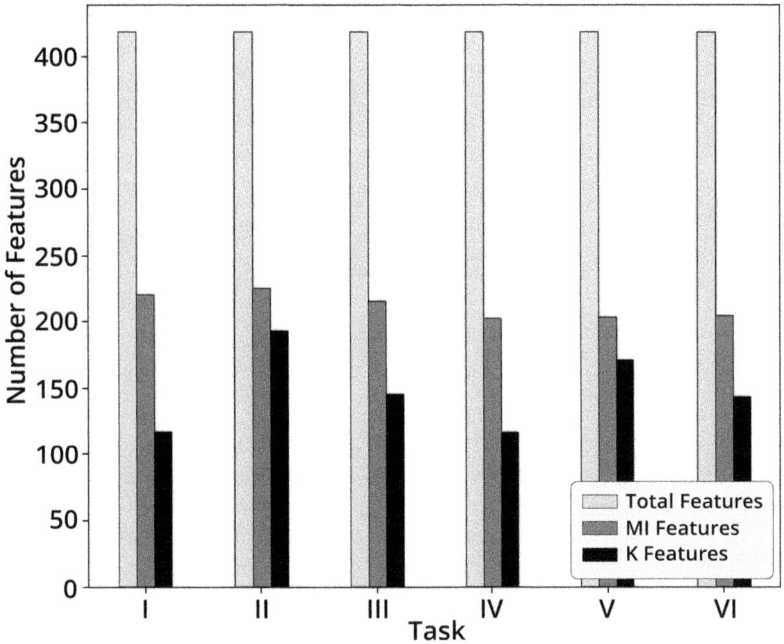

FIGURE 5.8 A comparison of the number of features that were selected by the first stage, i.e., mutual information, vs. the second stage, which was SFFS, for all six tasks.

stage, the SFFS algorithm sequentially explores various combinations of features without needing to select from a broader range, thus reducing the dimensionality. The proposed approach of feature selection is reliable in selecting the best subset of features from a diverse set of features. EEG recordings capture brain activity through multiple channels. Each channel records a unique signal, differing in amplitude, frequency, signal-to-noise ratio, and other characteristics. There's a chance that a single set of features from a few channels cannot capture all effective information for motor imagery classification. Employing a large number of features relevant to EEG signals can solve this problem but increases the computational cost due to the high dimensionality of the feature sets. Since our proposed approach uses multiple steps to reduce dimensionality to find an optimal number of features from a large pool of features, it is well-suited to handle large feature sets.

As shown in Figure 5.9, when locating the optimal set, the performance of Tasks I and III does not differ significantly for different sets of features, and several sets of features perform almost as well as the best optimal set. With the help of the proposed feature selection scheme, the most contributing channels were estimated by comparing the number of features that were selected from each channel to the maximum number of features that were selected by a channel for that specific classification task. The following mathematical expression was used to visualize the saliency maps:

$$\text{Channel Significance} = k_i \,/\, k_m \tag{5.6}$$

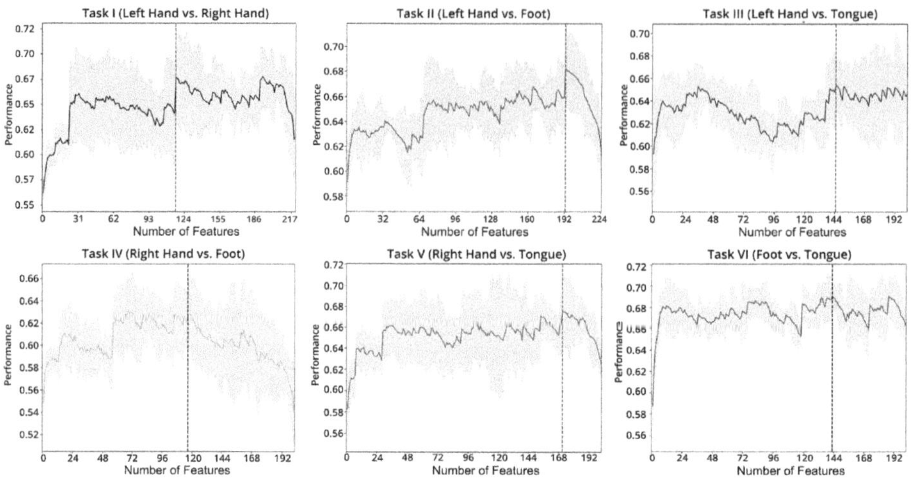

FIGURE 5.9 The optimal feature set is the final set of features that were selected for each classification task. The number of features chosen with respect to each channel implies the contribution of that channel to that specific task.

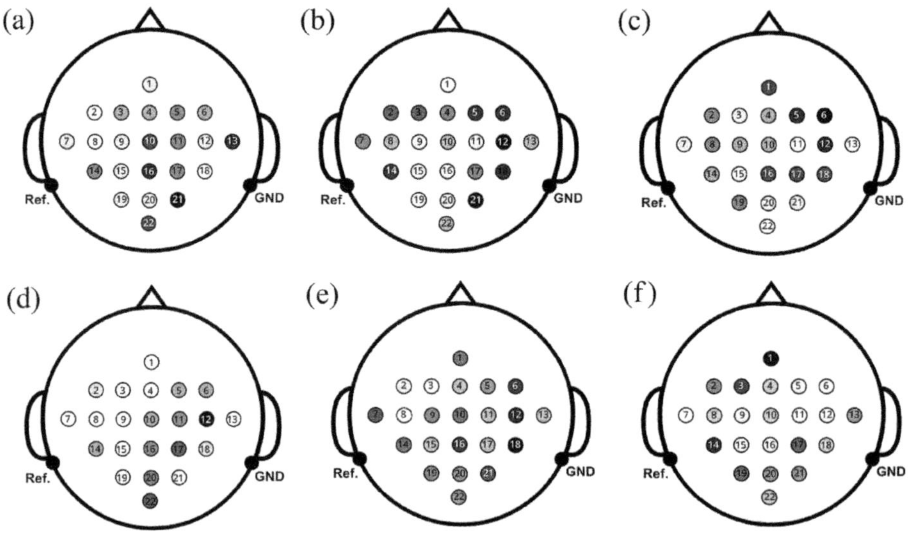

FIGURE 5.10 The optimal feature set is the final set of features that were selected for each classification task. The number of features that are selected with respect to each channel implies the contribution of that channel for that specific classification task. For one task, the frequency of some channels is higher than others, indicating their importance for that task. The number of selected features for each channel is shown on the heat map and also on the headset. The heat map depicts the priority of each value, with darker colors indicating higher priority, meaning a greater number of features are selected from that channel. The six heat maps and illustrated salient maps depict (a) Task I, (b) Task II, (c) Task III, (d) Task IV, (e) Task V, and (f) Task VI.

TABLE 5.6
The Comparison of the Proposed Work with the Existing Work in the Literature for Motor Imagery Classification Using EEG Signal

Study	Dataset	Feature	Classifier	Acc
Lawhern et al. (2018)	BCIC IV 2a	Time series	CNN	67.00
Luo et al. (2018)	BCIC IV 2a	Spatial frequency	RNN-GRU	73.56
Wang et al. (2018)	BCIC IV 2a	Calculated features	LSTM	70.77
Hassanpour et al. (2019)	BCIC IV 2a	Time series	SAE	71.08
Guo et al. (2020)	BCIC III 4a	FCCSP	LDA	82.01
Liu et al. (2017)	BCIC IV 2a	Calculated features	SRDA	70.20
Jiang et al. (2020)	BCIC (III 3a & 4a, IV-1)	CSP+LASSO	SVM	88.58
Yu et al. (2014)	BCIC III 4a	CAR, LAP, CSP+PCA	SVM	74.59
Malan and Sharma (2019)	BCIC (II-3 & IV-2b)	DTCWT+RNCA	SVM	80.70
Proposed work	BCIC IV 2a	Optimized features	FCNN	79.70

where k_i is the number of features selected for the *i*th channel, and k_m is the maximum number of features that are selected for the *m*th channel. We have plotted these estimated values of the spatial significance of each channel with a heat map and the location of channels on the scalp. All the saliency maps are shown in Figure 5.10, which indicates the significance of channels. For example, Figure 5.9b shows the highest number of features selected from the right side of the scalp and from the middle and front. Therefore, the information associated with the channel connected to its respective region exerts the most significant impact on classifying samples between the left hand and foot. Analysis of Table 5.4 reveals that Tasks II and IV achieved higher accuracy compared to other classification tasks. This suggests a potential correlation between higher accuracy and the ability to identify optimal channel priorities.

The comparison of the proposed work with existing literature for motor imagery classification is demonstrated in Table 5.6. As shown in Table 5.6, our model outperforms most of the reported results where the BCIC IV 2a benchmark dataset was used. The accuracy reported by Malan and Sharma (2019), Jiang et al. (2020), and Guo et al. (2020) was slightly higher than that of our model. This might be due to the use of different datasets that led to this kind of performance variation.

5.5 CONCLUSION

This work presents a hybrid feature selection technique to select features from a diverse set of EEG features and utilizes those selected features with a shallow neural network architecture for motor imagery classification. The proposed method can effectively discriminate between two different motor planning scenarios with an average accuracy of 79.70% and a standard deviation of 7.98%. The performance of our proposed model with different feature selection scenarios outperforms the existing state-of-the-art feature selection alternatives for EEG feature selection for motor imagery classification. These findings can potentially advance the development of different motor imagery applications using EEG signals.

REFERENCES

Al-Saegh, A., Dawwd, S. A., & Abdul-Jabbar, J. M. (2021). Deep learning for motor imagery EEG-based classification: A review. *Biomedical Signal Processing and Control*, *63*, 102172.

Ang, K. K., Chin, Z. Y., Zhang, H., & Guan, C. (2008). Filter bank common spatial pattern (FBCSP) in brain–computer interface. *2008 IEEE International Joint Conference on Neural Networks (IEEE World Congress on Computational Intelligence)*, Hong Kong, China, pp. 2390–2397, doi: 10.1109/IJCNN.2008.4634130.

Bhatti, M. H., Khan, J., Khan, M. U. G., Iqbal, R., Aloqaily, M., Jararweh, Y., & Gupta, B. (2019). Soft computing-based EEG classification by optimal feature selection and neural networks. *IEEE Transactions on Industrial Informatics*, *15*(10), 5747–5754.

Bishop, C. M. (1995). *Neural Networks for Pattern Recognition*. Oxford University Press, Oxford.

Blankertz, B., Tomioka, R., Lemm, S., Kawanabe, M., & Muller, K.-R. (2007). Optimizing spatial filters for robust EEG single-trial analysis. *IEEE Signal processing magazine*, *25*(1), 41–56.

Brunner, C., Leeb, R., Müller-Putz, G., Schlögl, A., & Pfurtscheller, G. (2008). BCI Competition 2008–Graz data set A. *Laboratory of Brain–Computer Interfaces*, *16*, 1–6.

Brunner, C., Naeem, M., Leeb, R., Graimann, B., & Pfurtscheller, G. (2007). Spatial filtering and selection of optimized components in four class motor imagery EEG data using independent components analysis. *Pattern Recognition Letters*, *28*(8), 957–964.

Chang, C.-Y., Hsu, S.-H., Pion-Tonachini, L., & Jung, T.-P. (2019). Evaluation of artifact subspace reconstruction for automatic artifact components removal in multi-channel EEG recordings. *IEEE Transactions on Biomedical Engineering*, *67*(4), 1114–1121.

Chen, X., Zhao, B., Wang, Y., Xu, S., & Gao, X. (2018). Control of a 7-DOF robotic arm system with an SSVEP-based BCI. *International Journal of Neural Systems*, *28*(08), 1850018.

Chin, Z. Y., Ang, K. K., Wang, C., Guan, C., & Zhang, H. (2009). Multi-class filter bank common spatial pattern for four-class motor imagery BCI. *2009 Annual International Conference of the IEEE Engineering in Medicine and Biology Society*, Minneapolis, MN, USA, pp. 571–574, doi: 10.1109/IEMBS.2009.5332383.

Corralejo, R., Hornero, R., & Álvarez, D. (2011). Feature selection using a genetic algorithm in a motor imagery-based brain computer interface. *2011 Annual International Conference of the IEEE Engineering in Medicine and Biology Society*, Boston, MA, USA, pp. 7703–7706, doi: 10.1109/IEMBS.2011.6091898.

Delorme, A., & Makeig, S. (2004). EEGLAB: An open source toolbox for analysis of single-trial EEG dynamics including independent component analysis. *Journal of Neuroscience Methods*, *134*(1), 9–21.

Graimann, B., Allison, B., & Pfurtscheller, G. (2010). Brain–computer interfaces: A gentle introduction. In: Graimann, B., Pfurtscheller, G., & Allison, B. (eds) *Brain–Computer Interfaces: Revolutionizing Human-Computer Interaction* (pp. 1–27). Springer, Berlin.

Guo, Y., Zhang, Y., Chen, Z., Liu, Y., & Chen, W. (2020). EEG classification by filter band component regularized common spatial pattern for motor imagery. *Biomedical Signal Processing and Control*, *59*, 101917.

Hsu, W.-Y., & Sun, Y.-N. (2009). EEG-based motor imagery analysis using weighted wavelet transform features. *Journal of Neuroscience Methods*, *176*(2), 310–318.

Hu, J., Xiao, D., & Mu, Z. (2009). Application of energy entropy in motor imagery EEG classification. *International Journal of Digital Content Technology and Its Applications*, *3*(2), 83–90.

Hassanpour, A., Moradikia, M., Adeli, H., Khayami, S. R., & Shamsinejadbabaki, P. (2019). A novel end-to-end deep learning scheme for classifying multi-class motor imagery electroencephalography signals. *Expert Systems*, *36*(6), e12494.

Jiang, J., Wang, C., Wu, J., Qin, W., Xu, M., & Yin, E. (2020). Temporal combination pattern optimization based on feature selection method for motor imagery BCIs. *Frontiers in Human Neuroscience*, *14*, 231.

Khorshidtalab, A., Salami, M.-J. E., & Hamedi, M. (2013). Robust classification of motor imagery EEG signals using statistical time–domain features. *Physiological Measurement*, *34*(11), 1563.

Kumar, S., Sharma, A., & Tsunoda, T. (2017). An improved discriminative filter bank selection approach for motor imagery EEG signal classification using mutual information. *BMC Bioinformatics*, *18*, 125–137.

Lawhern, V. J., Solon, A. J., Waytowich, N. R., Gordon, S. M., Hung, C. P., & Lance, B. J. (2018). EEGNet: a compact convolutional neural network for EEG-based brain–computer interfaces. *Journal of Neural Engineering*, *15*(5), 056013.

Liu, A., Chen, K., Liu, Q., Ai, Q., Xie, Y., & Chen, A. (2017). Feature selection for motor imagery EEG classification based on firefly algorithm and learning automata. *Sensors*, *17*(11), 2576.

Luo, T.-j., Zhou, C.-l., & Chao, F. (2018). Exploring spatial-frequency-sequential relationships for motor imagery classification with recurrent neural network. *BMC Bioinformatics*, *19*(1), 1–18.

Malan, N. S., & Sharma, S. (2019). Feature selection using regularized neighbourhood component analysis to enhance the classification performance of motor imagery signals. *Computers in Biology and Medicine*, *107*, 118–126.

Neuper, C., Scherer, R., Wriessnegger, S., & Pfurtscheller, G. (2009). Motor imagery and action observation: Modulation of sensorimotor brain rhythms during mental control of a brain–computer interface. *Clinical Neurophysiology*, *120*(2), 239–247.

Qiu, Z., Jin, J., Lam, H.-K., Zhang, Y., Wang, X., & Cichocki, A. (2016). Improved SFFS method for channel selection in motor imagery based BCI. *Neurocomputing*, *207*, 519–527.

Sanei, S. (2013). *Adaptive Processing of Brain Signals*. John Wiley & Sons, New York.

Sereshkeh, A. R., Trott, R., Bricout, A., & Chau, T. (2017). Online EEG classification of covert speech for brain–computer interfacing. *International Journal of Neural Systems*, *27*(08), 1750033.

Sun, C., & Mou, C. (2023). Survey on the research direction of EEG-based signal processing. *Frontiers in Neuroscience*, *17*, 1203059.

Tang, K.-S., Man, K.-F., Kwong, S., & He, Q. (1996). Genetic algorithms and their applications. *IEEE Signal Processing Magazine*, *13*(6), 22–37.

Team, K. (2022). Keras documentation: KerasTuner. Available: https://keras.io/keras_tuner/. [Accessed: 05-Feb-2022].

Thomas, K. P., Guan, C., Lau, C. T., Vinod, A. P., & Ang, K. K. (2009). A new discriminative common spatial pattern method for motor imagery brain–computer interfaces. *IEEE Transactions on Biomedical Engineering*, *56*(11), 2730–2733.

Uktveris, T., & Jusas, V. (2017). Application of convolutional neural networks to four-class motor imagery classification problem. *Information Technology and Control*, *46*(2), 260–273.

Wang, P., Jiang, A., Liu, X., Shang, J., & Zhang, L. (2018). LSTM-based EEG classification in motor imagery tasks. *IEEE Transactions on Neural Systems and Rehabilitation Engineering*, *26*(11), 2086–2095.

Wang, Y., Gao, S., & Gao, X. (2006). Common spatial pattern method for channel selelction in motor imagery based brain–computer interface. *2005 IEEE Engineering in Medicine and Biology 27th Annual Conference*, Shanghai, China, pp. 5392–5395, doi: 10.1109/IEMBS.2005.1615701.

Welch, P. (1967). The use of fast Fourier transform for the estimation of power spectra: A method based on time averaging over short, modified periodograms. *IEEE Transactions on Audio and Electroacoustics*, *15*(2), 70–73.

Xu, B.-G., & Song, A.-G. (2008). Pattern recognition of motor imagery EEG using wavelet transform. *Journal of Biomedical Science and Engineering*, *1*(1), 64.

Yu, X., Chum, P., & Sim, K.-B. (2014). Analysis the effect of PCA for feature reduction in non-stationary EEG based motor imagery of BCI system. *Optik*, *125*(3), 1498–1502.

Zhang, H., Zhou, Q.-Q., Chen, H., Hu, X.-Q., Li, W.-G., Bai, Y., Han, J.-X., Wang, Y., Liang, Z.-H., & Chen, D. (2023). The applied principles of EEG analysis methods in neuroscience and clinical neurology. *Military Medical Research*, *10*(1), 67.

6 Electroencephalogram-Based Driver Drowsiness Detection Using Entropy Features with Lightweight Deep Learning Model

Md. Anik Hasan, Mohammod Abdul Motin, Muhammad Sudipto Siam Dip, and G. K. M. Hasanuzzaman

6.1 INTRODUCTION

Drowsiness is a transitional state between wakefulness and sleep characterized by decreased alertness and cognitive function. In the state of drowsiness, the brain's electrical oscillations transition from a rapid and desynchronized pattern to a slower and more synchronized one [1]. It significantly impairs reaction time, posing a serious risk to road safety [2]. Distracted driving reduces the driver's attentiveness, concentration, and focus, contributing to 14%–20% of all traffic accidents worldwide [3]. Long-distance drivers are at an increased risk of drowsiness, which can be caused by sleep deprivation, monotonous driving, medical conditions, or drug or alcohol consumption. It presents a potential danger to the driver and other passengers, drivers of other vehicles, cyclists, and pedestrians in terms of the risk of injuries and deaths. An automatic drowsiness detection and monitoring system is necessary to alert the driver right before they begin to feel drowsy. Driver drowsiness detection systems are techniques that can monitor and determine a driver's alertness accurately. These systems can be used to develop real-time monitoring systems for a driver's drowsy state.

There are different paradigms of studies that have been of interest to researchers in recent decades, and these can be divided into three categories. Firstly, the vehicle-based system is detected by monitoring the variability in driving, such as motion, acceleration, and anomaly in movement. These methods are easy to implement and non-invasive, but they are affected by external factors. Secondly, the systems monitor the driver's behavioral patterns and predict judgment. Such systems can involve eye movement or facial expressions [4] and can be determined by data collected from external devices such as cameras. These methods are also non-invasive and can capture facial expressions and eye movements, but they require good lighting and camera quality. Thirdly, there are systems that observe physiological changes in the body. Different sensors can be attached

DOI: 10.1201/9781003611448-9

to the body for the acquisition of physiological signals such as photoplethysmography, electrooculogram (EOG), electrocardiogram (ECG), and electroencephalogram (EEG). The first two methods measure external physiological behaviors, which are the result of the different neurological states. They cannot capture the inherent information of the different brain states directly and are influenced by external factors that may lead to false classification results. However, the third method utilizes different physiological signals that directly carry the information content of the neurophysiological activities, considered a good indicator of drowsiness.

EEG-based frameworks for detecting drowsiness have become very popular in recent years. It is possible to produce more accurate and reliable methods with EEG [5]. Apart from performance, recent studies have also focused on reducing the number of channels and improving comfort in the real-time acquisition of EEG signals. The improvement of classification accuracy may vary across the segment length of EEG selected for the study. Different studies show that increasing or decreasing the segment length may significantly change model performance [6].

Chen et al. [7] proposed a novel network construction method and used graph theory to extract features for support vector machine (SVM) classification. Chai et al. [1] introduced entropy rate bound minimization analysis with independent component analysis to separate the source signal. Then, the multivariate auto-regressive model features are extracted and fed to the Bayesian neural network for classification. In [8], the author utilized different entropy, statistical, and frequency-domain features with SVM to classify drowsy and alert states. Kaur et al. [9] proposed the EMD-IMF method for feature extraction, explored different machine learning classifiers, and found that artificial neural network (ANN) outperformed other types of classifiers. Stancin et al. [10] exploited the ratio between different band powers as features input to an XGBoost classifier in detecting the drowsiness state. Chai et al. [11] proposed a novel EEG-based driver fatigue classification method by combining an AR modeling feature extractor and sparse-DBN classifier. Chaudhuri et al. [12] proposed a framework for alertness level detection using EEG source localization and chaotic entropy measures. Similarly, Subasi et al. [13] introduced a flexible analytic wavelet transform with different sub-band feature search spaces, and different boosting and classical models were trained with this feature to classify the vigilance space into two parts: fatigue and alert. Houshmand et al. [14] also proposed EEG, steering wheel, and vehicle motion data-based drowsiness alert systems employing a classifier tree-based learning system trained on statistical, entropy, and different sub-band features, respectively. An analysis was conducted on the power spectrum of EEG sub-bands. The results indicated a significant rise in the power of the Alpha and Theta bands during the transition from an alert to a drowsy state [15]. Budak et al. [16] suggested employing an ensemble majority voting approach with three deep networks. These networks were trained using different EEG data to classify the vigilance state into two distinct classes: awake and drowsy. Lan et al. [17] proposed a fusion of EEG and vehicle motion data. They extracted band ratio features from EEG, sample entropy, and standard deviation from vehicle information and selected the most important features using correlation analysis and classification with SVM. On the other hand, Li et al. [18] introduced a posterior probabilistic model based on an SVM. This model utilized the theta, alpha, and beta sub-bands of EEG data to convert the level of drowsiness into a value ranging from 0 to 1.

The recent progress in deep learning has created numerous possibilities in the domain of information processing and the identification of intricate patterns in signals. For example, Cui et al. [19] proposed a compact and interpretable convolutional neural network (CNN) to discover the intrinsic shared feature across subjects through the class activation map method using a single-channel signal. In another study, Cui et al. [20] used a separable convolution process for the spatial–temporal sequence of EEG signals. After speculating on the model, they found that it had learned meaningful biological features like alpha spindles, which are strong indicators of drowsiness. Similarly, Gao et al. [21] introduced a three-block CNN layer to extract temporal dependencies from the signal and then a fully connected layer to fuse these temporal features and realize drowsy state classification. Lee et al. [5] employed an LSTM-CNN network to recognize alert, sleep, and drowsy states with a multiple-segment size. To reveal the intrinsic inter-channel relation in a drowsy state and classify the fatigue neural state, Chen et al. [22] proposed a novel Self-Attentive Channel-Connectivity Capsule Network (SACC-CapsNet). Several pre-trained models, such as AlexNet, ResNet, and EEGNet, provide reliable performance when it comes to classification problems [23]. Specialized machine learning models have also performed well when it comes to driver drowsiness detection.

Although these methods have achieved notable success, they also have drawbacks, such as theoretical limitations and many constraints associated with classical algorithms, which restrict their practical implementation in real-world applications. Previous approaches utilized several feature extraction techniques in single- or multi-modal formats to detect drowsiness. Various techniques were utilized with multi-modal inputs, such as EEG+EOG [24], EEG+EOG+EMG [25], EEG+ECG+EOG+fNIRS [26], and EEG+EMG+respiration [27]. These approaches involved extracting features using distinct models, which were subsequently merged to yield the outcomes. While their main focus was on enhancing performance, the utilization of several process models to handle different types of input modalities increases both the size and complexity of the decision module. Consequently, this hampers real-time implementation. Furthermore, as previously stated, techniques that rely on EEG profiles are more accurate in their scope. An additional crucial factor is that the performance of the model relies on two key elements: the initial learning framework and the appropriate selection of the classification model. However, the most commonly used techniques involve the utilization of classical classification models such as SVM, logistic regression (LR), K-nearest neighbor (KNN), random forest, decision trees (DTs), CART, LAD-tree, and others. In addition to theoretical bottlenecks, these models frequently present challenges related to the curse of dimensionality, overfitting, and instability while handling large or small volumes of data. On the other hand, deep learning introduces a great degree of freedom because of multiple-layer structures but usually shows good learning ability. It functions as a black box, without a clear understanding of the reasons for greater classification performance.

Thus, it is essential to develop more accurate solutions by implementing new feature extraction methods, utilizing a more comprehensive version of learning models, or both, to enhance drowsiness detection and streamline human–machine interaction. This chapter addresses a fusion of multiple entropy features with a deep learning model, ensuring a robust and realistic implementation for classifying driver drowsiness states. The main contributions of this paper are as follows:

1. This work proposes an entropy-based deep learning model for driver drowsiness detection using EEG signals. It identifies the significant features from a pool of diverse feature sets. The importance of optimal channel selection was also investigated.
2. The performance of the proposed model was validated using publicly available datasets and compared with the existing state of studies in the literature. The robustness of the model was validated using subject-independent leave-one-out subject-wise cross-validation along with K-fold cross-validation.

The rest of the chapter is organized as follows: Section 6.2 describes the key attributes of the proposed framework. The experimental results are provided in Section 6.3, while the discussion based on the results is incorporated in Section 6.4. Finally, Section 6.5 concludes the work.

6.2 PROPOSED FRAMEWORK

The proposed framework is illustrated in Figure 6.1. The details of the proposed framework are described in this section, including the data acquisition procedure and pre-processing; feature extraction and theoretical framework; and classifier modeling and evaluation criteria.

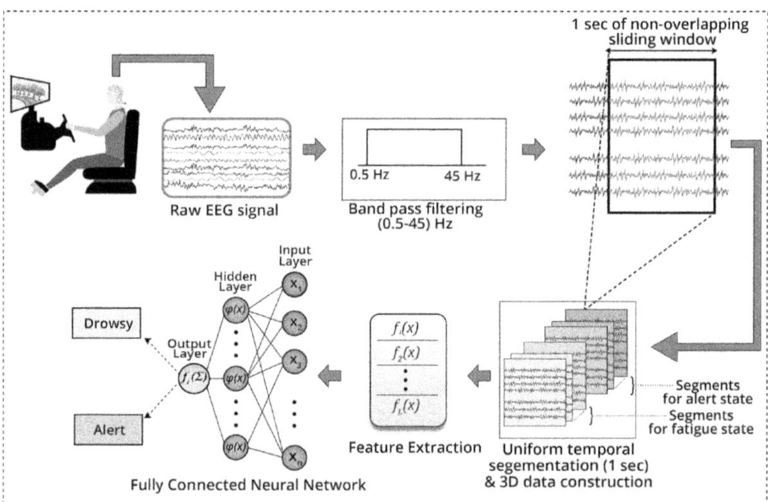

FIGURE 6.1 EEG data was acquired during alert and drowsy states. This data was pre-processed using a filter to remove noise and artifacts. We segmented the EEG recording using a non-overlapping sliding window of one second. Four different types of distinct feature sets were extracted to convert the data into a two-dimensional matrix. This data was trained and tested using leave-one-out cross-validation and five-fold cross-validation techniques for the detection of drowsiness or alert states.

6.2.1 Data Acquisition Procedure and Pre-Processing

In this study, we used a publicly available dataset published by Min et al. [28]. The dataset includes a total of 12 male young individuals in good health, with ages ranging from 19 to 24 years. The participants were instructed to abstain from stimulating activities, such as consuming alcohol or coffee. Before the commencement of the experiment, the participants were acquainted with the experimental protocol. EEG signals were recorded using a 32-channel electrode cap consisting of 30 active channels and 2 reference channels, with a sampling rate of 1,000 Hz; the electrodes of the cap were positioned following the international 10–20 system. The channel data was calibrated with respect to two mastoids, A_1 and A_2, which were electrically connected.

The EEG data was recorded in two phases. The first phase lasted 20 min; in the final 5 min of the first phase, the EEG signals were recorded and set as the alert state. During the second phase, continuous driving activity continued for 40–100 min, or until the subject's self-reported fatigue questionnaire showed that they were tired from driving. Both the Chalder Fatigue Scale and Li's Subjective Fatigue Scale were used to assess the fatigue state. In the last 5 min of the second stage, the EEG signals were recorded and classified as a drowsy state.

As illustrated in Figure 6.1, the raw EEG signals captured by the electrodes were passed through a band-pass filter to remove noise; the lower and higher cut-off frequencies of the band-pass filter were set to 0.5 and 45 Hz, respectively. Furthermore, the recorded session was segmented into 1-second epochs to produce approximately 300 epochs. As stated earlier, both the alert and drowsy states had a duration of five minutes for each individual, resulting in a total of 7,200 epochs, with 3,600 epochs from the alert state and 3,600 epochs from the drowsy state.

6.2.2 Feature Extraction

6.2.2.1 Statistical Feature

This study employed time-domain statistical features to characterize the EEG signals. Statistical features are computationally efficient and can effectively capture changes in the signal's amplitude, variability, and shape, often indicative of different vigilance states. The following ten statistical features were extracted from each EEG channel segment: mean, median, variance, standard deviation, skewness, kurtosis, first difference, normalized first difference, second difference, and normalized second difference.

6.2.2.2 Hjorth Parameters

Hjorth parameters, namely, Hjorth activity, mobility, and complexity, extract statistical properties from the EEG signal in the time domain. These parameters are features based on the variance of the signal's derivatives [29]. Hjorth activity measures the signal's variance, indicating the power of the EEG signal. A high activity value suggests a more powerful (greater amplitude) signal, which can indicate certain brain states or responses. Mobility is a measure of the standard deviation of the signal's power spectrum, reflecting the mean frequency or the rate of change in the signal. The complexity parameter measures how frequently the signal's frequency changes. It is a measure of the irregularity of a signal. It evaluates how closely the signal resembles a pure sine wave, with the value approaching 1 when it closely mirrors this sine waveform. If it deviates from the structured, predictable data, the complexity increases beyond unity. These parameters were chosen for their ability to capture information about the EEG signal's temporal dynamics and their utility in various EEG-based applications.

6.2.2.3 Band Power Feature

The energy of EEG is segmented into five distinct bands, each correlating with particular physiological states. These bands include delta (δ: 0.5–4 Hz), theta (θ: 4–8 Hz), alpha (α: 8–14 Hz), beta (β: 14–25 Hz), and gamma (γ: 25–45 Hz) frequencies. The Fast Fourier Transform was used to convert the signal from the time domain to the frequency domain. The power spectral density is calculated and integrated into distinct frequency bands to find the band power.

6.2.2.4 Entropy Feature

Entropy is a nonlinear parametric measure that quantifies the complexity of a time series. In this experiment, we employed various types of entropy measures, including approximate entropy, sample entropy, spectral entropy, and fuzzy entropy, to form a feature set of entropy that distinguishes the drowsy state from the alert state.

6.2.2.4.1 Approximate Entropy

Approximate entropy is a nonlinear dynamic parameter that can determine the regularity of signals and is well-suited for categorizing complex systems [30]. The equations for calculating approximate entropy are given below.

If the time series of the signal contains N samples, then $X = \left[\ x(1), x(2), \ldots..\ x(n)\right]$. The sequence can be rewritten into the following subsequence.

$$\begin{aligned} X(i) = &\quad \left[x(i), x(i+1), \ldots, x(i+m-1)\right], \\ &\qquad 1 \le i \le N-m+1 \\ X(j) = &\ \left[x(j), x(j+1), \ldots, x(j+m-1)\right], \\ &\qquad 1 \le j \le N-m+1 \end{aligned} \tag{6.1}$$

The distance between any two m-dimensional vectors $X(i)$ and $X(j)$ is

$$d\left|X(i), X(j)\right| = \max \mid x(i+k) - x(j+k) \tag{6.2}$$

Define a threshold where $r = k * SD$, ($r > 0$). For this experiment, k is set to 0.2, where SD denotes the standard deviation of the time series. num $\{d\left|X(i), X(j)\right| \le r$ is the number of pair of vector statistics that is less than threshold r, for each $1 \le i \le N-m+1$, which meets the condition of $d\left|X(i), X(j)\right| \le r$. The ratio of its total length to the vector is $C_i^m(r)$

$$C_i^m(r) = \frac{1}{N-m+1} * \text{num}\left\{d\left|X(i), X(j)\right| \le r\right\}$$
$$i = 1, \ldots, N-m+1, i \ne j \tag{6.3}$$

$C_i^{m+1}(r)$ can also be calculated similarly.

$$C_i^{m+1}(r) = \frac{1}{N-m+1} * \text{num}\left\{d\left|X(i), X(j)\right| \le r\right\}$$
$$i = 1, \ldots, N-m+1, i \ne j \tag{6.4}$$

$\text{AE}(m, r, N)$ is described as follows:

$$\text{AE}(m, r, N) = \frac{1}{N-m+1} \sum_{i=1}^{N-m+1} \log C_i^m(r) - \frac{1}{N-m} \sum_{i=1}^{N-m} \log C_i^{m+1}(r) \tag{6.5}$$

6.2.2.4.2 *Sample Entropy*

Sample entropy is a refined technique for calculating approximate entropy [31]. The formula for sample entropy is

$$\text{SamEn}(m, r, N) = -\ln \frac{H^{m+1}(r)}{H^m(r)} \tag{6.6}$$

Here, $H^m(r)$ is defined as

$$H^m(r) = \frac{1}{N-m} \sum_{i=1}^{N-m} C_i^m(r) \tag{6.7}$$

In this study, parameters are chosen as $m = 2$ and $r = 0.2 * SD$.

6.2.2.4.3 *Spectral Entropy*

The spectral entropy of a signal estimates its spectral power distribution. For a signal $x(n)$, the power spectrum is $S(m) = |X(m)|^2$, where $X(m)$ is the discrete Fourier transform of $x(n)$. The probability distribution $P(m)$ is then

$$P(m) = \frac{S(m)}{\sum_i S(i)} \tag{6.8}$$

The spectral entropy *SpEn* is

$$\text{SpEn} = -\sum_{m=1}^{N} P(m)\log_2 P(m) \tag{6.9}$$

6.2.2.4.4 Fuzzy Entropy

Fuzzy entropy has the ability to consistently provide reliable outcomes across many parameters and exhibit enhanced resistance to noise through the use of the well-defined fuzzy membership function, which is clearly shown in the studies [32].

6.2.3 Classifier Modeling and Evaluation Criteria

In this study, we employed a suite of classifiers: LR, linear discriminant analysis (LDA), KNN, SVM, ADABoost, DT, and an ANN to predict our target variable effectively. The ANN deep learning model had two fully connected hidden layers with 2,233 trainable parameters. This lightweight deep learning (LWDL) model enhances applicability and requires low computational power. The neural LWDL architecture was determined by manually tuning, utilizing different layer configurations and hyperparameters. The model selection process was based on optimizing performance metrics and the ability to handle specific patterns and complexities of our dataset. The model performance is evaluated through accuracy, sensitivity, specificity, and *F*1 score. The equations of performance parameters are given below:

$$\text{Accuracy} = \frac{\text{TP}+\text{TN}}{\text{TP}+\text{TN}+\text{FP}+\text{FN}} \tag{6.10}$$

$$\text{Sensitivity} = \frac{\text{TP}}{\text{TP}+\text{FN}} \tag{6.11}$$

$$\text{Specificity} = \frac{\text{TN}}{\text{FP}+\text{TN}} \tag{6.12}$$

$$\text{F1 score} = \frac{\text{TN}}{\text{TP}+0.5(\text{FP}+\text{FN})} \tag{6.13}$$

where TP, TN, FP, and FN describe the number of true positives, true negatives, false positives, and false negatives.

6.2.4 Key Brain Region Identification Criteria

The entire brain region is distributed into seven specific portions based on the spatial location of the EEG channel, as shown in Table 6.1. The region is denoted as R^x, where R is the specific brain portion and x is the number of regions. The total scalp is divided into seven specific regions to ensure comprehensive coverage. These spatial

TABLE 6.1
Different Brain Regions with Corresponding Electrodes

Brain Region	Electrode
R^1	FP1, FP2
R^2	F7, F3, Fz, F4, F8
R^3	FT7, FC3, FCz, FC4, FT8
R^4	T3, C3, Cz, C4, T4
R^5	TP7, CP3, CPZ, CP4, TP8
R^6	T5, P3, Pz, P4, T6
R^7	O1, OZ, O2

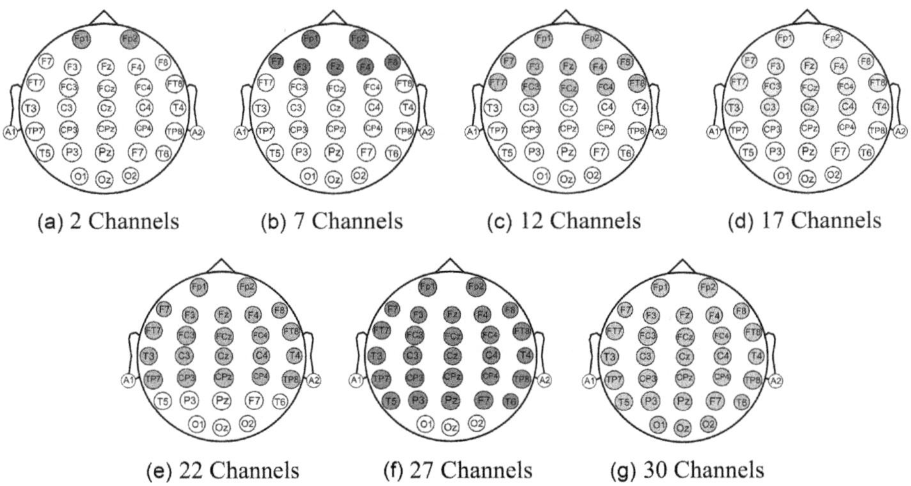

FIGURE 6.2 The seven regions of the brain and their corresponding channel locations are illustrated. The cumulative approach of channel selection and feature extraction, as well as the number of channels in each selected region, is depicted under each figure from (a–g).

regions are selected based on the channel locations from the forehead to the back of the scalp region, as shown in Figure 6.2.

To understand the brain regions R^x responsible for the drowsy neurophysiological state, a cumulative channel selection method is proposed. This algorithm selects the S^i electrodes based on the R^i regions to extract the entropy features φ^i and cumulatively incorporates the features for training and testing to calculate the results in terms of accuracy, as shown in Algorithm 6.1.

Algorithm 6.1: Highly contributing region identification with corresponding cumulative channel selection method

$$for\ i = 1\ to\ 7:$$

$$select\ S^i\ from\ R^i\ region$$

$$extract\ x^i = \varphi^i\left(S^i\right)$$

$$X = concat\left(X, x^i\right)$$

$$Train\ and\ test\ ANN\ by\ features\ X\ 5\ fold\ cross\ validation$$

$$calculate\ result$$

$$end$$

6.3 EXPERIMENTAL RESULTS

The dataset was divided into five sets for cross-validation, for training and measuring the performance of our proposed model. In this process, four sets of data were chosen to train the model, and one set was set aside for the evaluation of the model. The process continues until all the sets are used for training and testing separately. Each folded dataset contained the same data for both drowsy and alert classes.

We investigated the different classification model performances with four types of statistical parameters: the Hjorth parameter, the band power, and the entropy feature sets that were explained earlier. The performance was calculated in terms of mean accuracy in a five-fold cross-validation test. The performance of the other three features, other than entropy, is visible in Figure 6.3. The accuracies of these feature sets range from

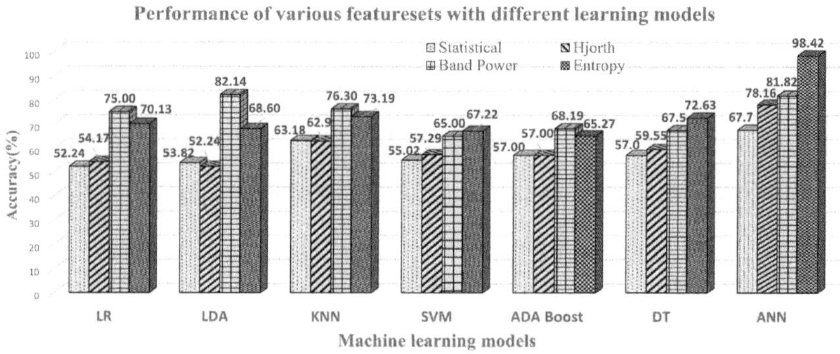

FIGURE 6.3 Visualization of the performance comparison of different machine learning models and the proposed model with the statistical, Hjorth, band power, and entropy feature sets. The performance is measured in terms of average accuracy in a five-fold cross-validation method. It depicts the overall comparison of corresponding models and the feature sets.

52.24% to 98.42% with different learning algorithms. The entropy feature sets combined with the LWDL model obtain the best mean accuracy of 98.42%. The highest accuracies for statistical and Hjorth parameters are achieved by LWDL, which are 67.70% and 78.16%, respectively. The performance of the band power feature was also consistent among the models. The range of accuracy in band power features is 65.00%–82.14%, which was achieved through SVM and LDA. The entropy feature set's accuracy ranges from 65.22% to 98.42%. It obtained the lowest and highest mean accuracy through the entropy+ ADABoost method and the entropy+ LWDL model method, respectively. In comparison, the performance of the entropy feature stands out significantly, especially when paired with the LWDL model. Combinations with superior performance indicate that the model effectively captures the underlying patterns in the data when using entropy features. It is suggested that selecting both the learning model and feature set is crucial for detecting the drowsy state.

Entropy-based biomarkers are further trained and validated through multiple learning algorithms for classifying drowsy and alert neurophysiological states. The performance of each dataset is measured over the folded dataset and reported as a mean value. The performance markers of the classifier model are presented in terms of sensitivity, specificity, accuracy, error rate, and *F*1 score (Table 6.2).

The LWDL classifier significantly outperformed other classifiers (Table 6.2), achieving the highest sensitivity of 98.47%, specificity of 98.37%, accuracy of 98.42%, and an *F*1 score of 98.44% with the lowest error rate of 1.58%. In comparison, the KNN and DT classifiers also performed well, with accuracies of 73.19% and 72.63%, respectively. LR, LDA, and SVM exhibited moderate performance, with accuracies of around 70%, while ADABoost showed the lowest accuracy of 65.27% and the highest error rate of 34.72%. These results indicate that entropy features can contain discriminative information for classifying drowsy and alert states with the help of the LWDL classifier. A comprehensive analysis of each fold's performance for the deep learning model is depicted in Figure 6.4, which displays the confusion matrix.

To further validate our entropy-based feature with the LWDL model, we performed the leave-one-out cross-validation method. In each iteration, one subject was left for validation, and the other 11 subjects were used to train the weights and biases.

TABLE 6.2
Classification Performance of Models Utilizing Entropy Features in Terms of Performance Markers

Classifier	Sensitivity (%)	Specificity (%)	Accuracy (%)	Error Rate (%)	*F*1 Score (%)
Logistic regression	70.06	70.21	70.13	29.87	70.54
LDA	68.02	69.21	68.60	31.40	69.09
KNN	73.18	73.20	73.19	26.80	73.48
SVM	67.35	67.08	67.22	32.77	67.53
ADABoost	65.47	65.07	65.27	34.72	65.56
Decision tree	72.21	73.09	72.63	27.36	73.19
LWDL	98.47	98.37	98.42	1.58	98.44

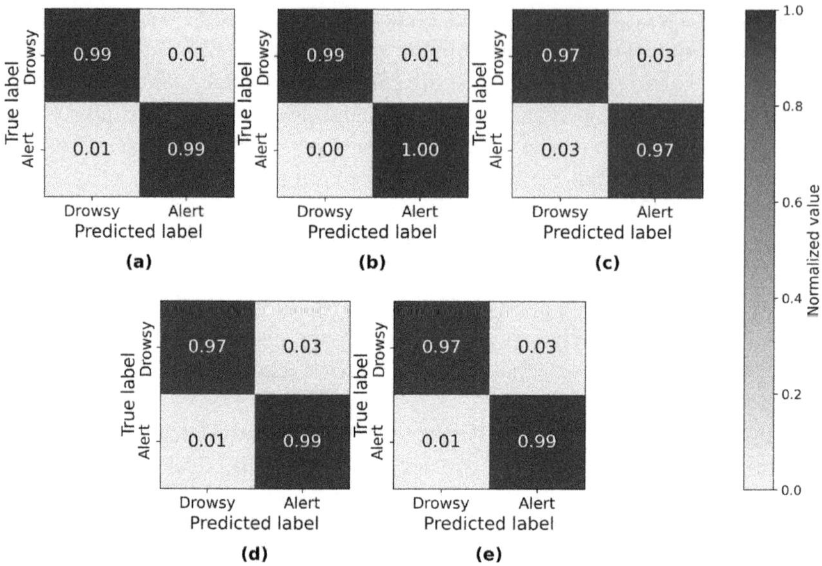

FIGURE 6.4 Confusion matrices for the proposed deep learning model using entropy-based feature sets. The figure displays the confusion matrices for each of the five-fold (a–e) cross-validation methods. Each matrix illustrates the true positive rate, false positive rate, true negative rate, and false negative rate in a normalized manner, demonstrating the model's performance and accuracy across different validation folds.

This validation technique demonstrates the robustness of our model and the entropy feature. Table 6.3 represents the performance in terms of accuracy, sensitivity, and specificity. We achieved the highest accuracy of 98.39% and the lowest of 79.96% in subjects 6 and 10. We obtained the average accuracy, sensitivity, and specificity of 91.93%, 91.45%, and 92.41%, respectively. Thus, these metrics demonstrate that our model's performance was robust, reliable, and generalized.

The proposed model includes four types of entropy features: AE, SE, PE, and FE. The performance impact of each entropy type was assessed by measuring accuracy metrics. Each feature was studied separately across various channels using the proposed model. Figure 6.5 illustrates that when evaluating the accuracy of all subjects using different combined entropies in a five-fold cross-validation, FE consistently outperforms other entropy features. FE achieves the highest median accuracy and maintains stable performance. AE demonstrates excellent accuracy and consistency, while SE offers a well-balanced performance. On the other hand, PE shows the least effective results, with greater inconsistency and the lowest average accuracy compared to the other types of entropy.

Identification of the brain region responsible for the drowsy state of the driver is crucial for the performance of the drowsiness detection system. In this experiment, we investigate various brain regions using entropy-based features and the proposed LWDL model. The utilization of entropy-based features enables capturing the dynamic changes present in EEG signals. The spatial distribution of brain activity varies according to different states and corresponding regions.

TABLE 6.3
Leave-One-Out Cross-Validation Performance for the Proposed Model

Subject	Accuracy (%)	Sensitivity (%)	Specificity (%)	F1 Score
Sub 1	94.19	92.42	95.96	93.89
Sub 2	92.00	91.48	92.53	91.98
Sub 3	88.23	89.21	87.26	88.22
Sub 4	95.69	97.55	93.84	95.66
Sub 5	96.21	97.46	94.97	96.20
Sub 6	98.39	98.51	98.28	98.39
Sub 7	86.00	82.03	89.98	85.82
Sub 8	91.97	90.88	93.06	91.37
Sub 9	93.88	91.65	96.12	93.60
Sub 10	79.96	77.26	82.66	79.90
Sub 11	93.92	93.72	94.12	93.92
Sub 12	92.69	95.25	90.14	92.66
Avg	91.93	91.45	92.41	91.80

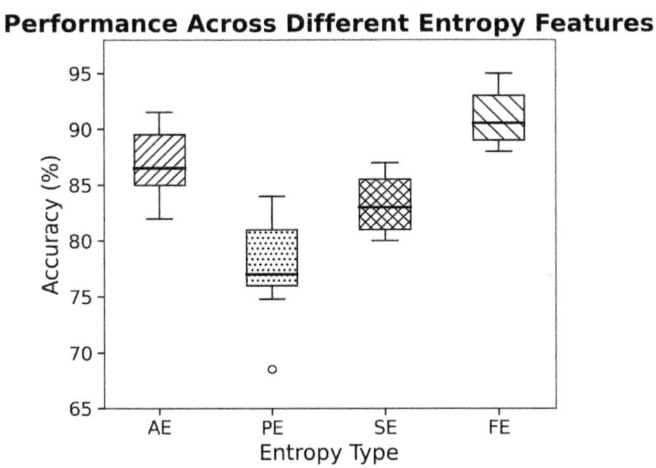

FIGURE 6.5 Illustrates performance across different entropy features in influencing the model's classification ability. Each box plot represents the accuracy distribution over all subjects in a five-fold cross-validation setup. FE achieves the most consistent and greatest accuracy, followed by AE and SE. PE exhibits the lowest performance with greater variability.

The result of Algorithm 6.1 is illustrated in Figure 6.6. It depicts that as the number of channels improves, the performance of the classification result increases. The performance metrics range: sensitivity between 69.06% and 98.2%, specificity between 64.58% and 98.83%, accuracy between 66.82% and 97.31%, and $F1$ score between 74.23% and 99.01%. An interesting trend found through this experiment was that after cumulatively extracting features from 17 electrodes, conserving the brain region of R^1– R^4, 96.02% accuracy, 95.26% sensitivity, 96.78% specificity, and 97.51% $F1$ score

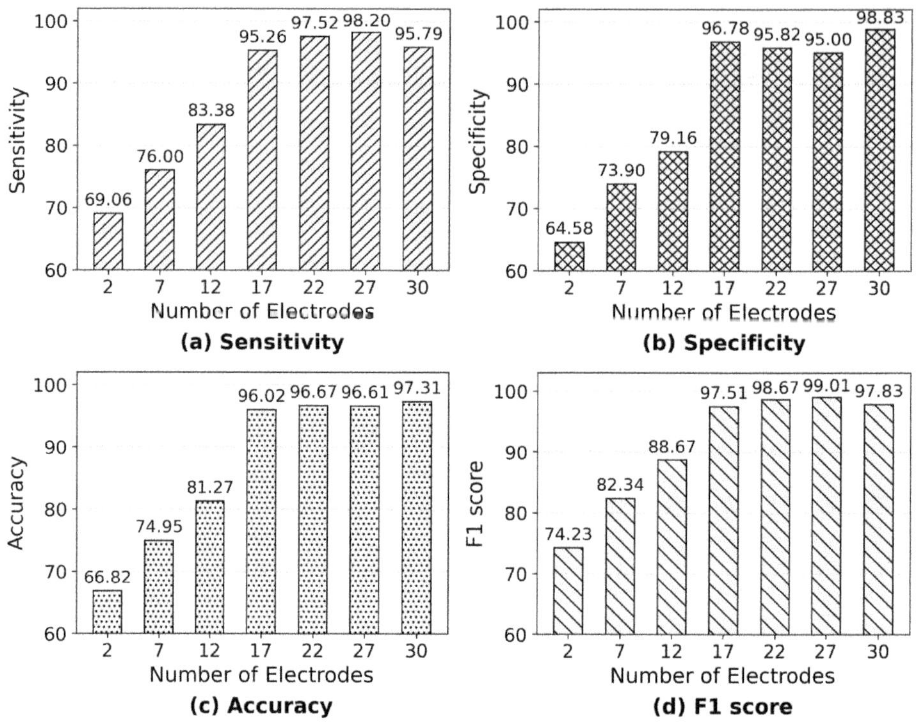

FIGURE 6.6 Performance of Algorithm 6.1 is illustrated in the bar plot in terms of sensitivity, specificity, accuracy, and $F1$ score. This graph shows an interesting trend of achieving close to maximum performance utilizing half of the channel information (17 electrodes) in categorizing the drowsy and alert states.

were obtained. This is almost the highest performance obtained with a total of 30 electrodes. Thus, almost all of the highest performances were obtained with around half of the electrode information.

6.4 DISCUSSION

In this study, we have developed a framework for the classification of a driver's alert and drowsy cognitive states while driving using EEG signals through multiple entropy feature space vectors with an LWDL model. The classification accuracy of 98.42%, with higher sensitivity and specificity of 98.47% and 98.37%, respectively, as shown in the results section of Table 6.2, shows the efficacy of our proposed method.

The performance of entropy features for the EEG signal's drowsy state is superior to that of any other features. The choice of each entropy feature is based on its definition and ability to distinguish randomness in signals. It extracts the signal's randomness and unpredictability in both the time and frequency domains. The four types of entropy features used in this work successfully represent the drowsy state through EEG signals. Both approximate entropy and sample entropy represent the complexity and irregularity of the signal. Here, sample entropy improves upon approximate entropy by avoiding the bias of counting self-similar patterns. During the awake and drowsy states, the EEG signal shows clear discrimination in complexity and regularity. Therefore, it satisfactorily represents the different states of the signal. To ensure accurate detection, both time-domain and frequency-domain entropy measures must be considered. The spectral entropy measures the complexity and irregularity in the frequency domain. It quantifies the power distribution uniformity across different frequency bands. In a drowsy and alert state, the power is distributed differently across the bands. Thus, the spectral entropy feature captures the differences between these states. This indicates the importance of the spectral entropy feature for the classification of different states. Fuzzy entropy finds the intrinsic patterns of similarity in real-world EEG signals. It shows resistance to noise and provides information on the transitions between different cognitive states. Fuzzy entropy outperforms the other entropy features (Figure 6.5), which may be based on its ability to handle noise artifacts in EEG signals. As EEG signals, which contain intrinsic neurophysiological drowsy state patterns, often require continuous separation from the artifacts and noise, fuzzy entropy outshines other entropy features. This feature makes it effective for the classification of different cognitive states.

The entropy feature captures more distinguishable information than other feature sets, which is visible in Figure 6.3. It shows that among the four types of features, the fourth feature set for the entropy features performs well. Among different classical machine learning classifiers, the LWDL model outperforms other classifiers with 98.42% average accuracy in five-fold cross-validation. The learning model's performance depends on two aspects: (1) the features that accurately represent the signal character and (2) the parameter configuration of the learning model. Our proposed LWDL model finds the intrinsic pattern in features for classifying different states. It shows the model's efficacy in finding the inherent pattern in features for the classification of different states.

Wearable technology with fewer electrodes should be used in future systems to deliver fatigue alerts since this will impact how convenient and pleasant a system is. In this chapter, we utilize cumulative channel selection, where the number of channels increases gradually from the frontal scalp region to the back of the head. The performance in Figure 6.6 indicates the top 17 electrodes that achieve higher classification accuracy, sensitivity, specificity, and $F1$ score. It is very close to the maximum performance obtained with 30 electrodes. The electrodes are shown in Table 6.4, namely, FP1, FP2, F7, F3, Fz, F4, F8, FT7, FC3, FCz, FC4, FT8, T3, C3, Cz, C4, and T4. Studies [33] stated that delta and alpha rhythms in the frontal and central regions build up as fatigue increases, while beta rhythm increases as alertness declines. In our study, the top 17 electrodes are located in the frontal and central regions of the cortex, and this shows that the frontal and central cortex areas are more responsible for detecting fatigue or alert states by analyzing the performance metrics.

This study aims to investigate the effectiveness of the recommended approach in the context of modality profile, type of feature, model, and learning ability. The study examines several existing two- and three-class classification approaches to accomplish this. Table 6.4 includes a two-class method that utilized the single EEG modality profile [1,8,9,13] and a three-class method with a multiple modalities profile [4,24,26]. It provides a concise overview of contemporary advancements in learning

TABLE 6.4
Comparison of Our Proposed Method in Signal, Feature, Classifier, and Accuracy

Method	Signal	Type of Feature	Classifier	Subject	State	Acc (%)
Chai et al. [2]	EEG	ICA-AR	BNN	43	Fatigue, alert	88.20
Khushaba et al. [24]	EEG, EOG	Fuzzy mutual-DWT	LDA, SVM, KNN, SVM, deep learning model	31	Drowsiness, Fatigue	95–97
Zhang et al. [4]	EEG, EOG, EMG	Approximate entropy	ANN	20	Nor, mild, ex-fat mood swing	96.50
Ahn et al. [26]	EEG, ECG, EOG, fNIRS	PSD-RPL	LDA	11	Rest, sleep-dep	84.50
Gangadharan and Vinod [8]	EEG	Sample entropy, envelope mean, Hjorth parameters, SD, spectral entropy (NSE)	SVM	18	Fat, alert	78.3
Kaur et al. [9]	EEG	EMD-IMF	ANN	8	Drowsy, awake	88.22
Stancin et al. [10]	EEG	Band power, multi-channel ratios	XGBoost	28	Alert, drowsy	86
Gao et al. [21]	EEG	Deep learning feature	ESTCNN	8	Fatigue, alert	97.37
Subasi et al. [13]	EEG	Flexible analytic wavelet transform	ANN, SVM	16	Fatigue, alert	97.10–97.90
Houshmand et al. [14]	EEG, lateral position, steering angle, vehicle dynamic	Statistical, entropy, PSD, band power	Classification tree	19	Drowsy, alert	87.55–91.31
Cui et al. [20]	EEG	Deep learning feature	Interpretable CNN	11	Drowsy, alert	78.35
Chen et al. [22]	EEG	Deep learning feature	SACC-CapsNet	31	Fatigue, alert	94.17/90.19
Lan et al. [17]	EEG, Vehicle motion information	Frequency band energy ration, sample entropy, standard deviation	SVM	16	Fatigue, alert	92.37
Proposed	EEG	Sample entropy, approximate entropy, spectral entropy, fuzzy entropy	ANN	12	Fatigue, alert	98.42

methodologies and their effectiveness, utilizing single and multiple modalities and typical models. The main emphasis of our work is on a single-modality profile. However, through a comparative analysis, we can gain `a deeper comprehension of and better understand the performance dependencies, the simplicity of the feature extraction framework, and the overall performance metrics. Many methods employed multiple modality profiles, such as EEG and EOG [23], EEG, EOG, and EMG [4], or EEG, ECG, EOG, and fNIRS [26] for feature extraction, as shown in Table 6.4, to increase classification performance. These methods required multiple computational frameworks for multiple modality signals. These processes introduce computational costs as well as theoretical bottlenecks. On the other hand, it is observed that the approaches employing typical shallow models such as SVM, KNN, LDA, K-SVM, and XGBoost are often known as weak classifiers due to their intrinsic theoretical bottlenecks. On the other hand, deep learning methods like CNN, BNN, and DBN exhibit good learning capability due to their layers and neurons. However, they introduce computational complexity, a lack of interpretation, and a lack of understanding regarding their performances. In contrast, our proposed solution is a simple, direct, well-explained, single-modality approach. The system employs a sophisticated learning approach based on distinct LWDL model criteria and achieves notable results that surpass previous systems. In comparison with the approaches mentioned in Table 6.4, our approach surpasses not only the single modality-based solutions such as [1,8–10,13,21] but also the multiple modality-based solutions mentioned in [14,17,24,26]. Further validation of the model using leave-one-out validation is also employed, as shown in Table 6.3, which achieves high classification accuracy, sensitivity, and specificity of an average of 91.93%, 91.45%, and 92.41%. This method verifies the robustness of our proposed model in real-world scenarios and has proven to be implementable.

Our research presents promising results, but several areas hold potential for further exploration and improvement. In future studies, a systematized and intelligent driver assistant can be designed and implemented for driver fatigue detection using the presented entropy-based feature extraction with an LWDL architecture framework. One key limitation of this study was that it was conducted offline, lacking real-time implementation and integration with actual vehicle systems. Real-time implementation with practical deployment can ensure timely intervention to prevent accidents. Furthermore, the dataset used primarily comprised male subjects. Studies show that demographic differences, such as age and gender, can influence driving behavior and the way brain patterns manifest during drowsiness [34]. Cross-cultural and cross-demographic validation is required to ensure robustness and generalizability.

6.5 CONCLUSION

In this study, we proposed a deep learning-based driver drowsiness detection system that leverages entropy-based features extracted from EEG signals. Our model demonstrated high effectiveness, achieving an average accuracy of 98.42%, a sensitivity of 98.47%, a specificity of 98.37%, and an $F1$ score of 98.44% using only four entropy features. We also investigated the importance of optimal channel selection. Furthermore, the model demonstrated an accuracy of 96.02%, a sensitivity of

95.26%, a specificity of 96.78%, and an $F1$ score of 97.51% while using almost half of the EEG channels, highlighting its efficiency in reducing the system's complexity without sacrificing performance. The ability of our approach to capture dynamic neurophysiological changes underscores its promise as a practical and reliable solution for enhancing road safety through real-time driver drowsiness detection.

REFERENCES

1. M. Steriade, D. A. McCormick, and T. J. Sejnowski, "Thalamocortical oscillations in the sleeping and aroused brain," *Science*, vol. 262, no. 5134, pp. 679–685, 1993, doi: 10.1126/science.8235588.
2. R. Chai et al., "Driver fatigue classification with independent component by entropy rate bound minimization analysis in an EEG-based system," *IEEE Journal of Biomedical and Health Informatics*, vol. 21, no. 3, pp. 715–724, 2016.
3. R. F. S. Job, A. Graham, C. Sakashita, and J. Hatfield, "Fatigue and road safety: Identifying crash involvement and addressing the problem within a safe systems approach," in P. A. Desmond (ed.), *The Handbook of Operator Fatigue*, CRC Press, London, 2017, pp. 349–363.
4. Y. Zhang and C. Hua, "Driver fatigue recognition based on facial expression analysis using local binary patterns," *Optik (Stuttg)*, vol. 126, no. 23, pp. 4501–4505, 2015.
5. M. M. Hasan, C. N. Watling, and G. S. Larue, "Validation and interpretation of a multimodal drowsiness detection system using explainable machine learning," *Computer Methods and Programs in Biomedicine*, vol. 243, p. 107925, 2024, doi: 10.1016/J.CMPB.2023.107925.
6. C. Lee and J. An, "LSTM-CNN model of drowsiness detection from multiple consciousness states acquired by EEG," *Expert Systems With Applications*, vol. 213, p. 119032, 2023.
7. H. Wang, C. Wu, T. Li, Y. He, P. Chen, and A. Bezerianos, "Driving fatigue classification based on fusion entropy analysis combining EOG and EEG," *IEEE Access*, vol. 7, pp. 61975–61986, 2019.
8. S. Gangadharan and A. P. Vinod, "Drowsiness detection using portable wireless EEG," *Computer Methods and Programs in Biomedicine*, vol. 214, p. 106535, 2022.
9. R. Kaur and K. Singh, "Drowsiness detection based on EEG signal analysis using EMD and trained neural network," *International Journal of Scientific Research*, vol. 10, pp. 157–161, 2013.
10. I. Stancin, N. Frid, M. Cifrek, and A. Jovic, "EEG signal multichannel frequency-domain ratio indices for drowsiness detection based on multicriteria optimization," *Sensors*, vol. 21, no. 20, p. 6932, 2021.
11. R. Chai et al., "Improving EEG-based driver fatigue classification using sparse-deep belief networks," *Frontiers in Neuroscience*, vol. 11, p. 234844, 2017, doi: 10.3389/fnins.2017.00103.
12. A. Chaudhuri and A. Routray, "Driver fatigue detection through chaotic entropy analysis of cortical sources obtained from scalp EEG signals," *IEEE Transactions on Intelligent Transportation Systems*, vol. 21, no. 1, pp. 185–198, 2019.
13. A. Subasi, A. Saikia, K. Bagedo, A. Singh, and A. Hazarika, "EEG-based driver fatigue detection using FAWT and multiboosting approaches," *IEEE Transactions on Industrial Informatics*, vol. 18, no. 10, pp. 6602–6609, 2022.
14. S. Houshmand, R. Kazemi, and H. Salmanzadeh, "An efficient approach for driver drowsiness detection at moderate drowsiness level based on electroencephalography signal and vehicle dynamics data," *Journal of Medical Signals and Sensors*, vol. 12, no. 4, pp. 294–305, 2022.

15. M. Awais, N. Badruddin, and M. Drieberg, "Driver drowsiness detection using EEG power spectrum analysis," in *2014 IEEE Region 10 Symposium*, IEEE, New York, 2014, pp. 244–247.
16. U. Budak, V. Bajaj, Y. Akbulut, O. Atila, and A. Sengur, "An effective hybrid model for EEG-based drowsiness detection," *IEEE Sensors Journal*, vol. 19, no. 17, pp. 7624–7631, 2019.
17. Z. Lan, J. Zhao, P. Liu, C. Zhang, N. Lyu, and L. Guo, "Driving fatigue detection based on fusion of EEG and vehicle motion information," *Biomedical Signal Processing and Control*, vol. 92, p. 106031, 2024.
18. G. Li, B.-L. Lee, and W.-Y. Chung, "Smartwatch-based wearable EEG system for driver drowsiness detection," *IEEE Sensors Journal*, vol. 15, no. 12, pp. 7169–7180, 2015.
19. J. Cui et al., "A compact and interpretable convolutional neural network for cross-subject driver drowsiness detection from single-channel EEG," *Methods*, vol. 202, pp. 173–184, 2022.
20. J. Cui, Z. Lan, O. Sourina, and W. Müller-Wittig, "EEG-based cross-subject driver drowsiness recognition with an interpretable convolutional neural network," *IEEE Transactions on Neural Networks and Learning Systems*, vol. 34, no. 10, pp. 7921–7933, 2022.
21. Z. Gao et al., "EEG-based spatio–temporal convolutional neural network for driver fatigue evaluation," *IEEE Transactions on Neural Networks and Learning Systems*, vol. 30, no. 9, pp. 2755–2763, 2019.
22. C. Chen, Z. Ji, Y. Sun, A. Bezerianos, N. Thakor, and H. Wang, "Self-attentive channel-connectivity capsule network for EEG-based driving fatigue detection," *IEEE Transactions on Neural Systems and Rehabilitation Engineering*, vol. 31, pp. 3152–3162, 2023.
23. V. J. Lawhern, A. J. Solon, N. R. Waytowich, S. M. Gordon, C. P. Hung, and B. J. Lance, "EEGNet: a compact convolutional neural network for EEG-based brain–computer interfaces," *Journal of Neural Engineering*, vol. 15, no. 5, p. 56013, 2018.
24. R. N. Khushaba, S. Kodagoda, S. Lal, and G. Dissanayake, "Driver drowsiness classification using fuzzy wavelet-packet-based feature-extraction algorithm," *IEEE Transactions on Biomedical Engineering*, vol. 58, no. 1, pp. 121–131, 2010.
25. C. Zhang, X. Wu, X. Zheng, and S. Yu, "Driver drowsiness detection using multichannel second order blind identifications," *IEEE Access*, vol. 7, pp. 11829–11843, 2019.
26. S. Ahn, T. Nguyen, H. Jang, J. G. Kim, and S. C. Jun, "Exploring neuro-physiological correlates of drivers' mental fatigue caused by sleep deprivation using simultaneous EEG, ECG, and fNIRS data," *Frontiers in Human Neuroscience*, vol. 10, p. 219, 2016.
27. R. Fu, H. Wang, and W. Zhao, "Dynamic driver fatigue detection using hidden Markov model in real driving condition," *Expert Systems With Applications*, vol. 63, pp. 397–411, 2016.
28. J. Min, P. Wang, and J. Hu, "Driver fatigue detection through multiple entropy fusion analysis in an EEG-based system," *PLoS One*, vol. 12, no. 12, p. e0188756, 2017.
29. M. J. Francis, M. P. Keran, R. Chetan, and B. N. Krupa, "EEG-controlled robot navigation using Hjorth parameters and Welch-PSD," *International Journal of Intelligent Engineering & Systems*, vol. 14, no. 4, p. 231, 2021.
30. C. Zhang, H. Wang, and R. Fu, "Automated detection of driver fatigue based on entropy and complexity measures," *IEEE Transactions on Intelligent Transportation Systems*, vol. 15, no. 1, pp. 168–177, 2013.
31. J. S. Richman and J. R. Moorman, "Physiological time-series analysis using approximate entropy and sample entropy," *American Journal of Physiology-Heart and Circulatory Physiology*, vol. 278, no. 6, pp. H2039–H2049, 2000.
32. J. Xiang et al., "The detection of epileptic seizure signals based on fuzzy entropy," *Journal of Neuroscience Methods*, vol. 243, pp. 18–25, 2015.

33. F. Barwick, P. Arnett, and S. Slobounov, "EEG correlates of fatigue during administration of a neuropsychological test battery," *Clinical Neurophysiology*, vol. 123, no. 2, pp. 278–284, 2012.
34. N. Rhodes and K. Pivik, "Age and gender differences in risky driving: The roles of positive affect and risk perception," *Accident Analysis and Prevention*, vol. 43, no. 3, pp. 923–931, 2011.

Section IV

Seizure Detection and Classification Using Time-Frequency Features

7 From Signals to Automated System

Seizure Detection Using Time–Frequency EEG Features: An Experimental Investigation

Abhishek Sathyendran, Ankith P. Shetty, Roopa B. Hegde, and G. R. Pradyumna

7.1 INTRODUCTION

Epilepsy, a neurological disorder characterised by recurrent and unpredictable seizures, affects millions worldwide, with an estimated 5 million people diagnosed each year, according to the WHO. These seizures result from abnormal electrical activity in the brain, often detectable through electroencephalogram (EEG) recordings. Accurate detection and diagnosis are essential for effective management and treatment. Accurate and timely detection of seizures is essential for enabling early intervention and improving long-term treatment outcomes. Traditionally, epilepsy diagnosis involves neurologists manually analysing EEG recordings for seizure patterns. Still, this method is time-consuming, susceptible to human error, and may not capture subtle or complex patterns in the data. Moreover, manual analysis requires expert knowledge and consistency, which can be challenging in regions with limited access to experts.

The advent of machine learning (ML) has opened new avenues for developing automated methods for epileptic seizure detection using EEG signals. By training algorithms on large EEG datasets, these models can learn to identify features indicative of epileptic activity, aiming to create reliable, real-time systems that assist clinicians in diagnosing and managing epilepsy more effectively. Recent advancements in deep learning (DL) and artificial intelligence have driven research in this field, leading to the development of systems capable of handling high-dimensional EEG data while improving seizure classification accuracy. Hence, developing automated systems using ML and DL has transformed this field, providing promising real-time and precise detection solutions. Recent studies have explored various ML and DL models, from support vector machines (SVMs) and random forests (RFs) to advanced

DOI: 10.1201/9781003611448-11

convolutional neural networks and long short-term memory networks. These models have improved accuracy by utilising diverse feature sets extracted from EEG data, such as time-domain, frequency-domain, and wavelet-based features. Despite these advancements, challenges in this domain persist, including model generalisation, computational complexity, and the need for extensive, diverse datasets. The variability of EEG signals among individuals, differences in seizure types, and noise in real-world scenarios further complicate model performance, necessitating robust pre-processing and feature selection techniques. Hence, to overcome these issues, researchers have proposed hybrid models integrating ML and DL techniques. Additionally, integrating real-time detection systems with platforms like Raspberry Pi shows potential for practical applications in clinical and home environments. Edge computing solutions and embedded artificial intelligence systems enable efficient real-time classification, making seizure detection systems more reliable.

This study aims to advance EEG-based seizure detection by developing a robust, real-time system using an ensemble of ML models tailored to manage the variability in patient data and seizure patterns effectively. We aim to enhance model interpretability and improve detection accuracy by integrating an ensemble model with a Raspberry Pi for better clinical decision-making.

7.2 LITERATURE SURVEY

Many researchers have proposed various methods for detecting epileptic seizures from EEG signals using ML algorithms. A recent survey paper (Farooq et al. 2023) detailed the articles published in 2022, including those from MDPI, IEEE Xplore, Wiley, Elsevier, ACM, and Springer. This paper reviewed the feature selection and classification process of EEG signals into seizure and non-seizure. An experimental study (Kunekar et al. 2024) focused on both ML and DL methods for seizure detection, thereby comparing both methods. This study used the UCI-Epileptic Seizure Recognition dataset and found that the long short-term memory algorithm performed better by yielding 97% accuracy than the other ML algorithms. Another study conducted by Tran et al. (2022) employed SVM, k-nearest neighbour (k-NN), decision tree (DT), and RF for the detection of epileptic seizures. This study utilised statistical features extracted from the original EEG data using discrete wavelet transform analysis and binary particle swarm optimiser for feature selection, thereby obtaining the highest accuracy of 98.4% with the SVM classifier. A public EEG dataset from the University of Bonn was used to train and evaluate the ML models. A similar study (Nahzat and Yaganoglu 2021) was carried out to compare the performances of k-NN, RF, SVM, artificial neural network, and DT using principal component analysis for feature reduction. RF classifiers without principal component analysis produced an accuracy of around 99%. A group of researchers employed SVM (Dhondiyal and Dimri 2022) and a combination of SVM and gradient-boosting DT (He et al. 2022) for automatically detecting epileptic seizures, obtaining an accuracy of around 82% and 90%, respectively. Features used for training any ML algorithm affect the algorithm's performance. To select appropriate features, the fuzzy-based seizure detection approach (Qureshi et al. 2022) showed an accuracy range of 98%–100% for various seizure types. This study included time-domain, frequency-domain, and non-linear features. A study (Wang et al. 2024) focused on developing RF and SVM for automated seizure detection, utilising state-space features extracted from EEG signals. These traditional models laid the groundwork for more sophisticated approaches, such as the multi-input deep feature learning network proposed by Sun et al. (2024), which enhanced accuracy by combining temporal and spatial features from EEG data. Intongkum et al. (2024) proposed a system that integrates various human monitoring systems, including posture and seizure detection, using a combination of received signal strength and ML models. This highlights the trend of incorporating multiple physiological signals to improve detection robustness.

Feature extraction and selection are crucial for the performance of ML models in seizure detection. Traditional methods rely on handcrafted features like time-domain, frequency-domain, and non-linear characteristics (Xiong et al. 2023). Recent studies have shifted towards automated feature extraction using models. For instance, Hasan et al. (2023) demonstrated the effectiveness of ResNest in learning complex patterns from EEG data. Another study (Palanisamy and Rengaraj, 2024) addressed the detection of anxiety-based epileptic seizures, utilising augmented data and DL techniques to achieve high predictive accuracy. This study emphasises the need for tailored models to differentiate between seizures and related disorders.

The quality and diversity of datasets are pivotal for training reliable seizure detection models. The CHB-MIT and Bonn EEG datasets are among the most used in research (Xiong et al. 2023; Wei and Mooney, 2023). However, these datasets have size and patient diversity limitations, and a study (Intongkum et al. 2024) highlighted the need for larger, more representative datasets to improve model generalisation. Several studies have also utilised novel datasets. For example, Thomas et al. (2024) leveraged a unique stereo-EEG dataset, providing a more comprehensive view of seizure activity. Such datasets can capture the spatial dynamics of seizures, offering valuable insights for model development. While most studies report high accuracy, several limitations persist. The unpredictability of epileptic seizures and noise in EEG data pose significant challenges (Sun et al. 2024).

Additionally, many ML models suffer from overfitting due to small, homogeneous datasets. An experimental analysis (Hasan et al. 2023) reported that despite high accuracy in training, the ResNet model underperformed on unseen data, highlighting the need for better generalisation techniques. Paediatric-specific models are another area of concern. Wei and Mooney (2023) emphasised the inadequacy of adult-based models for detecting seizures in children, suggesting the development of paediatric-specific models to address the unique patterns of childhood epilepsy. Real-time implementation of seizure detection systems is crucial for their clinical utility. A group of researchers (Wang et al. 2024) developed a system integrating ML models into a Raspberry Pi, enabling real-time monitoring and alerting caregivers upon seizure detection. Similarly, another study (Thomas et al. 2024) focused on a stereo-EEG-based seizure matching system, which can considerably reduce the time required for diagnosis in clinical settings. Xiong et al. (2023) explored the fusion of features from multiple datasets to improve the robustness of seizure detection algorithms. This approach, though computationally intensive, can enhance the reliability of real-time systems, making them more adaptable to different clinical scenarios.

Many studies rely on limited datasets that do not capture the full spectrum of epileptic activity (Hasan et al. 2023). Model generalisation is another challenge because models trained on specific datasets often fail to generalise to new data due to overfitting and lack of diversity in training samples. Computational complexity and resource limitations are also challenges in EEG signal analysis. While accurate DL models are computationally intensive, they are challenging to deploy in resource-constrained environments. This work explores the application of conventional ML algorithms for epileptic seizure detection from EEG signals, addressing challenges such as computational complexity and resource limitations. It reviews various approaches and techniques proposed to overcome these challenges and evaluates the performance metrics used to assess the efficacy of ML models in seizure detection. The significance of this research lies in enhancing hardware setups for epilepsy diagnosis, providing clinicians with automated tools for real-time seizure detection by deploying the ML model to hardware setups, leading to timely interventions and improved patient outcomes. Advancements in this field hold promise for enhancing healthcare for individuals with epilepsy.

7.3 METHODOLOGY

In the proposed experimental study, various ML models were trained to detect epileptic seizures. Different stages of the epileptic seizure detection process are shown in Figure 7.1. This section details the various stages in the workflow, from data collection to the development of the detection system.

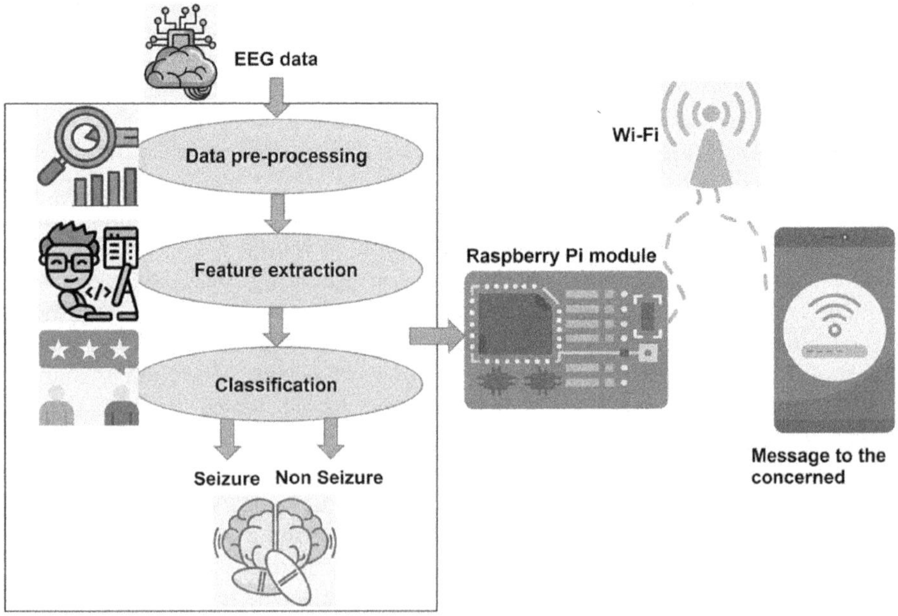

FIGURE 7.1 Workflow of the proposed seizure detection system.

7.3.1 EEG Data

Electroencephalography signals are recorded from electrodes placed on the patient's scalp under controlled conditions. This study obtained the necessary EEG data from the CHB-MIT Scalp EEG Database (Guttag 2022). This repository includes a variety of EEG samples collected at Children's Hospital Boston. Subjects were monitored over several days after discontinuing antiseizure medications to document their seizure activity and determine their suitability for surgical treatment. The dataset comprises recordings organised into 23 cases from 22 individuals (5 males, aged 3–22 years, and 17 females, aged 1.5–19 years). Most recordings contain 23 EEG channels, and the international 10–20 system was used for electrode placement and nomenclature.

7.3.2 Development of the Decision Support System

The development stages of an automated decision support system are shown in Figure 7.1, and this section provides details of the different stages involved in the present work.

7.3.2.1 EEG Data Pre-processing

Data pre-processing encompasses a series of techniques to prepare a dataset before further analysis. The primary objective is to enhance the quality and suitability of raw EEG data for efficient classification. Moreover, it enhances the readability of the data, making it more accessible for ML algorithms to interpret and establish consistency in the analysis. The raw data files obtained from the database are in .edf format and converted to .CSV format for easy handling. Furthermore, the sensor output values are in microvolts, which are too small for the feature extraction stage, and are converted to volts. The human brain primarily generates 1–100 Hz electrical signals. This range encompasses various brain rhythms associated with specific cognitive functions (Tortora and Derrickson, 2017). However, noise is an undesired signal that enters the EEG data during acquisition. A band-pass filter removes noise with a lower cut-off frequency of 1 Hz and an upper cut-off frequency of 50 Hz. Slow drifts, originating from movement artefacts, electrode instability, or power line interference, constitute frequencies below 1 Hz.

In comparison, high-frequency noise from muscle activity or electrical noise from the recording environment constitutes frequencies above 50 Hz. This study employs a Butterworth filter of order six after several experiments. The general expression of the Butterworth filter is given in Equation 7.1.

$$H(z) = \frac{q_0 + q_1 z^{-1} + \ldots + q_A z^{-A}}{1 + p_1 z^{-1} + \ldots + p_B z^{-B}} \tag{7.1}$$

where p and q are the feedback and feedforward coefficients, respectively, and B and A are the orders of the feedback and feedforward parts, respectively. For the filter specifications, the Butterworth band-pass filter details are as follows. Select a sampling frequency (as given in the database) equal to 256 Hz. Normalise the lower and

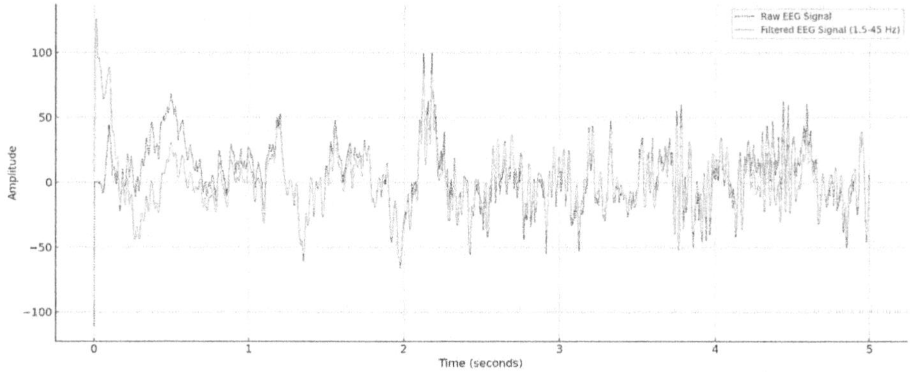

FIGURE 7.2 Raw and filtered EEG signals.

upper cut-off frequencies, dividing them by half of the sampling frequency (Nyquist frequency) as shown in Equation 7.2.

$$w_l = \frac{1}{256/2} = 0.0078125 \quad \text{and} \quad w_h = \frac{50}{256/2} = 0.390625 \tag{7.2}$$

where w_l and w_h are the lower cut-off and upper cut-off frequencies.

Compute the analogue filter parameters called poles of the transfer function. The circle radius is determined by the geometric mean of the lower and upper cut-off frequencies as given in Equation 7.3.

$$\text{Radius} = \sqrt{w_l \, . \, w_h} = 0.0551981648 \tag{7.3}$$

Obtain digital filter coefficients by applying bilinear transformation to each pole using the relation in Equation 7.4.

$$S = \frac{2}{T}\frac{1-z^{-1}}{1+z^{-1}} \tag{7.4}$$

where T is the sampling period = 1/126 = 3.9×10^{-3}.

Compute filter coefficients p and q using the transfer function given in Equation 7.1. Use the filter coefficients to filter the EEG signal.

Comparison of raw and filtered EEG signals is shown in Figure 7.2. It is evident that the filter effectively suppresses high-frequency noise and low-frequency drifts, preserving the EEG components required for detecting seizure cases.

7.3.2.2 Feature: Time–Frequency Domain

Feature extraction is a vital step in decision-making using ML. This is because it contributes to enhancing the efficiency of ML models. In the present study, ML models are trained separately using different types of feature sets: time-domain features, frequency-domain features, autoregressive (AR) model features, and wavelet-based features, as listed in Table 7.1, to evaluate the performance of the models. Time-domain features are statistical measures such as mean, standard deviation, variance, skewness, kurtosis, energy, etc. used to identify basic patterns such as overall activity and overall variability of EEG signals. However, the brain functions at different frequencies during normal and seizure conditions. This can be obtained by frequency-domain features, namely power spectral densities (PSD). It is a fundamental technique in understanding the frequency distribution of EEG signals, which represent the brain's electrical activity over time. This provides the intensity of the signal at different frequency components that are present in the signal. Hence, PSD is vital in identifying epileptic seizures from the EEG signal. During an epileptic seizure, there is often a pronounced increase in the power of specific frequency bands, particularly in the ranges of 4–30 Hz.

Additionally, features extracted from PSD, such as spectral entropy, relative power, absolute power, and peak frequency, are used to train ML models. The relative power value increases during seizure activity compared to non-seizure periods. Similarly, absolute power also increases during seizure activity. Spectral entropy, which measures the irregularity of the PSD, may decrease during seizure activity. The peak frequency of the PSD may become more pronounced during seizure events, reflecting changes in dominant oscillatory patterns. Furthermore, AR model features are the coefficients obtained by fitting an AR model that represents the relation between current and past values of the EEG signal, thereby

TABLE 7.1
List of Main Features

Feature Type	Feature Name	Description
Time domain	Mean, standard deviation, skewness, kurtosis, energy	Sensitive to changes in the electrical activity of the brain. An increase in mean and standard deviation with high kurtosis indicates seizures
Frequency domain	Delta power, theta power, alpha power, beta power, total power	Power at frequency bands reveals the brain's functional state. Different types of seizures may exhibit characteristic patterns in these power levels
AR model	AR coefficients	Represents the influence of the ith previous observation on the current value and captures temporal dependencies. Magnitude indicates temporal dependency and sign indicates frequency shift in EEG signals
Wavelet-based	Mean, standard deviation, min and max values, energy, entropy, and statistical moments	Provides both time and frequency localisation. Highlights average amplitude and its variability at different scales, capturing transient patterns and changes in frequency content. These features reflect spikes, sharp waves, or bursts and shape of wavelet coefficients

predicting the progression of seizures. Finally, wavelet features are derived by taking wavelet transforms of EEG signals that decompose them into different frequency scales, thereby aiding in detecting non-stationary and transient features such as spikes and seizures. The main features used for training the classifiers are listed in Table 7.1.

7.3.2.3 Classification

In the proposed experimental study, the performance of three classifiers, namely RF, SVM, and k-NN, is evaluated for different sets of features: time-domain features, frequency-domain features, AR features, and wavelet features. Performances are evaluated using accuracy, precision rate, recall rate, and F1-score. All the classifiers are trained and assessed using time-domain features, followed by frequency-domain, AR model features, and wavelet features. It was found that RF classifiers with all the features outperformed the other feature combinations. The best results are obtained using RF classifiers. Technical details are shown in Figure 7.3.

In the experimental study, the 'TreeBagger' method with 100 DTs is used with the 'Ginni index' as the splitting criterion. Furthermore, the best split at each node is selected, and weighted voting is employed to classify the EEG into normal and seizure. The SVM classifier is trained using a radial basis kernel, and the results are recorded. Similarly, the performance of k-NN is recorded with $k=3$ and the Euclidean distance measure. Finally, the RF classifier results are found to be better than the other two; hence, it is used in the automated system.

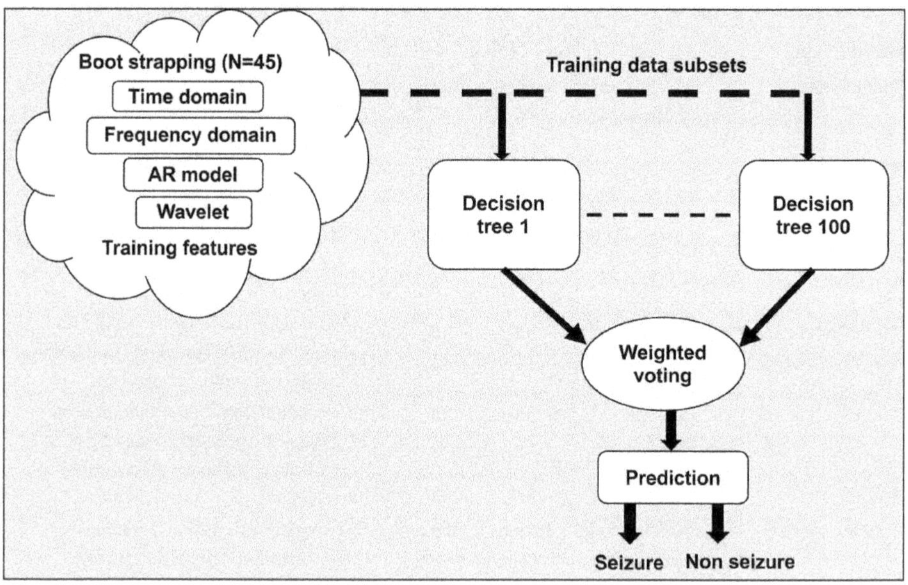

FIGURE 7.3 Random forest classifier.

7.3.2.4 Automated Seizure Detection System

The RF classifier is thoroughly trained to create a computerised detection system, and the trained model is imported to Raspberry Pi 3B+. It has built-in dual-band (2.4 and 5 GHz) Wi-Fi 802.11ac and Bluetooth 4.2/BLE, providing wireless connectivity options for networking and peripherals. The idea is to streamline EEG monitoring in tertiary healthcare centres by providing real-time updates to caregivers or doctors. Hence, a provision is made to send an alert message to the attending doctor or caregiver only if a seizure is detected. If the EEG signal is normal, alerting messages are not sent, so normal cases do not need to wait for further examination. Such a system can enhance the diagnosis when integrated with real-time EEG acquisition instruments.

7.4 RESULTS AND DISCUSSION

This section demonstrates the results obtained during experimentation. After pre-processing, the raw dataset is converted from microvolts to volts and passed through a band-pass filter. Further, the features are extracted from the filtered signals to train the classifiers. Among the three classifiers, RF classifiers provided the highest accuracy when trained using all features. The performances of the classifiers in terms of accuracy, precision rate, recall rate, and F1-score are provided in this section. Initially, the classifier is trained using time-domain features only. Further, the remaining feature sets are added one after the other and evaluated for classifier performance. The confusion matrix for each step is shown in Figure 7.4. It can be observed that the classifier performed better when all the features were combined.

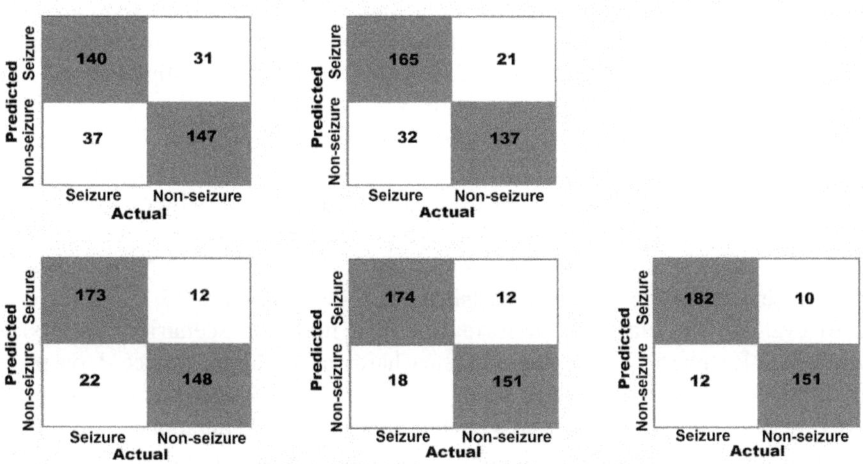

FIGURE 7.4 Confusion matrix of RF classifier: (a) time-domain features, (b) additional derivatives of time-domain features, (c) frequency-domain features, (d) AR model features, and (e) wavelet features.

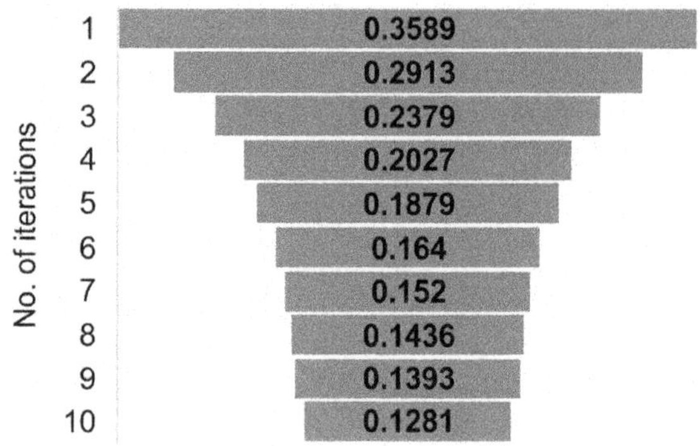

FIGURE 7.5 Out-of-bag error for ten iterations of the RF classifier.

To design a reliable RF model, the out-of-bag error is computed to estimate the number of trees. Figure 7.5 shows the gradual decrease in the error with an increase in the number of trees. The values show an apparent reduction in the error from 0.3589 to 0.1281, indicating an improvement in the model's performance.

In the proposed approach, model performances are evaluated by adding the feature sets one after the other. Finally, it is observed that performances are better after adding all thirty features. Hence, scores of the features are also extracted and analysed as shown in Figure 7.6. It is observed that many features contributed well to distinguishing standard and seizure EEG signals. However, PSD of the frequency range of 8–12 Hz marked a significant contribution with the highest score of 3.6279. During the experimental study, RF model performances are evaluated by dropping a few features whose ranks are below 0.8. This is selected empirically to obtain maximum accuracy.

Performance metrics for RF classifiers are tabulated in Table 7.2. It is evident that combinations of features are well suited for detecting seizures from EEG data. Similar experiments are carried out using SVM and k-NN classifiers, and their performances are recorded. It is observed that both classifiers performed well. However, the performance of the SVM classifier is better than that of k-NN. RF classifiers perform better than the other two classifiers as shown in Figure 7.7.

To evaluate the classifier performance in a real-time scenario, the model is imported to Raspberry Pi, and the necessary hardware setup is connected to a printed circuit board (PCB), as shown in Figure 7.8. During EEG signal acquisition, if seizures are detected, an alert message is sent to the concerned personnel. Additionally, light indicators are provided in the hardware setup: a green light to indicate normal EEG and a red light to indicate predicted seizures.

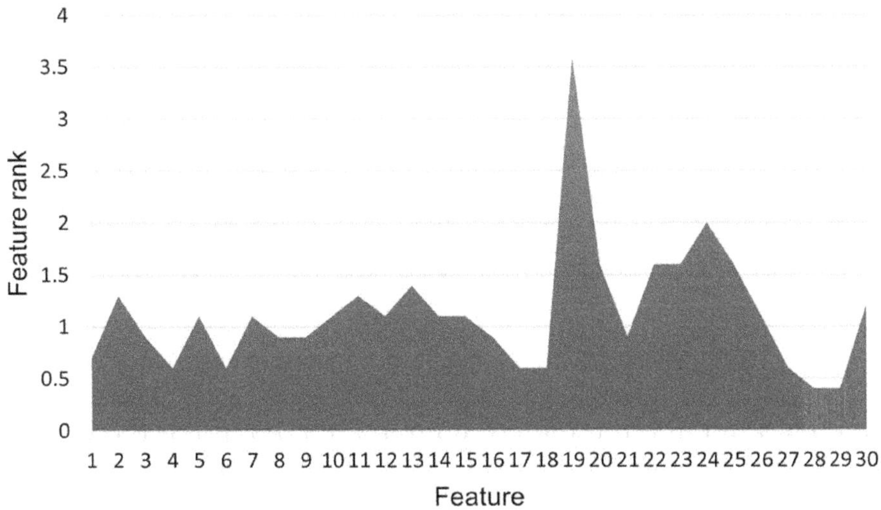

FIGURE 7.6 Scores of the features obtained from the RF classifier.

TABLE 7.2
Performance of RF Classifier

Performance/ Feature Set	Accuracy	Precision Rate	Recall Rate	F1-Score
Time domain	0.81	0.83	0.80	0.81
Addition time domain	0.85	0.87	0.81	0.84
Frequency domain	0.90	0.93	0.87	0.90
AR model	0.92	0.93	0.89	0.91
Wavelet-based	0.94	0.94	0.95	0.94

7.5 CONCLUSION

The proposed experimental study demonstrated the effectiveness of ML models, mainly RF, in detecting seizures from EEG signals. Various features such as time-domain, frequency-domain, AR model, and wavelet-based features are extracted, and the performance of the classifier models is evaluated by adding the feature sets one after the

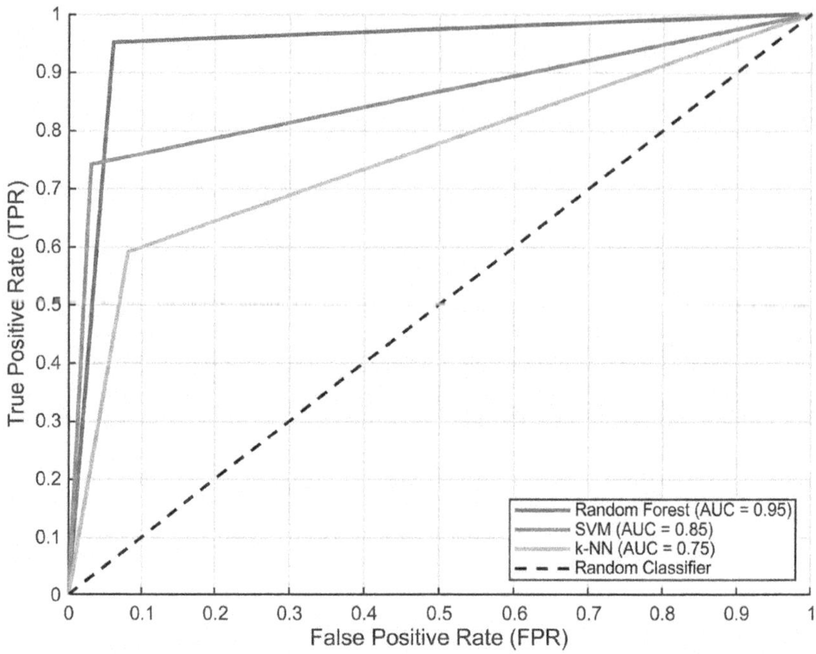

FIGURE 7.7 Comparison of the classifiers.

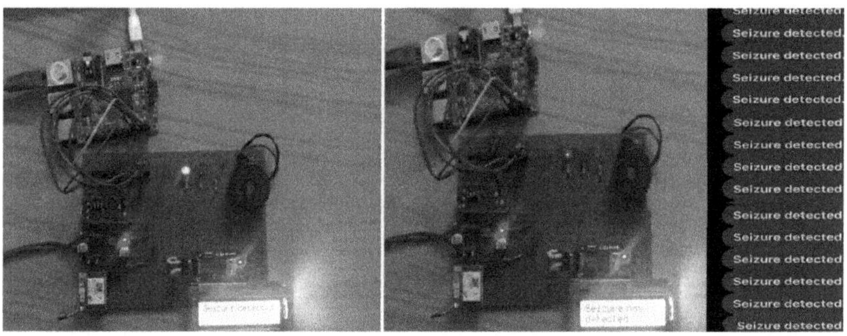

FIGURE 7.8 Real-time decision support seizure detection system.

other. It was found that RF classifiers achieved a maximum accuracy of 94% when trained using all the features. Further, a hardware system is developed to detect seizures in real time and to send an alert message to the concerned if seizures are detected. Integrating this automated detection system into the EEG acquisition system enhances and improves the efficiency of EEG signal analysis. The deployment of a Raspberry Pi with real-time alerting capabilities highlights its practical utility for continuous monitoring and decision support. Future work will focus on integrating DL techniques to improve the model's performance and robustness. Additionally, the dataset should be expanded to include various types of seizures.

REFERENCES

C. Intongkum, K. Sengchuai, and A. Booranawong, "A system for monitoring human postures, seizures, and falls from bed using radio and surface electromyography signals," *Bulletin of Electrical Engineering and Informatics*, vol. 13, no. 4, pp. 2517–2526, Aug. 2024. doi: 10.11591/eei.v13i4.7241.

G. J. Tortora and B. H. Derrickson, "*Principles of Anatomy and Physiology: Global Edition*", Wiley, 15th Edition, 2017.

J. Guttag, "CHB-MIT Scalp EEG Database," 2022, https://physionet.org/content/chbmit/1.0.0/ [Accessed: (23/08/2023)].

J. He, L. Yang, D. Liu, and Z. Song, "Automatic recognition of high-density epileptic EEG using support vector machine and gradient boosting decision tree," *Brain Science*, vol. 5, no. 12, p. 1197, 2022. doi: 10.3390/brainsci12091197. PMID: 36138933; PMCID: PMC9497056.

J. Thomas et al., "Development of a stereo-EEG based seizure matching system for clinical decision making in epilepsy surgery," *Journal of Neural Engineering*, vol. 21, no. 5, p. 1088, Aug. 2024. doi: 10.1088/1741-2552/ad7323.

K. K. Palanisamy and A. Rengaraj, "Detection of anxiety-based epileptic seizures in EEG signals using fuzzy features and parrot optimization-tuned LSTM," *Brain Sciences*, vol. 14, no. 8, p. 848, Aug. 2024. doi: 10.3390/brainsci14080848.

L. V. Tran, H. M. Tran, T. M. Le, T. T. M. Huynh, H. T. Tran, and S. V. T. Dao, "Application of machine learning in epileptic seizure detection," *Diagnostics*, vol. 12, no. 11, p. 2879, Nov. 2022. doi: 10.3390/diagnostics12112879.

L. Wei and C. Mooney, "Investigating the Need for Pediatric-Specific Machine Learning Approaches for Seizure Detection in EEG," in *2023 11th International Conference on Bioinformatics and Computational Biology (ICBCB)*, Hangzhou, China: IEEE, Apr. 2023, pp. 57–63. doi: 10.1109/ICBCB57893.2023.10246719.

M. B. Qureshi, M. A. Qureshi, M. S. Qureshi, and J. Gwak, "Fuzzybased automatic epileptic seizure detection framework," *Computers, Materials & Continua*, vol. 70, no. 3, pp. 5601–5630, 2022. doi: 10.32604/cmc.2022.020348.

M. M. Hasan, S. Rahman, A. Sarkar, F. Khan, and A. Seum, "Different methods of EEG signal analysis using power spectral density, ChronoNet and ResNest," *European Journal of Electrical Engineering and Computer Science*, vol. 7, no. 5, pp. 20–27, Oct. 2023. doi: 10.24018/ejece.2023.7.5.551.

M. S. Farooq, A. Zulfiqar, and S. Riaz, "Epileptic seizure detection using machine learning: taxonomy, opportunities, and challenges," *Diagnostics*, vol. 13, no. 6, p. 1058, 2023. doi: 10.3390/diagnostics13061058.

P. Kunekar, M. K. Gupta, and P. Gaur, "Detection of epileptic seizure in EEG signals using machine learning and deep learning techniques," *Journal of Engineering Applied Science*, vol. 71, no. 21, p. 21, 2024. doi: 10.1186/s44147-023-00353-y.

Q. Sun, Y. Liu, and S. Li, "Automatic seizure detection using multiinput deep feature learning networks for EEG signals," *Journal of Sensors*, vol. 2024, pp. 1–15, Feb. 2024. doi: 10.1155/2024/8835396.

S. A. Dhondiyal and S. C. Dimri, "Detection of Epilepsy Disease Using Support Vector Machine Based on EEG Signals," in *2022 OPJU International Technology Conference on Emerging Technologies for Sustainable Development (OTCON)*, Raigarh, Chhattisgarh, India: IEEE, Feb. 2023, pp. 1–6. doi: 10.1109/OTCON56053.2023.10114028.

S. Nahzat and M. Yaganoglu, "Classification of epileptic seizure dataset using different machine learning algorithms and PCA feature reduction technique," *Journal of Investigations on Engineering and Technology*, vol. 4, no. 2, pp. 47–60, 2021.

Y. Xiong et al., “Seizure detection algorithm based on fusion of spatio-temporal network constructed with dispersion index,” *Biomedical Signal Processing and Control*, vol. 79, p. 104155, Jan. 2023. doi: 10.1016/j.bspc.2022.104155.

Z. Wang, M. R. Sperling, D. Wyeth, and A. Guez, “Automated seizure detection based on state-space model identification,” *Sensors*, vol. 24, no. 6, p. 1902, Mar. 2024. doi: 10.3390/s24061902.

8 Deep Learning-Based Epileptic Seizure Classification in Neonates Using STFT-Transformed EEG Signals

Saif Al-Jumaili, Salam Alyassri, Ahmed A. Mohammed, Adil Deniz Duru, and Osman Nuri Uçan

8.1 INTRODUCTION

Interpreting the brain signals of a newborn baby requires specialized and experienced doctors; therefore, the number of specialists is limited. Early detection of the disease helps the doctor begin treatment and reduce the risk of worsening the condition. Usually, epileptic seizures are clear in adults, but in newborns, the seizures are more complex and similar to the symptoms of other diseases, such as hypoxic–ischemic encephalopathy. Therefore, the use of artificial intelligence (AI) represents a breakthrough in diagnosing disease because of its high diagnostic ability and speed of completion. Accordingly, one can benefit from the ability of AI to diagnose diseases. It has shown high potential for diagnosing epileptic seizures in several different ways, including machine learning (Alazzawı, Aljumaili et al. 2024) and deep learning (DL) (Al-Jumaili, Duru et al. 2021; Al-Jumaili, Al-Jumaili et al. 2022; Al-Jumaili, Duru et al. 2023), among the methods used in diagnosing disease through brain signals (Al-Jumaili, Al-Azzawi et al. 2021).

Indeed, the action of these neurons, triggered by different nerve stimuli, depends on the nervous system's responses (Al-Azzawi, Al-Jumaili et al. 2023, Saif 2022, 2023a, 2023b). In a normal state, the body is affected by electrical shocks resulting from a defect in brain functions, which manifests as epileptic seizures (Zazzaro, Cuomo et al. 2021), interpreted as epilepsy. These are temporary brain signals accompanied by symptoms that occur periodically or non-periodically. Based on the International League Against Epilepsy, electroencephalogram (EEG) signals are a reference tool for seizure diagnosis (Al-Azzawi, Al-Jumaili et al. 2022; Al-Azzawi, Al-Jumaili et al. 2023).

DOI: 10.1201/9781003611448-12

Therefore, it is considered a golden window for brain signals and represents indicative data for brain activities under internal and external conditions (Al-Jumaili, Duru et al. 2023). Moreover, the number of electrodes in the brain is one of the factors that most significantly affects the collection of signals from the human brain. Since the size of a child's head is small, the number of electrodes is limited compared to adults, making the diagnosis of head diseases in children difficult. Therefore, developing algorithms in the health sector that can diagnose neurological diseases with high accuracy is a challenge for researchers.

Many publicly available datasets have been created to help researchers in diagnosing epileptic seizures. Each dataset has different features, such as the University of Bonn (Andrzejak, Lehnertz et al. 2001), the TUH EEG Seizure Corpus (Shah, Von Weltin et al. 2018), and CHB-MIT Scalp EEG (Gupta, Singh et al. 2018). All these datasets are stored in different cloud storage, but they primarily represent adult subjects and are used by many researchers. Therefore, in this chapter, we study neonatal seizures due to the difficulties in classifying and diagnosing the disease using AI. The use of AI has significantly impacted the diagnosis of epilepsy through various techniques, including numerous machine learning algorithms based on iterative methods, such as support vector machines (SVM), k-nearest neighbors, and others (Ali, Al-Jumaili et al. 2022), or those relying on gradient descent mechanisms such as DL (Al-Jumaili, Al-Azzawi et al. 2023).

In the literature, many studies have been conducted to classify epileptic seizures. One of these studies was conducted by Chavan and Desai (2023). They analyzed EEG signals from the CHB-MIT dataset to diagnose focal and non-focal classes. Researchers depend on enhancing human learning (HLO) to benefit from its role in simplifying the model, in addition to enhancing feature extraction. With a feature extraction focus, in another study by (Ullah, Hussain et al. 2018), the research was supposed to be transferred using the ordinary convolutional neural network (CNN) to a one-dimensional neural network. The aim was to classify epilepsy patients whose data are registered in the University of Bonn dataset. The results show that memory savings reached 61%.

In a study by Abbasi, Rashad et al. (2019), there is work on recurrent neural networks used in epilepsy classification. Long short-term memory (LSTM) is a specific type of recurrent neural networks used. The development involves using two in the memory layers of LSTM, and the results of autoregressive moving average classification are compared with the SVM classifier, demonstrating how much the proposed model surpasses the SVM classifier. With the advancement of LSTM, research has explored the use of convolutional LSTM to describe epileptic seizures in relation to the classification of brain tumor type. The dataset is available on the UCI website and was collected from people with three eye conditions, while normal people only had two (Liu, Huang et al. 2020). The results obtained by the convolutional LSTM were more accurate than those obtained by the deep CNN. The models, including the LSTM within them, are proposed by Qiu, Yan et al. (2023); the result is the DARLNet model, which also uses CNN and ResNet. This model serves as a diagnostic tool for epileptic seizures by being trained on the Boone EEG signal records. The researchers show how accurate this model is in detecting five categories of epileptic seizures.

Mathematical calculations also aid in building the complete classifier system by re-transferring the EEG data to another domain, such as the Fourier–Bessel series expansion-based empirical wavelet transform (Anuragi, Sisodia et al. 2021). After applying the proposed method, Fourier–Bessel series expansion-based empirical wavelet transform, to the CHB-MIT dataset, the segments were of different lengths, which helped define the results and develop the work of the classifier. In Amiri, Aghaeinia et al. (2023), there is an application of the sparse common sCSP and adaptive FSST. FSST's role is to clean the time–frequency transformation of EEG from noise, while sCSP aids in selecting valuable channels, and the proposed model has the ability to reduce the data and offer only the most important channels.

On the other side, there are papers describing the effect of developing new DL models to help in classification. This was done by Dissanayake, Fernando et al. (2021), through implementing the geometric DL and training it on EEG signals. The research outcomes show the ability to allocate the seizures in the brain by constructing the brain's graphs by applying geometric DL on EEG data. The given model presents a novel method for detecting epileptic seizures using EEG signals, data augmentation, integrating feature extraction, and a DL architecture optimized with a well-constructed hybrid algorithm. Various extraction features (e.g., power spectral density, relative amplitude, multiple kernel-weighted Mel frequency cepstral coefficients, logarithmic band power, tonal power ratio, spectral entropy, and weighted one-dimensional local binary pattern) were extracted to improve the discriminative ability of the proposed model. To systematically tackle the issue of variability in EEG signals, data augmentation systems are employed (both for vertical flipping and circular shifting of EEG data) to ensure high structural reliability while expanding the obtained data size. It's known that the classification phase employs a deep maxout network (DMN), which has a robust approximation of activation functions and highly boosted feature learning. To ensure effective exploration and exploitation, training of the DMN is conducted using a novel dual algorithm involving the Taylor series for theoretical error reduction and the Henry gas solubility optimization. This integration increases the convergence speed, solves optimizations of multiple tasks, and avoids local minima. The given model proves excellent performance on the TUEP dataset, attaining accuracy, specificity rates, and sensitivity of 93.6%, 93.4%, and 94.7%, respectively, and 93.2%, 92.4%, and 89.2% on the CHB-MIT dataset. Compared to existing available models, the Taylor Henry gas solubility optimization-based DMN steadily outperforms many metrics, revealing its reliability for scalable and competent computerized seizure recognition in the medical field (Jaffino, Sundaram et al. 2022).

8.1.1 Motivation

Hybrid techniques are an effective answer since, eventually, single techniques for EEG seizure categorization might not yield satisfactory findings. Classifying newborn seizures within a millisecond or less is a significant problem, which is exacerbated by several aspects:

1. The possibility of diagnosing epileptic seizures in newborns is difficult and requires time, great effort, and the best specialists in diagnosing seizures in children.

2. The presence of signal artifacts greatly affects the accuracy of diagnosis, prompting scientists to develop a method that can diagnose the disease with high accuracy.

8.1.2 Contributions

We summarize the most important contributions made in this chapter as follows:

1. Developing a method that can diagnose epileptic seizures in newborns.
2. Developing a method that can convert the EEG signal into an image while preserving the signal pattern within a time difference, clarifying the effect on the brain.
3. Developing a new method that can extract features from images using multiple inputs to obtain the best features for each class, followed by the fusion of features.
4. Developing a method that has the ability to classify epilepsy with high accuracy and reduce time consumption for training.
5. To our knowledge, this is the first neonatal study that can classify epileptic seizures using spectrogram images with feature fusion.

8.2 MATERIALS AND METHODS

8.2.1 Dataset

To construct and improve frameworks that ultimately result in a reliable assessment of their models' efficiency, researchers and academics entirely depend on datasets. Seventy-nine newborn records and three specialists' comments on seizures from an EEG dataset were used in the current study (Stevenson, Tapani et al. 2019). With a postmenstrual age of 35–45 weeks, every record relates to a single neonate (subject). Due to their immaturity, these kinds of seizures are not the same as those in older children and adults. The newborn infant dataset was recorded with 19 electrodes distributed according to the global standard 10–20, and the dataset was made accessible to everyone. A spare electrode was positioned in the midline as a reference point. Initially, the data signals were arranged into an 18-channel "double banana montage" (Fp2-F4, F4-C4, C4-P4, P4-O2, Fp1-F3, F3-C3, C3-P3, P3-O1, Fp2-F8, F8-T4, T4-T6, T6-O2, Fp1-F7, F7-T3, T3-T5, T5-O1, Fz-Cz, Cz-Pz) as illustrated in Figure 8.1 with a sampling rate of 256 Hz.

This included an average record duration of one hour for each subject; thus, in the end, the neonatal EEG data had a total duration of 111.9 h, with a median of 74 min. In terms of diagnosis, the three experts agreed on the following two groups. The data gathered showed that 40 newborns had a seizure confirmed by all three experts (infants No. 1, 4, 5, 7, 9, 11, 13, 14, 15, 16, 17, 19, 20, 21, 22, 25, 31, 34, 36, 38, 39, 40, 41, 44, 47, 50, 51, 52, 62, 63, 66, 67, 69, 71, 73, 75, 76, 77, 78, 79), whereas 22 were seizure-free (infants No. 3, 10, 18, 27, 28, 29, 30, 32, 35, 37, 42, 45, 48, 49, 53, 55, 57, 58, 59, 60, 70, 72), the scientists said. Finally, the remaining 17 newborns were affected by one or two expert decisions confirming the possibility of epilepsy (infants Nos. 2, 6, 8, 12, 23, 24, 26, 33, 43, 46, 54, 56, 61, 64, 65, 68, 74). Table 8.1 presents a comprehensive summary of the groups, classified according to the clinical

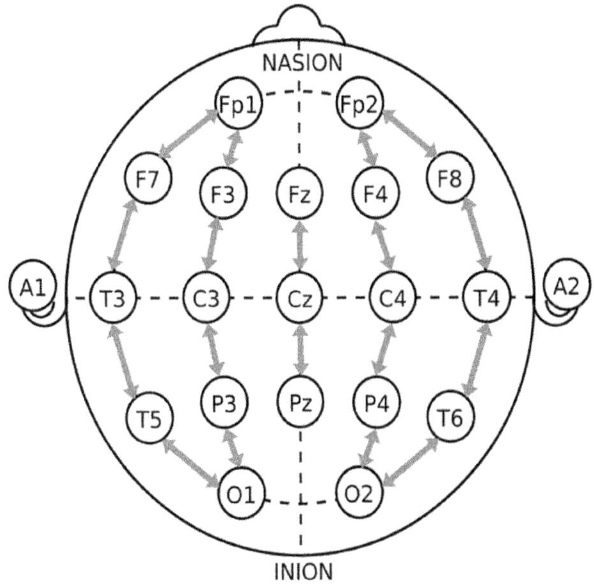

FIGURE 8.1 EEG signal distribution using the 10–20 system (Stevenson, Tapani et al. 2019).

TABLE 8.1
Provides a Detailed Overview of the Groupings

Group	Description	Infants Number	Seizures Annotated
EXP12_17	The collected data demonstrated that 17 neonates were impacted by one or two expert assessments affirming the potential for seizures	2, 6, 8, 12, 23, 24, 26, 33, 43, 46, 54, 56, 61, 64, 65, 68, 74	46
EXP0_22	The collected data indicated that 22 neonates experienced a seizure verified by all three specialists	3, 10, 18, 27, 28, 29, 30, 32, 35, 37, 42, 45, 48, 49, 53, 55, 57, 58, 59, 60, 70, 72	45
EXP3_40	The collected data indicated that 40 neonates had seizures corroborated by all three specialists	1, 4, 5, 7, 9, 11, 13, 14, 15, 16, 17, 19, 20, 21, 22, 25, 31, 34, 36, 38, 39, 40, 41, 44, 47, 50, 51, 52, 62, 63, 66, 67, 69, 71, 73, 75, 76, 77, 78, 79	53

diagnosis performed by three independent specialists. These specialists assessed all children and verified the existence or nonexistence of epileptic seizures, thus confirming the validity and precision of the diagnostic criteria used.

8.3 METHODOLOGY

This section explains the method used in this chapter. The method is divided into two stages. The first stage involves converting brain signals into images using the short-time Fourier transform (STFT) to convert each channel into an image. Given that the duration of the signal recorded from the brain was more than an hour for each patient, converting every second into an image would generate a very large number of images. Therefore, by converting every 200 ms into images, we can increase the extracted information, while reducing the total number of images. The total number of images extracted for all classes was 52,708 images (21,000 for EXP0_22, 17,895 for EXP3_40, and 13,813 for EXP12_17). Although this is still a large number, we selected a group of random images for each class and divided them into two parts for feature extraction and training. The second stage involves the image classification method, which was carried out using a new method that features fusion DL. Features were extracted from the images for each class separately using several pre-trained algorithms. These features were then combined and input into the classification system to classify the diseases. Figure 8.2 illustrates the method comprehensively.

8.3.1 SHORT-TIME FOURIER TRANSFORM

Transforming the domain of any function can be accomplished by different mathematical approaches. One of them is integral transformation, which uses a kernel function, as shown in Equation 8.1 (Sneddon 1995):

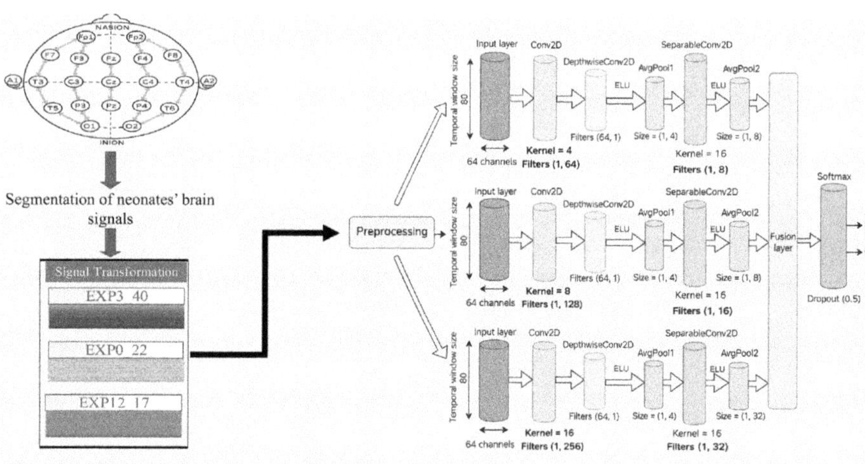

FIGURE 8.2 A comprehensive overview of the method used in this chapter, where the signal was divided into small segments, and then these parts were converted to spectrogram images, which were used as input to algorithms to extract the features.

$$X(\alpha) = \int_{0}^{\infty} x(t) K(\alpha, t) d(t) \tag{8.1}$$

where $K(\alpha, t)$ is a known function in the α and t domains.

If $x(t)$ is differentiable in the t domain, then the operand ($s = d/d(t)$) will create a new domain, and the kernel function will change to e^{-st}. Thus, Equation 8.1 can be rewritten as Equation 8.2 (Spiegel, 1965):

$$X(s) = \int_{0}^{\infty} e^{-st} x(t) d(t) \tag{8.2}$$

This transformation is called as the Laplace transform, with the notation $X(s) = L[x(t)]$. Conversely, the original signal $x(t)$ can be reconstructed from the Laplace transformed signal $X(s)$ by taking the inverse Laplace transform, as shown in Equation 8.3 (Dyke and Dyke 2001):

$$L^{-1}\{X(s)\} = x(t) \tag{8.3}$$

Assuming that $s = k + i\omega$, we have $ds = i\, d\omega$ and the integration limits change from $(0, \infty)$ to $(k - i\infty, k + i\infty)$. $x(t)$ is calculated as in Equation 8.4 (Dyke and Dyke 2001):

$$x(t) = \frac{1}{2\pi i} \int_{k-i\infty}^{k+i\infty} e^{st} X(s) ds \tag{8.4}$$

In special cases when $s = i\omega$, the result is given in Equation 8.5:

$$x(t) = \frac{1}{2\pi} \int_{-\infty}^{\infty} e^{i\omega t} X(\omega) d\omega \tag{8.5}$$

So,

$$X(\omega) = \int_{-\infty}^{\infty} e^{-i\omega t} x(t) dt \tag{8.6}$$

$X(\omega)$ is called the Fourier transform of $x(t)$ and (ω) is called the frequency domain. Modern calculations are carried out via computers, which work in discrete time, necessitating the derivation of the discrete form of the Fourier transform, shown in Equation 8.7, called the discrete Fourier transform (DFT) (Selesnick and Schuller 2001):

$$X(k)=\sum_{n=0}^{N-1}x(n)W_N^{kn}, \quad k=0,\ldots..,N-1 \tag{8.7}$$

whereas

$$W_n=e^{-i2\pi/N}=\cos\left(\frac{2\pi}{N}\right)-\sin\left(\frac{2\pi}{N}\right) \tag{8.8}$$

With the few obvious frequency component signals, the DFT is an essential transformation that reduces the data for processing. However, with periodic signals like brain signals, analysis can be less informative depending on the frequency. The solution comes from segmenting these signals quickly within a short time to introduce features depending on the shape of the signals during this windowing period. This analysis is called frequency–time transforming (Baba 2012).

To accumulate accurate features, the aim is to make the time window as narrow as possible, leading to the terminology of STFT. The slice can be shortened with samples that consist of a wide frequency, and the reason for this is the relationship between time and frequency. The figures resulting from taking two dimensions, one for time and another for frequency, are called a spectrogram.

From Equation 8.7, the width of the periodic DFT (N) is one parameter used to estimate the accuracy of the transformation. With STFT, the sampling time will be short, so the width of the sampling time is another parameter controlling the transformation results.

The sampling will be done through convolution with a shifting window equation, as shown in (Equation 8.9):

$$r(n)=x(n)\otimes W_n(n-k.S_n) \tag{8.9}$$

From Equation 8.9, there are other transform control operators such as the amount of time shifting (S_n) and the shape of the (W_n). As an explanatory example for the sample of the signal in the frequency and time domain, Figure 8.3a and b shows the parameters of STFT, in which a chirp waveform was selected as W_n.

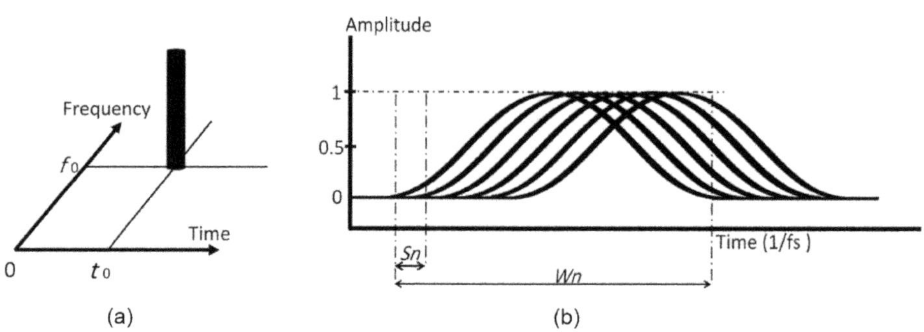

FIGURE 8.3 (a) Time–frequency sample. (b) STFT parameters.

As noted above, the STFT has the following parameters that control the result of the transformation:

1. W_n wave shape form.
2. W_n time width.
3. S_n time delay.
4. N components for DFT size.

8.3.2 Feature Extraction

The classification of epilepsy is one of the most important aspects related to neurological degeneration in the brain, as DL is currently being used to classify epileptic seizures. This section will explain the method used to extract features from images generated using STFT. We used three different models to obtain the best features, fuse them, and then use them for training and classification.

8.3.3 ResNet

The ResNet model is one of the best models developed recently because of its ability to extract features from images (Lu and Triesch 2019). The ResNet family consists of many models, all of which use residual blocks. The number in the model indicates the number of layers used in the structure of the model, which in turn affects its depth, for example, ResNet16, ResNet18, ResNet34, ResNet50, ResNet101, ResNet110, ResNet152, ResNet164, and ResNet1202 (Al-Jumaili, Al-Jumaili et al. 2022). On the other hand, increasing the depth of the model negatively affects its performance, leading to degradation and vanishing gradients. To avoid this problem, the ResNet model uses the method of adding a skip connection, which prevents loss of information when the model has several layers

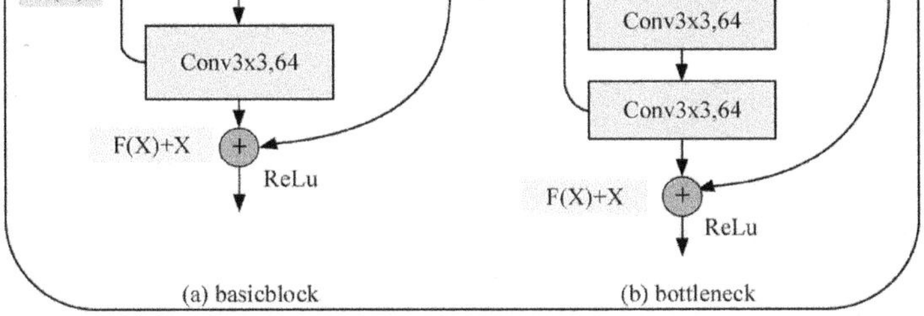

FIGURE 8.4 Residual blocks architecture used in ResNet18 and ResNet50 (Memari, Shekaramiz et al. 2024).

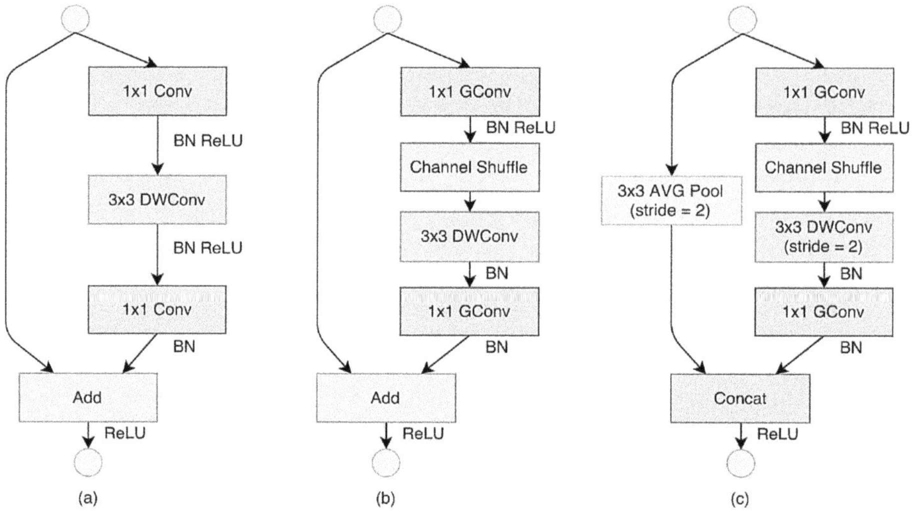

FIGURE 8.5 ShuffleNet network architecture (Zhang, Zhou et al. 2018).

(He, Zhang et al. 2016). Figure 8.4 shows the method used in the residual blocks, where there are two convolutional layers in 4a on the left. In addition to using 3×3 kernels, the spatial dimensions are maintained. The ResNet18 architecture uses a technique on the right side that uses skip connections by connecting the input data to the outputs. As shown in Figure 8.4-b, the input is passed to convolutional networks with different kernel sizes of 1×1 and 3×3. The correct method is to skip the connection that connects the inputs of the unit through an addition operation with the outputs of the left path.

8.3.4 ShuffleNet

ShuffleNet is a convolutional neutral network primarily designed for mobile devices but also partially for computers. The design uses two novel operations ("pointwise group" convolutions along with channel shuffling) to reduce processing costs while maintaining precision. ShuffleNet is a CNN design with exceptionally significant computational effectiveness and low computational consumption. Figure 8.5 illustrates the shuffle block.

8.3.5 Evaluation Metrics

To evaluate the results obtained, we used various performance evaluation techniques to investigate the model individually, relying on the results obtained using the confusion matrix after validation. The content of the confusion matrix was used to include entries to measure the following criteria: accuracy, precision, recall, and F1 score. The equations used to calculate these metrics are shown in Table 8.2.

TABLE 8.2
Equations Used for Measuring the Results

Equation	Number
$\text{Accuracy} = \frac{TP + TN}{TP + FP + TN + FN}$	(1)
$\text{Sensitivity} = \frac{TP}{TP + FN}$	(2)
$\text{Precision}(PPV) = \frac{TP}{TP + FP}$	(3)
$\text{F1} - \text{Score} = \frac{2 * TP}{2 * TP + FP + FN}$	(4)

8.4 RESULTS AND DISCUSSION

To classify satisfactorily and build a CNN model from scratch, the model needs to be trained on a set of images. Therefore, the number of images must be large to obtain good training for the model. Using pre-trained algorithms is one of the best quick solutions because of their ability to save time because they are trained on a different set of shapes and images. In addition to the accuracy of the classification, the strength of the architecture used in the structure of the models is also essential to obtain the best features from the image. In the proposed method, features were extracted using pre-trained models from images of each class individually, and then these features were combined and used as input for classification. The images were divided into 70% for training and 30% for testing. The results obtained by using the input images will be calculated using pre-trained models for training and testing. Table 8.3 shows the results obtained from the confusion matrix.

TABLE 8.3
Confusion Matrix Results Based on Three Different Types of Pre-Trained Models

CNN Name	Classes Name	Seizure Name	Predicted Class		
ResNet18	Actual Class	EXP0_22	3,917	0	53
		EXP3_40	0	5,959	181
		EXP12_17	226	342	5,134
ResNet50	Actual Class	EXP0_22	4,091	0	55
		EXP3_40	0	6,172	340
		EXP12_17	53	128	4,973
ShuffleNet	Actual Class	EXP0_22	4,101	2	248
		EXP3_40	0	6,017	262
		EXP12_17	43	281	4,858

The ResNet18 model performs well, with accuracy ranging from 94.93% to 98.24% across many datasets. However, the accuracy fluctuated, reaching 99% in the EXP0_22 dataset and falling to 90% in the EXP12_17 dataset, indicating that the model is less effective in some cases. Regarding recall, the model largely maintained a value ranging from 95% to 96%, which indicates its effectiveness in positive cases. The F1 rates indicate about 95.33%, which indicates that the model's performance is fairly acceptable. The ShuffleNet model shows performance similar to ResNet18, with accuracy ranging between 94.73% and 98.15%. This model shows relatively stable accuracy at around 94% on average and a recall between 90% and 99%. F1 rates are around 95%, reflecting a good balance between precision and recall, although the overall performance is not up to the level of ResNet50. In the last case, for the ShuffleNet model, the results were similar to the ResNet18 model, where the accuracy was between 94.73% and 98.15%. From the accuracy obtained, the model shows relative stability, while for precision, the value was constant at 94%. As for recall, the results ranged between 90% and 99%. The results for the last F1 score criterion varied between 92 and 97. In the general evaluation, the performance of this model was acceptable. The presented Receiver Operating Characteristic (ROC) curves demonstrate the performance of three well-known deep learning models, ShuffleNet, ResNet18, and ResNet50, on seizure recognition tasks. Each subplot links to one of those architectures, showing the true-positive rate in contrast to the false-positive rate during experimental setups for the model. The ResNet (including both ResNet18 and ResNet50) operates residual connections to address the vanishing gradient problem, providing deeper training. Since ResNet50 consists of 50 layers compared to ResNet18, which has 18 layers, it favorably captures more complex structures but characteristically requires extra computational resources. On the other hand, ShuffleNet is typically designed for efficient resource-constrained systems, for instance, mobile devices, which use pointwise cluster convolution and channel shuffling to achieve higher performance with minimized computational environments.

As shown in Figure 8.6, all the experiments and tested models accomplish high area under the curve values, approaching 1.0, demonstrating outstanding

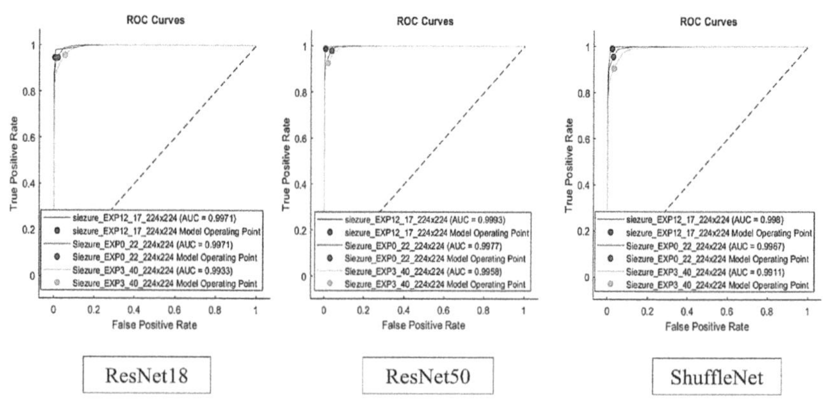

FIGURE 8.6 ROC curves demonstrate the performance of three well-known deep learning models.

classification accuracy for seizure recognition. ResNet18 reveals strong performance with relatively minor complexity compared to ResNet50, making it appropriate when computational efficiency is critical. Moreover, ResNet50 displays comparable performance, if not slightly greater, benefiting from its deeper design to detect complicated patterns in the given data. Meanwhile, ShuffleNet provides competitive performance that is ideal for various scenarios with limited hardware capabilities while prioritizing efficiency. The high values of area under the curve across all models highlight the effectiveness of these architectures in this particular task, even if the selection between them would rely on available resources and application necessities.

The results obtained using the three models shown in Table 8.4, in terms of accuracy, were the highest for the ResNet50 model, with a rate exceeding 97%, compared to the other models, where the accuracy was close but significantly lower. As for other criteria, the ResNet50 model obtained the highest results. Finally, to verify the results obtained using feature fusion, we compared our results with previous studies classifying epilepsy in newborns. Table 8.5 shows that our results were better than all previous studies. This confirms that the proposed method has the potential to classify epileptic seizures in newborns, as well as the possibility of using this method in clinical applications. Based on the results obtained, the fusion method provided very good results compared to the number of images used in training and testing. Seizure diagnostic technology represents a great opportunity to enhance the monitoring and treatment of seizures in infants. These new methods can contribute significantly to the possibility of diagnosing epilepsy in children and thus help in treatment faster compared to manual diagnosis, which may have a higher error rate than diagnosis using a computer. In addition to the time factor, manual diagnosis requires considerably more time compared to automatic diagnosis.

TABLE 8.4
Classification Results Using the Three Models and Applying 5 Cross-Validation

Optimizer	Model	Metrices	EXP0_22	EXP3_40	EXP12_17	Average
Sgdm	ResNet18	Accuracy	98.24	96.69	94.93	96.62
		Precision	99	97	90	95.33
		Recall	95	95	96	95.33
		F1 score	97	96	93	95.33
	ResNet50	Accuracy	99.32	97.04	96.36	97.57
		Precision	99	95	96	96.66
		Recall	99	98	93	96.66
		F1 score	99	96	95	96.66
	ShuffleNet	Accuracy	98.15	96.55	94.73	96.47
		Precision	94	96	94	94.66
		Recall	99	96	90	95
		F1 score	97	96	92	95

TABLE 8.5
Comparison of the Results Obtained with Several Previous Studies Classifying Epileptic Seizures

References	Features	Classifiers	Sen %	Spec %	Acc %
Acharya, Oh et al. (2018)	Raw signal	CNN	95	90	88
Li, Karmakar et al. (2018)	Fuzzy entropy and distribution entropy	QDA	90	92	92
Ocak (2009)	Approximate entropy	Thresholding	N.A.	N.A.	73
Tawfik, Youssef et al. (2016)	Weighted permutation entropy	SVM	N.A.	N.A.	91
Tawfik, Youssef et al. (2016)	Intrinsic mode functions	Random forest	91	93	91
Cao, Guo et al. (2017)	STFT	CNN	N.A.	N.A.	80
Gill, Fatima et al. (2015)	Hybrid feature set	GMM	87	86	86
Acharya, Hagiwara et al. (2018)	Multiple	CNN	95	90	88
Chen, Wan et al. (2017)	Multiple	SVM	N.A.	N.A.	92.30
Selvakumari, Mahalakshmi et al. (2019)	Multiple of entropy	SVM	95.7	96.55	95.63
Raghu and Sriraam (2018)	Multiple	SVM	97.6	94.4	96.1
Zahra, Kanwal et al. (2017)	EMD	ANN	N.A.	N.A.	87.2
Kavitha, Ashok et al. (2022)	DWT	k-NN	N.A.	N.A.	91.67
Tuncer (2021)	Statistical features	Bi-LSTM	N.A.	N.A.	96
Khan, Khan et al. (2021)	Statistical features	LSTM	N.A.	N.A.	96
Elakkiya (2021)	Deep CNN	SVM	N.A.	N.A.	92.3
Ryu, Back et al. (2021)	Principal component analysis	Cos-KNN	N.A.	N.A.	68.14
Tapani, Nevalainen et al. (2022)	Statistical features	SVM	N.A.	N.A.	93
Tanveer, Khan et al. (2021)	Segmentation	CNN	N.A.	N.A.	95.6
Webb, Kauppila et al. (2021)	Artifact detection algorithm	Residual deep neural network	N.A.	N.A.	95
Mumenin, Biswas et al. (2023)	Multiple	Gradient boosting	93.38	77.52	93.38
Ours	ResNet18&50, ShuffleNet	SoftMax	96.66	96.66	97.57

8.5 CONCLUSION

AI is one of the most recent developments in diagnosing diseases. It has the ability to diagnose diseases with high accuracy. As for epilepsy, diagnosing it in children is considered one of the most complex matters because of the difficulty in determining the disease due to the small size of the child's brain and the similarity of the disease to several other conditions. In addition, seizures do not appear clearly in the brain's diagram, so scientists resort to methods that can accurately diagnose the disease in the early stages. Therefore, in this chapter, epileptic seizures were classified using EEG signals, where the signals were converted into images using the STFT algorithm, which divides the signal into small parts lasting 1 second and then converts them into histogram images. Three types of pre-trained algorithms were applied to fuse the features extracted individually for each class. Based on the data obtained by applying the three models, the best model, which is clearly superior in all criteria compared to the other models, is ResNet50. It achieved an accuracy of about 99.32%. For both ResNet18 and ShuffleNet, the results were close but at a lower level, as the accuracy ranged between 94.73% and 98.24%. The ResNet18 model showed an improvement of 4% over the previous one, but in general, the accuracy in the three cases was acceptable. The discrepancy in class classification (EXP0_22, EXP3_40, and EXP12_17) shows that the distribution of images plays a vital role in the accuracy of the final classification of the model, as the ResNet50 model has the ability to adapt to changes in the number of images and thus demonstrates the possibility and effectiveness of the model in diagnosing epileptic seizures in children.

REFERENCES

Abbasi, M. U., A. Rashad, A. Basalamah and M. Tariq (2019). "Detection of epilepsy seizures in neo-natal EEG using LSTM architecture." *IEEE Access* 7: 179074–179085.

Acharya, U. R., Y. Hagiwara and H. Adeli (2018). "Automated seizure prediction." *Epilepsy & Behavior* 88: 251–261.

Acharya, U. R., S. L. Oh, Y. Hagiwara, J. H. Tan and H. Adeli (2018). "Deep convolutional neural network for the automated detection and diagnosis of seizure using EEG signals." *Computers in Biology and Medicine* 100: 270–278.

Al-Azzawi, A., S. Al-Jumaili, A. D. Duru, O. Bayat, S. Kurnaz and O. N. Uçan (2023). Pseudopapilledema diagnosis based on a hybrid approach using deep transfer learning. *2023 7th International Symposium on Multidisciplinary Studies and Innovative Technologies (ISMSIT)*, IEEE, New York.

Al-Azzawi, A., S. Al-Jumaili, A. D. Duru, D. G. Duru and O. N. Uçan (2023). "Evaluation of deep transfer learning methodologies on the COVID-19 radiographic chest images." *Traitement du Signal* 40(2): 407–420.

Al-Azzawi, A. H. A. L., S. Al-Jumaili, A. A. Ibrahim and A. D. Duru (2022). "Classification of epileptic seizure features from scalp electrical measurements using KNN and SVM based on Fourier transform." *AIP Conference Proceedings* 2499: 020003.

Al-Jumaili, S. (2022). "Classification of COVID-19 omicron variant using hybrid deep transfer learning based on X-Ray images chest." *Aurum Journal of Health Sciences* 4(3): 153–165.

Al-Jumaili, S., A. Al-Azzawi, A. D. Duru and A. A. Ibrahim (2021). Covid-19 X-ray image classification using SVM based on local binary pattern. *2021 5th International Symposium on Multidisciplinary Studies and Innovative Technologies (ISMSIT)*, IEEE, New York.

Al-Jumaili, S., A. Al-Azzawi, O. N. Uçan and A. D. Duru (2023). Classification of the level of Alzheimer's disease using anatomical magnetic resonance images based on a novel deep learning structure. In Chaki, J. (ed.), *Diagnosis of Neurological Disorders Based on Deep Learning Techniques*. CRC Press, Boca Raton, FL, pp. 29–46.

Al-Jumaili, S., A. Al-Jumaili, S. Alyassri, A. D. Duru and O. N. Uçan (2022). Recent advances on convolutional architectures in medical applications: Classical or Quantum? *2022 International Symposium on Multidisciplinary Studies and Innovative Technologies (ISMSIT)*, IEEE, New York.

Al-Jumaili, S., A. D. Duru, A. A. Ibrahim and O. N. Uçan (2023). "Investigation of epileptic seizure signatures classification in EEG using supervised machine learning algorithms." *Traitement du Signal* 40(1): 43–54.

Al-Jumaili, S., A. D. Duru and O. N. Uçan (2021). Covid-19 Ultrasound image classification using SVM based on kernels deduced from Convolutional neural network. *2021 5th International Symposium on Multidisciplinary Studies and Innovative Technologies (ISMSIT)*, IEEE, New York.

Al-Jumaili, S., D. G. Duru, B. Ucan, O. N. Uçan and A. D. Duru (2022). "Classification of Covid-19 effected CT images using a hybrid approach based on deep transfer learning and machine learning." *Research Square*. DOI: 10.21203/rs.3.rs-1541093/v1.

Alazzawı, A., S. Aljumaili, A. D. Duru, O. N. Uçan, O. Bayat, P. J. Coelho and I. M. Pires (2024). "Schizophrenia diagnosis based on diverse epoch size resting-state EEG using machine learning." *PeerJ Computer Science* 10: e2170.

Ali, R., S. Al-Jumaili, A. D. Duru, O. N. Uçan, A. Boyaci and D. G. Duru (2022). Classification of brain tumors using MRI images based on convolutional neural network and supervised machine learning algorithms. *2022 International Symposium on Multidisciplinary Studies and Innovative Technologies (ISMSIT)*, IEEE, New York.

Amiri, M., H. Aghaeinia and H. R. Amindavar (2023). "Automatic epileptic seizure detection in EEG signals using sparse common spatial pattern and adaptive short-time Fourier transform-based synchrosqueezing transform." *Biomedical Signal Processing and Control* 79: 104022.

Andrzejak, R. G., K. Lehnertz, F. Mormann, C. Rieke, P. David and C. E. Elger (2001). "Indications of nonlinear deterministic and finite-dimensional structures in time series of brain electrical activity: Dependence on recording region and brain state." *Physical Review E* 64(6): 061907.

Anuragi, A., D. S. Sisodia and R. B. Pachori (2021). "Automated FBSE-EWT based learning framework for detection of epileptic seizures using time-segmented EEG signals." *Computers in Biology and Medicine* 136: 104708.

Baba, T. (2012). "Time-frequency analysis using short time Fourier transform." *The Open Acoustics Journal* 5(1): 32–38.

Cao, Y., Y. Guo, H. Yu and X. Yu (2017). Epileptic seizure auto-detection using deep learning method. *2017 4th International Conference on Systems and Informatics (ICSAI)*, IEEE, New York.

Chavan, P. A. and S. Desai (2023). "Effective epileptic seizure detection by classifying focal and non-focal EEG signals using human learning optimization-based hidden Markov model." *Biomedical Signal Processing and Control* 83: 104682.

Chen, D., S. Wan, J. Xiang and F. S. Bao (2017). "A high-performance seizure detection algorithm based on Discrete Wavelet Transform (DWT) and EEG." *PLoS One* 12(3): e0173138.

Dissanayake, T., T. Fernando, S. Denman, S. Sridharan and C. Fookes (2021). "Geometric deep learning for subject independent epileptic seizure prediction using scalp EEG signals." *IEEE Journal of Biomedical and Health Informatics* 26(2): 527–538.

Dyke, P. P. and P. Dyke (2001). *An introduction to Laplace transforms and Fourier Series*. Springer, Berlin.

Elakkiya, R. (2021). "Machine learning based intelligent automated neonatal epileptic seizure detection." *Journal of Intelligent & Fuzzy Systems* 40(5): 8847–8855.

Gill, A. F., S. A. Fatima, M. U. Akram, S. G. Khawaja and S. E. Awan (2015). Analysis of EEG signals for detection of epileptic seizure using hybrid feature set. *Theory and Applications of Applied Electromagnetics: APPEIC 2014*, Springer, Berlin.

Gupta, A., P. Singh and M. Karlekar (2018). "A novel signal modeling approach for classification of seizure and seizure-free EEG signals." *IEEE Transactions on Neural Systems and Rehabilitation Engineering* 26(5): 925–935.

He, K., X. Zhang, S. Ren and J. Sun (2016). Deep residual learning for image recognition. *Proceedings of the IEEE Conference on Computer vision and Pattern Recognition*, IEEE, New York.

Jaffino, G., M. Sundaram and J. P. Jose (2022). "Weighted 1D-local binary pattern features and Taylor-Henry gas solubility optimization based Deep Maxout network for discovering epileptic seizure using EEG." *Digital Signal Processing* 122: 103349.

Kavitha, K., S. Ashok, A. L. Imoize, S. Ojo, K. S. Selvan, T. A. Ahanger and M. Alhassan (2022). "On the use of wavelet domain and machine learning for the analysis of epileptic seizure detection from EEG signals." *Journal of Healthcare Engineering* 2022(1): 8928021.

Khan, P., Y. Khan, S. Kumar, M. S. Khan and A. H. Gandomi (2021). "HVD-LSTM based recognition of epileptic seizures and normal human activity." *Computers in Biology and Medicine* 136: 104684.

Li, P., C. Karmakar, J. Yearwood, S. Venkatesh, M. Palaniswami and C. Liu (2018). "Detection of epileptic seizure based on entropy analysis of short-term EEG." *PLoS One* 13(3): e0193691.

Liu, Y., Y.-X. Huang, X. Zhang, W. Qi, J. Guo, Y. Hu, L. Zhang and H. Su (2020). "Deep C-LSTM neural network for epileptic seizure and tumor detection using high-dimension EEG signals." *IEEE Access* 8: 37495–37504.

Lu, D. and J. Triesch (2019). "Residual deep convolutional neural network for eeg signal classification in epilepsy." arXiv preprint arXiv:1903.08100.

Memari, M., M. Shekaramiz, M. A. Masoum and A. C. Seibi (2024). "Data fusion and ensemble learning for advanced anomaly detection using multi-spectral RGB and thermal imaging of small wind turbine blades." *Energies* 17(3): 673.

Mumenin, K. M., P. Biswas, M. A.-M. Khan, A. S. Alammary and A.-A. Nahid (2023). "A modified aquila-based optimized XGBoost framework for detecting probable seizure status in neonates." *Sensors* 23(16): 7037.

Ocak, H. (2009). "Automatic detection of epileptic seizures in EEG using discrete wavelet transform and approximate entropy." *Expert Systems with Applications* 36(2): 2027–2036.

Qiu, X., F. Yan and H. Liu (2023). "A difference attention ResNet-LSTM network for epileptic seizure detection using EEG signal." *Biomedical Signal Processing and Control* 83: 104652.

Raghu, S. and N. Sriraam (2018). "Classification of focal and non-focal EEG signals using neighborhood component analysis and machine learning algorithms." *Expert Systems with Applications* 113: 18–32.

Ryu, S., S. Back, S. Lee, H. Seo, C. Park, K. Lee and D.-S. Kim (2021). "Pilot study of a single-channel EEG seizure detection algorithm using machine learning." *Child's Nervous System* 37(7): 2239–2244.

Saif, A.-J. (2022). "Classification of COVID-19 omicron variant using hybrid deep transfer learning based on X-Ray images chest." *Aurum Journal of Health Sciences* 4(3): 153–165.

Saif, A.-J. (2023a). "ParkinsonNet: Classification Parkinson's disease model based on novel deep learning structure." *AURUM Journal of Engineering Systems and Architecture* 7(2): 259–276.

Saif, A.-J. (2023b). "Efficient mental arithmetic classification using approximate entropy features and machine learning classifiers." *Aurum Journal of Health Sciences* 5(3): 109–120.

Selesnick, I. W. and G. Schuller (2001). The discrete Fourier transform. In Rao, K. R. and Yip, P. C. (ed.) *The Transform and Data Compression Handbook.* CRC Press, Boca Raton, FL, pp. 37–74.

Selvakumari, R. S., M. Mahalakshmi and P. Prashalee (2019). "Patient-specific seizure detection method using hybrid classifier with optimized electrodes." *Journal of Medical Systems* 43(5): 121.

Shah, V., E. Von Weltin, S. Lopez, J. R. McHugh, L. Veloso, M. Golmohammadi, I. Obeid and J. Picone (2018). "The temple university hospital seizure detection corpus." *Frontiers in Neuroinformatics* 12: 83.

Sneddon, I. N. (1995). *Fourier Transforms.* Courier Corporation, New York.

Spiegel, M. R. (1965). *Laplace Transforms.* McGraw-Hill, New York.

Stevenson, N. J., K. Tapani, L. Lauronen and S. Vanhatalo (2019). "A dataset of neonatal EEG recordings with seizure annotations." *Scientific Data* 6(1): 1–8.

Tanveer, M. A., M. J. Khan, H. Sajid and N. Naseer (2021). "Convolutional neural networks ensemble model for neonatal seizure detection." *Journal of Neuroscience Methods* 358: 109197.

Tapani, K. T., P. Nevalainen, S. Vanhatalo and N. J. Stevenson (2022). "Validating an SVM-based neonatal seizure detection algorithm for generalizability, non-inferiority and clinical efficacy." *Computers in Biology and Medicine* 145: 105399.

Tawfik, N. S., S. M. Youssef and M. Kholief (2016). "A hybrid automated detection of epileptic seizures in EEG records." *Computers & Electrical Engineering* 53: 177–190.

Tuncer, T. (2021). "A new stable nonlinear textural feature extraction method based EEG signal classification method using substitution Box of the Hamsi hash function: Hamsi pattern." *Applied Acoustics* 172: 107607.

Ullah, I., M. Hussain and H. Aboalsamh (2018). "An automated system for epilepsy detection using EEG brain signals based on deep learning approach." *Expert Systems with Applications* 107: 61–71.

Webb, L., M. Kauppila, J. A. Roberts, S. Vanhatalo and N. J. Stevenson (2021). "Automated detection of artefacts in neonatal EEG with residual neural networks." *Computer Methods and Programs in Biomedicine* 208: 106194.

Zahra, A., N. Kanwal, N. ur Rehman, S. Ehsan and K. D. McDonald-Maier (2017). "Seizure detection from EEG signals using multivariate empirical mode decomposition." *Computers in Biology and Medicine* 88: 132–141.

Zazzaro, G., S. Cuomo, A. Martone, R. V. Montaquila, G. Toraldo and L. Pavone (2021). "Eeg signal analysis for epileptic seizures detection by applying data mining techniques." *Internet of Things* 14: 100048.

Zhang, X., X. Zhou, M. Lin and J. Sun (2018). Shufflenet: An extremely efficient convolutional neural network for mobile devices. *Proceedings of the IEEE Conference on Computer Vision and Pattern Recognition*, Salt Lake City, UT.

9 E-PRESTO
Epileptic PREictal State Detection Using Time Series MOdelling

Priya Shree, Manvir Bhatia, Tapan Kumar Gandhi, and Piyush Swami

9.1 INTRODUCTION

Epilepsy is one of the most prevalent neurological disorders, characterized by spontaneous, recurrent seizures caused by abnormal, excessive, and synchronous neural firing in the brain. It affects individuals of all ages, with a global prevalence of approximately 1% (Elger et al. 2007). The clinical manifestations of seizures vary widely, including involuntary motor movements, muscle twitches, altered consciousness, dizziness, blackouts, and, in some cases, complete loss of awareness. For many patients, antiepileptic drugs are effective in managing symptoms. However, a substantial portion of the epileptic population either does not respond to pharmacological treatment or develops drug resistance over time. Moreover, side effects of prolonged medication usage can contribute to additional neurological and systemic complications (Corne and Moghim 2014). In light of these challenges, seizure forecasting systems have been proposed as a proactive solution to mitigate the risk of seizure-related injury and improve patient autonomy. An automated system capable of predicting seizures in real time can inform the timely administration of rescue medication, trigger neuromodulation devices, or alert caregivers and patients to take precautionary measures. Such predictive technologies are particularly valuable for patients with drug-resistant epilepsy or those experiencing infrequent but severe episodes.

Historically, seizures were considered unpredictable events—a phenomenon often described as a 'bolt from the blue'—attributed to sudden, stochastic changes in neural activity. As a result, early research efforts focused primarily on seizure detection rather than prediction (Swami et al. 2016a; Swami et al. 2024a; Pontes et al. 2024). Early studies using intracranial electroencephalography (EEG) identified changes in brain network connectivity and synchronization patterns occurring seconds to minutes before seizure onset. This perception began to shift in the 1990s, when studies demonstrated that seizures are often preceded by subtle but measurable physiological changes. For instance, researchers observed increased neuronal

DOI: 10.1201/9781003611448-13

firing rates, shifts in spectral power, and evolving network dynamics within the seizure onset zone, suggesting that the brain undergoes a measurable transition into a seizure-prone state (Sumsky and Greenfield 2022). These also include increased cerebral blood flow, alterations in heart rate, and fluctuations in blood oxygenation, suggesting the presence of a transitional preictal state before seizure onset (Viglione and Walsh 1975). The identification of this preictal phase catalysed a new wave of research aimed at developing reliable biomarkers and computational tools to anticipate seizures. One of the earliest seizure prediction studies was conducted by Viglione and Walsh in 1975 using linear analytical measures to identify precursors to absence seizures. Subsequent work by Rogowski, Gath, and Bental (1981) applied auto-regressive (AR) modelling to EEG signals, detecting changes up to 6 s prior to seizure onset (Salant, Gath, and Henriksen 1998; Siegel, Mirsky, and Grady 1982). Siegel et al. analysed spike frequency and bilateral spiking rates, finding that seizure-related changes often occurred within one minute before the clinical event (Iasemidis et al. 1990). However, such findings were not universally reproducible. For example, failure to observe these patterns in a more comprehensive dataset was reported, challenging the reliability of spike rate-based prediction (Martinerie et al. 1998). The emergence of nonlinear and dynamical systems theory in the 1980s provided a new conceptual framework for understanding brain activity. Given that the human brain is a high-dimensional, nonlinear dynamical system, researchers began exploring nonlinear features to model seizure transitions. Measures such as the largest Lyapunov exponent (Le Van Quyen et al. 1999), correlation dimension, correlation density, and phase coherence were introduced to better characterize the chaotic and complex behaviour of epileptic brain networks (Andrzejak et al. 2003; Hagemann et al. 2021; Kreuz et al. 2003). Iasemidis et al. in 1990 demonstrated that the largest Lyapunov exponent values tend to decrease in the last minute before seizure onset and shows a decrease in the chaoticity of the system (Le Van Quyen et al. 1999). Le Van Quyen and colleagues further advanced this by using nonlinear similarity measures and moving window dynamics to successfully predict seizures in both animal and human models (Hagemann et al. 2021; Kreuz et al. 2003). Among nonlinear features, phase synchronization has emerged as one of the most promising metrics for seizure prediction. It quantifies the consistency in the instantaneous phase of two oscillating signals—such as those from EEG channels—and serves as a measure of functional connectivity. A study by Andrzejak et al. (2003) revealed that a decrease in phase synchronization precedes seizure onset, possibly due to a breakdown in global neural coordination across cortical areas (Tarassenko et al. 2003). This reduction in coherence provides a window of opportunity to identify the preictal state with high temporal precision. However, seizure prediction based on nonlinear measures has its limitations. In 2005, Mormann's group conducted a comprehensive evaluation of both linear and nonlinear metrics using surrogate time series analysis (Mormann et al. 2005). Their study found that many nonlinear methods failed to generalize and often performed no better than simpler linear approaches. Moreover, earlier studies often lacked rigorous statistical validation, failing to report false-positive rates (FPRs) or test their algorithms on independent data.

To address these limitations, recent studies have turned to machine learning and time series modelling. The linear features, such as Hjorth parameters, wavelet coefficients, and statistical moments, were used to construct a high-dimensional feature matrix, which was then fed into classifiers, including support vector machines and artificial neural networks (Parhi et al. 2011). Their system achieved low sensitivity between 70% and 80% with high inter-patient variability. Corne and Moghim in 2014 introduced the Advanced Seizure Prediction via Preictal Re-labelling framework, which dynamically redefined preictal boundaries to improve prediction accuracy and model selection (LeCun et al. 2009). Despite these advances, many machine learning approaches suffer from drawbacks such as the need for extensive labelled data, risk of overfitting, and limited physiological interpretability. Moreover, generalization across patients remains a major challenge, given the heterogeneity in seizure onset zones, EEG patterns, and brain dynamics.

In this work, we present E-PRESTO—Epileptic PREictal State detection using Time series mOdelling—a hybrid system that combines the strengths of nonlinear phase synchronization analysis and time series modelling via auto-regressive integrated moving average (ARIMA) simulation error. Phase synchronization is employed to detect the preictal state, while simulation error from AR modelling is used to detect the transition into the ictal phase. The system is trained on a short segment of interictal EEG, making it well-suited for real-time, low-supervision deployment. We evaluate the system on EEG recordings from six patients with focal and generalized epilepsy, and report its sensitivity, FPR, and prediction horizon in terms relevant to clinical translation.

Key contributions of this work include:

- **Low supervision requirements** and **short training durations**, suitable for online monitoring,
- **Complementary modelling of preictal and ictal phases**, leading to comprehensive seizure event coverage,
- **Patient-specific adaptation**, avoiding the need for large-scale cohort training,
- **Physiological interpretability**, enhancing clinical acceptance and trust in the system's outputs.

In the sections that follow, we detail the development and evaluation of the proposed E-PRESTO system. Section 9.2 presents the methodology, including dataset characteristics, data segmentation and preprocessing steps, and the computational strategies used for seizure prediction and detection. Phase synchronization is introduced as a nonlinear measure for identifying preictal states, while ARIMA-based simulation error is employed for ictal detection. Section 9.3 reports the results obtained across patient recordings. The implications and limitations of these findings are discussed in Section 9.4. Finally, Section 9.5 concludes the study with future directions for enhancing the clinical viability and adaptability of the E-PRESTO framework.

9.2 MATERIALS AND METHODS

9.2.1 Datasets

The dataset used for this project was collected at the Neurology and Sleep Centre in New Delhi, India. It consisted of scalp recordings of six epilepsy patients. Two patients experienced focal epilepsy while the rest experienced generalized epilepsy. The data was acquired at a sampling frequency of 200 Hz. The length of the recordings varied from 9 to 32 min. The data was acquired from 18 channels using a unipolar montage. The electrodes were placed according to the standard 10–20 international electrode placement system (Homan 1988), ensuring uniform coverage of the entire scalp.

9.2.2 Data Segmentation

The data from each patient was divided into windows to perform preprocessing and feature extraction. The length of the window was chosen to be 1000 samples (5 s). The window was verified for its weak sense stationarity by plotting the mean and standard deviation of several windows and observing that both measures remained almost constant over a 5-second time span. Figures 9.1 and 9.2 show examples of mean and standard deviation, respectively, across segments.

9.2.3 Data Preprocessing

The dataset was preprocessed before extracting any information from it. The preprocessing helped remove artefacts that may have been acquired along with the data during recording. The preprocessing step was applied to each extracted

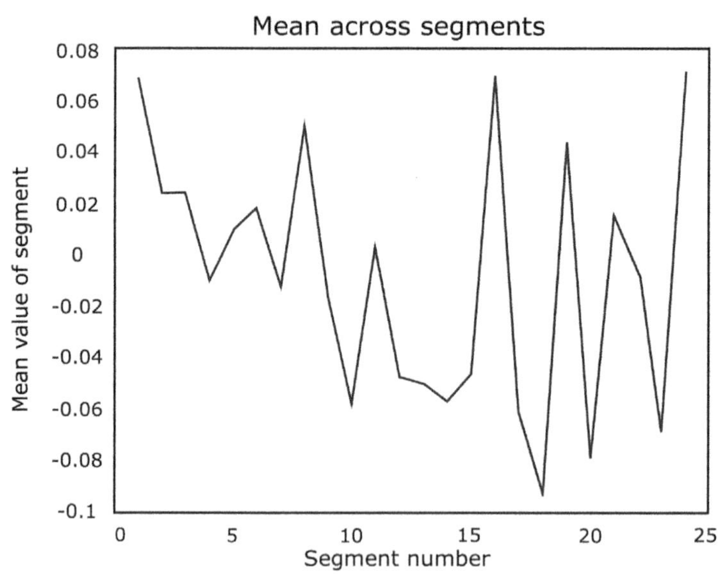

FIGURE 9.1 An example of the mean across segments.

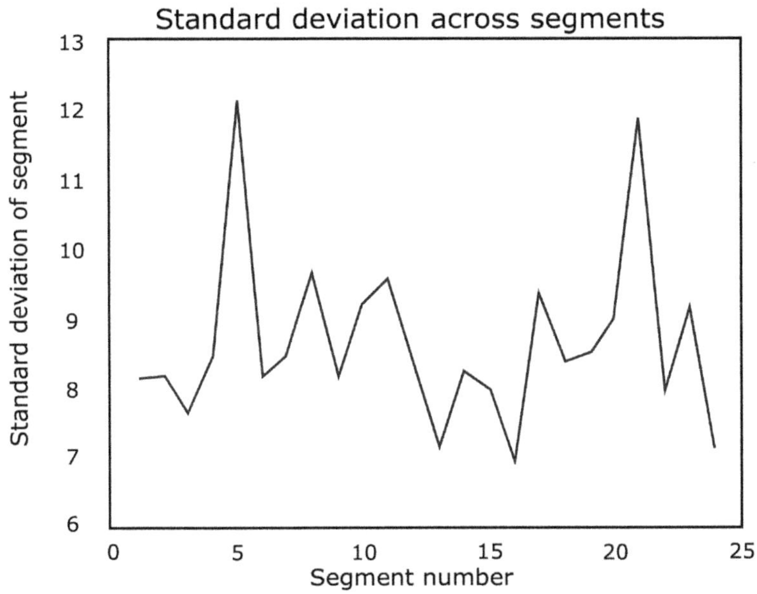

FIGURE 9.2 An example of the standard deviation across segments.

window of 1,000 samples. A discrete wavelet transform was performed on the data in each window to obtain wavelet coefficients corresponding to different frequency bands. The frequency bands above 45 Hz and below 0.5 Hz were rejected as they mostly comprised high-frequency noise and artefacts.

9.2.4 Phase Synchronization for Distinguishing Preictal from Interictal State

One of the biggest challenges in the prediction of epileptic seizures is to find features that can distinguish the preictal state from the interictal state. Many features have been used by various researchers to distinguish between the two states. However, very few features have been reported to be capable of distinguishing between the two states. The feature used in this study to distinguish between preictal and interictal states was phase synchronization (Andrzejak et al. 2003).

Phase synchronization is defined as the locking of the phases of two oscillating signals $A(t)$ and $B(t)$. This measure determines whether the phases of two oscillating signals are in coherence or not. However, it is not easy to find the phase variables for brain signals due to their highly chaotic and dynamic nature. Therefore, an analytic signal approach is followed, which defines the instantaneous phase of an oscillating signal as

$$\varphi = \tan^{-1}\frac{A_h(t)}{A(t)} \tag{9.1}$$

where $A_h(t)$ is the Hilbert transform of the signal given by

$$A_h(t) = \frac{1}{\pi}\text{ p.v.}\left(\int_{-\infty}^{+\infty} \frac{s(\tau)}{s(t-\tau)}\right) \tag{9.2}$$

where p.v. corresponds to Cauchy's principle value.

Phase synchronization tells us about the phase relationship between two signals acquired through different EEG channels. The mean value of phase synchronization between two channels is computed for the entire window length using the formula:

$$R = \frac{\sum_{j=1}^{N} \left| e^{i[\varnothing_A(t_j) - \varnothing_B(t_j)]} \right|}{N} \tag{9.3}$$

where N is the length of the window.

Φ_A and Φ_B are the instantaneous phases of the signals A and B.

The mean value of phase synchronization (R) lies between 0 and 1. If there is no phase locking between the signals acquired from the channels, the value of R turns out to be 0, whereas $R = 1$ indicates perfect phase synchronization between the two channels. Phase synchronization is a bivariate measure, i.e., it requires signals from two channels to compute the measure. The number of combinations possible for 18 channels is 18Combination2, which equals 153. However, we don't require as many computations of phase synchronization values, as neighbouring channels have more synchronization compared to the farther ones. To reduce computation time without compromising the sensitivity of this measure, we defined 35 channel combinations based on the proximity of electrodes and computed phase synchronization values for these channel pairs.

Several studies have shown that there is a decline in the value of R when an epileptic seizure is imminent (Hagemann et al. 2021). Thus, phase synchronization can be used as a measure to distinguish between preictal and interictal states. The preictal states have been observed to exhibit lower values of phase synchronization than the interictal state (Hagemann et al. 2021). A window is said to be in a preictal state if the mean phase synchronization value is some 'r' standard deviations below the mean phase synchronization value of the baseline interictal state data, i.e., if $R \leq m -' r'\sigma$, the state was preictal; else if $R > m -' r'\sigma$, the state was interictal. To calculate the mean phase synchronization for each channel combination, the data was demeaned for both windows for which phase synchronization was to be computed. Then, the phase synchronization between the windows from the two channels was computed using the formula:

$$\varphi(n) = \tan^{-1} \frac{H_1(n) * D_2(n) - H_2(n) * D_2(n)}{D_1(n) * D_2(n) - H_1(n) * H_2(n)} \tag{9.4}$$

where H_1 and H_2 are the imaginary parts of the Hilbert transform, D_1 and D_2 are demeaned signals from the window, and n varies from 1 to the length of the window (1,000 samples). The mean values for each channel combination were computed using Equation 9.3. This was followed by splitting each of the continuous EEG recordings in the dataset into two parts: the training data and the testing data. The length of the training data was kept relatively short (2–5 min) based on the length of the recording. The training data included interictal recordings and were used to compute the baseline for mean phase synchronization.

The standard deviation values were also calculated for the windows that constitute the training data. The remaining data constituted the testing data. Thus, a window was classified to be in a preictal state if its mean phase synchronization was '*r*' standard deviations below the interictal phase synchronization. To obtain the optimum value of parameter '*r*', the value of *r* was varied from 0 to 6 standard deviations, the step size being 0.5. The value of *r*, which yielded better values of sensitivity and specificity in the validation part of the training data, was chosen for that dataset (Kreuz et al. 2003; Mormann et al. 2005). Then the simulation error from AR models for distinguishing ictal from preictal and interictal states was calculated. The ARIMA model was used in this study to detect the occurrence of seizures. ARIMA models can be used to model time series data. These models can predict the future value of the time series or can simulate the time series with the same persistence structure as the original. The time series from patients was modelled as an ARIMA model. Each patient's recording was divided into a window of 1,000 samples, as mentioned before. A suitable model order was chosen to fit the data in each window. The fitted model was then used to simulate the data in each window. Studies have shown that the error in prediction increases when there are spikes or seizures in the window for which the data is being predicted. Similarly, we found an increase in the simulation error for windows that contain epileptic seizures or spikes. When a model is fitted to the data, there is an inherent assumption of stationarity of the data. However, when a seizure approaches, the stationarity of the signal is compromised, resulting in sudden changes in the time series. This results in high simulation errors for windows having epileptic seizures or spikes. Based on these observations, we attempted to model the seizures using an ARIMA model (Tarassenko et al. 2003).

The first step in fitting a model to the time series data is identifying the model order for the ARIMA process. The Ljung–Box Pierce test and Akaike's Information Criterion (AIC) and Bayesian Information Criterion (BIC) were used to decide the model order. The model with the minimum values of AIC and BIC was chosen, as it displayed no serial autocorrelation among the residuals (Parhi et al. 2011; LeCun et al. 2009). Table 9.1 shows various AIC and BIC measures for different AR models. After settling on an order of 11 for the AR model (the AR model of order 11 had the minimum AIC and BIC values), we tried various moving average (MA) and integration orders. Table 9.2 represents the AIC and BIC measures for different ARIMA models.

Based on the values of the AIC and BIC measures, we finally selected a model order of 11 for the AR model and 0 for the MA order (Swapna et al. 2013). The next step after model order selection was to estimate the model using the model's specifications and the original data. Once the model was calculated, the simulation error was computed for the entire window. The length of the simulation error variable was equal to the size of the window (1,000 samples). The error was averaged for each window and was used to compute the threshold for classifying a window as ictal or non-ictal (Quiroga et al. 1995). The training part of the data had interictal recordings

TABLE 9.1
Order of Auto-Regressive (AR) Model with Its Akaike's Information Criterion (AIC) and Bayesian Information Criterion (BIC) Measures for Segment of Length 100 Samples (5 s)

Order of Auto-Regressive (AR) Model (*p,d,q*)[a]	Akaike's Information Criterion (AIC) Measure	Bayesian Information Criterion (BIC) Measure
(8,0,0)	304	328
(9,0,0)	302	328
(10,0,0)	303	331
(11,0,0)	297	328
(12,0,0)	298	331
(13,0,0)	299	336
(14,0,0)	300	339

[a] *p* indicates the auto-regressive order, *d* indicates the differential order, and *q* indicates the moving average order.

TABLE 9.2
Order of Auto-Regressive Integrated Moving Average (ARIMA) Model with Its Akaike's Information Criterion (AIC) and Bayesian Information Criterion (BIC) Measures for Segment of Length 100 Samples (5 s)

Order of Auto-Regressive Integrated Moving Average (ARIMA) Model (*p,d,q*)[a]	Akaike's Information Criterion (AIC) Measure	Bayesian Information Criterion (BIC) Measure
(11,0,0)	297	328
(11,1,0)	298	329
(11,0,1)	300	329
(11,1,1)	298	332

[a] *p* indicates the auto-regressive order, *d* indicates the differential order, and *q* indicates the moving average order.

for the patient. The error for these training windows would be less compared to the windows that would have ictal state data. Thus, a threshold was set to classify the window as ictal if the error increased above that threshold. The threshold was chosen using the formula: Threshold = $\gamma *$ average error in the interictal windows, where γ is the scalar which is chosen based on the performance of the model on the validation part of the dataset (Rickus et al. 2010). Any window having an average error above the threshold was classified as an ictal window. The windows having artefacts also had large simulation errors. However, it was observed that their errors were significantly lower than those for windows with epileptic seizures or spikes. The threshold was kept high enough to avoid classifying artefacts as ictal windows.

The automated system developed for prediction and early detection of epileptic seizures in this work can be summarized into nine steps. For clarification and ease of replication, the pseudo code is provided in Table 9.3.

TABLE 9.3
The Pseudo Code of the Automated System for Prediction and Early Detection of Epileptic Seizures in this Work

```
BEGIN
## Step 1: Load EEG signals
eeg_data = load_eeg_signals()

## Step 2: Select initial interictal segment as training data
training_data = select_interictal_segment(eeg_data)

## Step 3: Divide training data into windows of 1000 samples
training_windows = divide_into_windows(training_data,
window_size=1000)

## Step 4: Process each training window
simulation_errors = []
phase_sync_values = []

for window in training_windows:
    preprocessed = apply_dwt(window)
    simulated = simulate_ar_model(preprocessed, order=11)

    error = compute_simulation_error(preprocessed, simulated)
    simulation_errors.append(error)

    phase_sync = compute_phase_synchronization(preprocessed,
channel_pairs=35)
    phase_sync_values.append(phase_sync)

## Step 5: Compute mean and std of phase synchronization
mean_phase_sync = compute_mean(phase_sync_values)
std_phase_sync = compute_std(phase_sync_values)

## Step 6: Compute mean simulation error
mean_sim_error = compute_mean(simulation_errors)

## Step 7: Set thresholds
r = predefined_constant()
threshold_sim_error = mean_sim_error
threshold_phase_sync = mean_phase_sync - r * std_phase_sync

## Step 8: Evaluate testing windows
testing_windows = get_testing_windows(eeg_data)

for test_window in testing_windows:
    preprocessed_test = apply_dwt(test_window)
```

```
    simulated_test = simulate_ar_model(preprocessed_test, order=11)

    test_error = compute_simulation_error(preprocessed_test,
simulated_test)
    test_phase_sync = compute_phase_synchronization(preprocessed_test,
channel_pairs=35)

    if test_error > threshold_sim_error:
       label_window(test_window, label="ICTAL")
    else:
     for pair in test_phase_sync:
       if test_phase_sync[pair] <= threshold_phase_sync:
         label_channels(pair, label="PREICTAL")
     else:
         label_window(test_window, label="INTERICTAL")

## Step 9: Monitor synchronization drops
sync_drop_tracker = track_sync_drops(test_phase_sync)

if sync_drop_tracker.has_persistent_drop(min_consecutive_windows=3):
  raise_alarm("Impending Seizure")

END
```

9.2.5 Performance Metrics

The following performance metrics were used to analyse the performance of the developed automated seizure prediction system:

9.2.5.1 Sensitivity

It is defined as the ratio of the number of seizures predicted by the system to the number of seizures that occurred in the recording. It can be formulated as:

$$\text{Sensitivity} = \frac{\text{Number of seizures predicted}}{\text{Number of seizures in the recording}} \tag{9.5}$$

9.2.5.2 False-Positive Rate

It is defined as the ratio of false predictions made by the system (i.e., the number of times an alarm was raised for an impending seizure, but no seizure occurred) to the duration of the recording (Viglione and Walsh 1975). It can be simply represented as follows:

$$\text{False Positive Rate} = \frac{\text{Number of false alarms}}{\text{Duration of recording}} \tag{9.6}$$

The unit of FPR is time^{-1}. Mormann et al. (2005) have mentioned that the denominator in the formula for FPR for seizure prediction must account for the time during which the system was in the warning state, i.e., the alarm had already been raised (Elger et al. 2007). During that period, no further false alarms can occur, as the system is already in a warning state (Pontes et al. 2024). The prediction horizon is defined as the period during which a seizure is expected to occur after an alarm has been raised. The formulation for FPR by Pontes et al. (2024) considers the prediction horizon. They have formulated an effective FPR as follows:

$$\text{FPR} = \frac{\text{Number of false predictions}}{\text{Duration of recording} - \text{Number of seizures} * \text{Prediction Horizon}} \tag{9.7}$$

This was used for the calculation of FPR.

9.3 RESULTS

The performance of the proposed E-PRESTO system was evaluated on scalp EEG recordings from six patients diagnosed with either generalized or focal epilepsy. Table 9.4 presents the performance metrics, including sensitivity, FPR, and mean prediction time for each subject. The system achieved an average sensitivity of 85% across the dataset, with four out of six analysed subjects reaching 100% seizure prediction accuracy. Notably, subject 2 exhibited a sensitivity of 80% with an FPR of 0 min^{-1}, indicating high specificity in this case. The lowest sensitivity was observed in subject 4, where only 50% of seizures were correctly predicted. The FPR averaged to 0.1278 min^{-1}, with the best case at 0 and the highest at 0.222 min^{-1}. These values were calculated using a prediction horizon-adjusted FPR formulation (Equation 9.7), as recommended in a prior study (Pontes et al. 2024). The mean prediction horizon—defined as the interval between seizure prediction and clinical onset—was approximately 2 min and 56 s. In some cases, such as subject 3, the prediction time was extended to 5 min, suggesting potential for early intervention.

The simulation error computed from AR modelling showed a marked increase in the presence of ictal activity. Out of 15 total seizures in the dataset, 12 seizures (80%) were successfully detected using AR model-based simulation error, confirming its utility in ictal state identification.

TABLE 9.4
Performance Metrics for Various Recordings in the Dataset

Subject Number	Sensitivity (%)	False-Positive Rate (min^{-1})	Mean Prediction Time
1	100	0.222	1 min 58 s
2	80	0	1 min 50 s
3	100	0.069	5 min
4	50	0.138	4 mins 52 s
5	100	0.132	2 mins 30 s
6	80	0.206	1 min

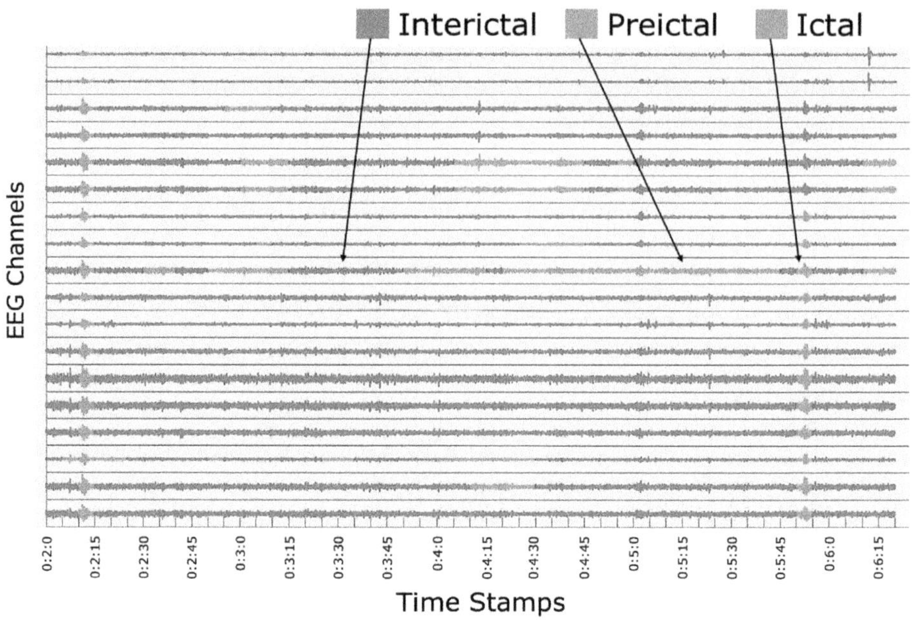

FIGURE 9.3 Automated system plot for a patient with generalized epilepsy.

For the preictal phase classification, phase synchronization analysis was applied to 35 carefully selected channel pairs. A consistent decline in mean synchronization was observed 30–35 windows (approximately 2.5–3 min) prior to seizure onset in subjects. This validated the selection of phase synchronization as a physiologically meaningful feature for preictal state identification.

Figures 9.3 and 9.4 illustrate example outputs from the system for patients with generalized and focal epilepsy, respectively, showing temporal markers of synchronization drop and ictal detection based on simulation error.

9.4 DISCUSSIONS

The results obtained from the E-PRESTO system highlight the promise of a hybrid, lightweight framework for automated seizure prediction and detection. The integration of nonlinear phase synchronization measures for preictal state identification and AR modelling for ictal detection enables complementary tracking of dynamic changes in brain states.

The high sensitivity values in several subjects suggest that the system is capable of learning subject-specific patterns from minimal interictal state data. More importantly, using only 2–5 min of interictal recording for training reduces the burden of large-scale data collection, making the system adaptable for real-time deployment in clinical or ambulatory settings.

The decline in phase synchronization preceding seizures, as observed in multiple subjects, aligns with the findings reported in previous studies (Andrzejak et al.

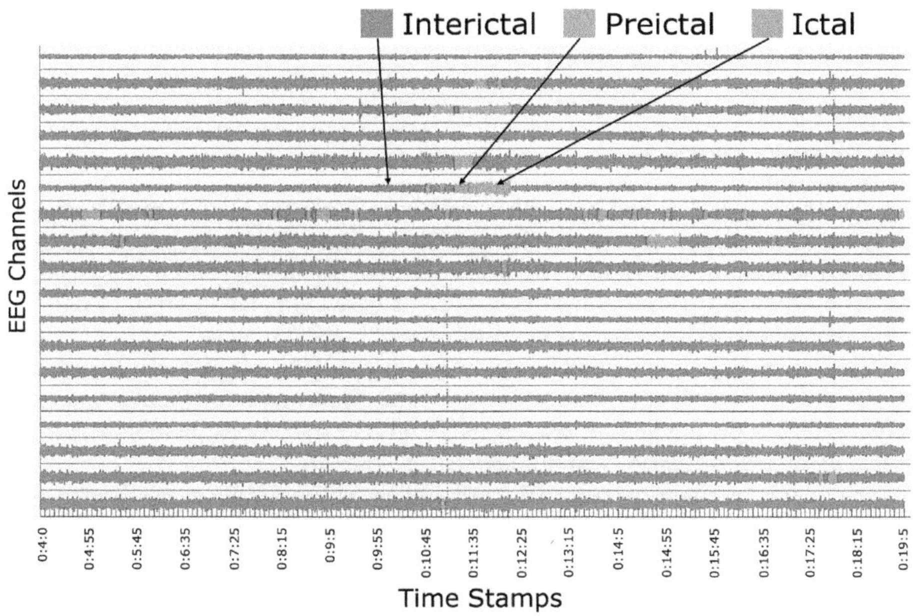

FIGURE 9.4 Automated system plot for a patient with focal epilepsy.

2003; Hagemann et al. 2021) that associate desynchronization with the loss of coordinated neuronal activity during the transition into seizures. This further reinforces the relevance of bivariate dynamical measures over traditional univariate signal statistics. The simulation error derived from ARIMA modelling functioned effectively as an ictal marker, particularly in cases with pronounced spikes or sudden shifts in the EEG. The model's assumption of weak stationarity fails during ictal transitions, resulting in high simulation errors that serve as a discriminative feature. This mechanism is both interpretable and computationally efficient.

Nevertheless, the performance variance observed in some subjects, such as subject 4 with 50% sensitivity, underscores the individual variability in seizure dynamics. This suggests that phase synchronization and AR modelling, while effective, may need to be complemented with additional biomarkers such as EEG entropy, cross-frequency coupling, or even non-EEG features (e.g., heart rate and electrodermal activity) to account for patient heterogeneity. Another advantage of the proposed method is its low computational overhead. The system does not rely on deep learning or high-dimensional feature spaces and can be implemented on wearable or embedded systems for on-device seizure forecasting. However, limitations include a small patient cohort and short recording durations. Future studies should aim to validate the model on longer recordings across diverse epilepsy subtypes. Since most existing publicly available datasets lack this information, our research group is also working on creating new 'annotated' datasets, like Swami et al. (2016b), but with continuous EEG signals consisting of preictal, interictal, and ictal states. The existing dataset (Swami et al. 2016b) could also be used to train an artificial intelligence-based classifier. The future scope also includes the real-time implementation of an amplification unit (Swami et al. 2024b) with feedback to raise warnings of impending seizures.

9.5 CONCLUSIONS

This work presented E-PRESTO, an automated system for seizure prediction and early detection, designed around low-supervision signal processing techniques and nonlinear dynamical measures. The system combines phase synchronization analysis for identifying preictal states with ARIMA-based simulation error modelling, achieving robust performance on multi-subject scalp EEG signals.

The system demonstrated an average sensitivity of 85%, a low FPR of 0.1278 min^{-1}, and an average prediction time of nearly 3 min. The seizure prediction approach requires only a few minutes of baseline interictal state data for calibration, making it viable for clinical scenarios with limited data availability.

The findings support the feasibility of building patient-adaptive, low-latency seizure prediction systems using classical time series models and nonlinear features. The hybrid nature of E-PRESTO enables both seizure forecasting and detection, forming a closed-loop foundation for potential integration into wearable periprosthetic or neurostimulation devices. Future extensions of this work will focus on real-time deployment (Beniczky et al. 2025), model personalization, and integration with multimodal physiological sensing platforms (Swami et al. 2024b, Beniczky et al. 2025).

ACKNOWLEDGEMENT

The authors would like to thank Mrs. Anu Mol, Senior Clinician at the Neurology and Sleep Centre, Hauz Khas, New Delhi, India, for her assistance in the annotation of the datasets.

REFERENCES

Andrzejak, R. G., P. David, K. Lehnertz, C. E. Elger, F. Mormann, and T. Kreuz. "Epileptic Seizures Are Preceded by a Decrease in Synchronization." *Epilepsy Research* 53, no. 3 (2003): 173–185.

Beniczky, S., P. Ryvlin, H. Stefan, T. Tomson, H. de Boer, and E. Perucca. "Artificial Intelligence, Digital Technology, and Mobile Health in Epilepsy." *Epilepsy* 66, no. S3 (2025): 1–3.

Corne, D. W., and N. Moghim. "Predicting Epileptic Seizures in Advance." *PLoS One* 9 (2014): e99334.

Elger, C. E., F. Mormann, R. G. Andrzejak, and K. Lehnertz. "Seizure Prediction: The Long and Winding Road." *Brain* 130 (2007): 314–333.

Hagemann, A., A. G. Boye, P. Wollstadt, V. Priesemann, C. G. Benar, and F. Mormann. "Assessing Criticality in Pre-Seizure Single-Neuron Activity of Human Epileptic Cortex." *PLoS Computational Biology* 17, no. 3 (2021): e1008773.

Homan, R. W. "The 10–20 Electrode System and Cerebral Location." *American Journal of EEG Technology* 28, no. 4 (1988): 269–279.

Iasemidis, L. D., H. P. Zaveri, W. J. Williams, and J. C. Sackellares. "Phase Space Topography and the Lyapunov Exponent of Electrocorticograms in Partial Seizures." *Brain Topography* 2, no. 3 (1990): 187–201.

Kreuz, T., C. Rieke, P. David, C. E. Elger, K. Lehnertz, F. Mormann, and R. G. Andrzejak. "Automated Detection of a Pre-Seizure State Based on a Decrease in Synchronization in Intracranial EEG Recordings." *Physical Review E* 67, no. 2 (2003): 021912.

Le Van Quyen, M., M. Baulac, S. Clemenceau, F. J. Varela, J. Martinerie, and C. Adam. "Anticipating Epileptic Seizures in Real Time by a Nonlinear Analysis of Similarity Between EEG Recordings." *Neuroreport* 13, no. 10 (1999): 2149–2155.

LeCun, Y., R. Kuzniecky, P. Mirowski, and D. Madhavan. "Classification of Patterns of EEG Synchronization for Seizure Prediction." *Clinical Neurophysiology* 120, no. 11 (2009): 1927–1940.

Martinerie, J., C. Adam, M. Le Van Quyen, M. Baulac, S. Clemenceau, B. Renault, and F. J. Varela. "Epileptic Seizures Can Be Anticipated by Nonlinear Analysis." *Nature Medicine* 4 (1998): 1173–1176.

Mormann, F., T. Kreuz, C. Rieke, R. G. Andrzejak, A. Kraskov, P. David, C. E. Elger, and K. Lehnertz. "On the Predictability of Epileptic Seizures." *Clinical Neurophysiology* 116, no. 3 (2005): 569–587.

Parhi, K. K., T. Neto, Y. Park, and L. Luo. "Seizure Prediction with Spectral Power of EEG Using Cost-Sensitive Support Vector Machines." *Epilepsia* 52, no. 10 (2011): 1761–1770.

Pontes, E. D., D. F. Silva, D. Rodrigues, I. N. da Silva, M. C. Oliveira, F. M. de Assis, and J. P. Papa. "Concept-Drifts Adaptation for Machine Learning EEG Epilepsy Seizure Prediction." *Scientific Reports* 14, no. 1 (2024): 8204.

Quiroga, R. Q., L. Romanelli, O. A. Rosso, S. Blanco, and H. Garcia. "Stationarity of the EEG Series." *IEEE Engineering in Medicine and Biology Magazine* 14, no. 4 (1995): 395–399.

Rickus, J., R. Worth, P. P. Irazoqui, P. Rajdev, and M. P. Ward. "Real-Time Seizure Prediction from Local Field Potentials Using an Adaptive Wiener Algorithm." *Computers in Biology and Medicine* 40, no. 1 (2010): 97–108.

Rogowski, Z., I. Gath, and E. Bental. "On the Prediction of Epileptic Seizures." *Biological Cybernetics* 42, no. 1 (1981): 9–15.

Salant, Y., I. Gath, and O. Henriksen. "Prediction of Epileptic Seizures from Two-Channel EEG." *Medical & Biological Engineering & Computing* 36, no. 5 (1998): 549–556.

Siegel, A., A. F. Mirsky, and C. L. Grady. "Prediction of Spike-Wave Bursts in Absence Epilepsy by EEG Power-Spectrum Signals." *Epilepsia* 23 (1982): 47–60.

Sumsky, S., and L. J. Greenfield Jr. "Network Analysis of Preictal iEEG Reveals Changes in Network Structure Preceding Seizure Onset." *Scientific Reports* 12, no. 1 (2022): 12526.

Swami, P., B. K. Panigrahi, and T. K. Gandhi. *A Device for Automated Diagnosis of Epilepsy*. Indian Patent IN500189. Application No. 201611043166. Filed 2016, Published January 17, 2024a.

Swami, P., B. K. Panigrahi, S. Nara, M. Bhatia, and T. K. Gandhi. "EEG Epilepsy Datasets." *Research Gate* 2016b. https://doi.org/10.13140/RG.2.2.14280.32006.

Swami, P., J. Maheshwari, M. Kumar, and M. Bhatia. "Evolutionary Transfer Optimization-Based Approach for Automated Ictal Pattern Recognition Using Brain Signals." *Frontiers in Human Neuroscience* 18 (2024b): 1386168.

Swami, P., T. K. Gandhi, B. K. Panigrahi, M. Tripathi, and S. Anand. "A Novel Robust Diagnostic Model to Detect Seizures in Electroencephalography." *Expert Systems with Applications* 56 (2016a): 116–130.

Swapna, G., R. J. Martis, J. S. Suri, U. R. Acharya, and S. V. Sree. "Automated EEG Analysis of Epilepsy: A Review." *Knowledge-Based Systems* 45 (2013): 147–165.

Tarassenko, L., P. E. McSharry, and L. A. Smith. "Comparison of Predictability of Epileptic Seizures by a Linear and a Nonlinear Method." *IEEE Transactions on Biomedical Engineering* 50, no. 5 (2003): 628–633.

Viglione, S. S., and G. O. Walsh. "Epileptic Seizure Prediction." *Electroencephalography and Clinical Neurophysiology* 39, no. 4 (1975): 435–436.

10 Sliding Window-Based Epileptic Seizure Detection Using Classifier Fusion and TQWT with Statistical Features

Durga Siva Teja Behara, Anirudh Kumar, Manvir Bhatia, Tapan Kumar Gandhi, Bijaya Ketan Panigrahi, and Piyush Swami

10.1 INTRODUCTION

The International League Against Epilepsy defines an epileptic seizure (also called fits) as a transient occurrence of signs and/or symptoms due to abnormally excessive or synchronous neuronal activities in the brain (Moshé et al. 2015). The enduring tendency of the brain to produce fits is called epilepsy. It is one of the major chronic neurogenic disorders, affecting approximately 1%–2% of the global population (Moshé et al. 2015). The most used method for diagnosing epilepsy is electroencephalography (EEG) (Gajic et al. 2015). Neuro-clinicians usually depend on their visual search skills for inspecting the EEG signals to diagnose epilepsy. This method, however, proves to be very tedious and challenging. It can even lead to errors, which can result in misdiagnosis (Gandhi et al. 2010, 2015). A promising solution to this problem is the application of an expert system for detecting epileptic seizures (Swami et al. 2018).

The existing expert systems are mostly based on feature extraction techniques. This typically involves analysing EEG signals in the time domain, frequency domain, and time–frequency domain (Tzallas et al. 2012). In the past, various time-domain approaches (such as fractional linear prediction (Tzallas et al. 2012) and empirical mode decomposition (Swami et al. 2015a) and frequency-domain approaches such as the fast Fourier Transform (Tzallas et al. 2012) have been proposed for the detection of epileptic seizures. Various time-frequency-domain approaches have also been used for feature extraction. The discrete wavelet transform (DWT) has been among the most sought-after time-frequency-domain analysis techniques used for the

DOI: 10.1201/9781003611448-14

decomposition of EEG signals (Gandhi, Panigrahi, and Anand 2011; Li et al. 2016; Swami et al. 2017). However, it is well known that the DWT suffers from shift variance (Chen 2014; Swami et al. 2016a). This results in poor denoising of signals and even affects the classification of signals in the later stage (Swami et al. 2017, 2019). The tunable-Q wavelet transform (TQWT) technique overcomes the aforementioned drawbacks with high computational efficiency (Selesnick 2011) and has been utilized in this work to effectively denoise EEG signals.

After signal decomposition, various features such as energy (Gandhi et al. 2010), standard deviation (SD) (Swami et al. 2014; Gajic et al. 2015; Swami et al. 2015b), entropy (Gandhi et al. 2012; Swami et al. 2016a), Lyapunov exponents (Sahu et al. 2010), Hurst exponents (Acharya et al. 2013), and correlation dimension (Sahu et al. 2010) have been extracted for characterizing the epileptic seizures. These features were later used as inputs to the machine learning classifier(s). This work extracted a novel set of features using TQWT and a sliding window. The first feature set used in this study was the SD measures of the TQWT wavelet coefficients. The second feature set consisted of the centred correntropy (CCE) values, which were also calculated from the extracted wavelet coefficients. The third feature set included the sliding window features formed by concatenating lower (mean values) and higher-order (skewness and kurtosis) statistical parameters.

The extracted features were provided as inputs to the classifier(s) to detect epileptic seizures present in EEG signals. Some of the classification methods that have been successfully demonstrated in the literature for the detection of epileptic seizures include support vector machines (SVMs) (Gandhi, Panigrahi, and Anand 2011; Swami et al. 2014; Tiwari et al. 2017; Swami et al. 2015a), k-nearest neighbour (Tzallas et al. 2012), artificial neural networks (Gandhi et al. 2010; Swami et al. 2017), fuzzy Sugeno (Swami et al. 2016c), and Gaussian mixture models (Tzallas et al. 2012). Nevertheless, to the best of our knowledge, most existing expert systems in the literature achieve high efficiency only with a very high percentage of training data (≥ 50%) (Tzallas et al. 2012; Acharya et al. 2013; Chen 2014). In this study, the robustness of the expert system was validated using different amounts of training data, ranging from 10% to 90%. A fusion of three different classifiers, viz., decision tree (DT), naïve Bayes (NB), and least-squares SVM (LS-SVM), was applied to detect epileptic seizures.

The next section in this chapter illustrates the systematic methodology followed to develop the proposed expert system. Section 10.3 describes the classification results obtained from the system. The discussion based on the results is provided in Section 10.4. Lastly, Section 10.5 of this chapter concludes the study by briefly mentioning the major findings and their future scope.

10.2 MATERIALS AND METHODS

10.2.1 Datasets

The EEG datasets from two different databases were used in this study. The first database considered was publicly available from the University of Bonn (UoB), Germany (Andrzejak et al. 2001). This database has become a standard benchmark for the development and validation of seizure detection systems for epilepsy. Its datasets were acquired from five epilepsy patients at a sampling rate of 173.61 Hz and then divided into three sets (denoted by letters C, D, and E). Each set contained segments of EEG signals lasting 23.6 s. Both sets C and D contained segments of EEG signals acquired during a non-seizure condition. The segments in set C were acquired from the hippocampal region opposite to the site of the epileptogenic zone, while the segments in set D were acquired from the epileptogenic zone. The EEG segments in set E were recorded while the patients were experiencing epileptic seizures. Since the objective of this study was to classify segments of non-seizure and epileptic signals, a combination of sets C and D versus E was considered for classification using this database.

The second EEG dataset used in this study was obtained from our database (Gandhi et al. 2010, 2012; Swami et al. 2016a, 2017). It contained EEG signals from 21 epilepsy patients, collected using a Grass Telefactor Twin3 16-channel scalp EEG recording machine. The international 10–20 system was used for the placement of gold-plated EEG electrodes (Moshé et al. 2015), and the sampling rate was set to 400 Hz during acquisition. The data was recorded in the presence of an experienced neurophysiologist. Later, the recorded data were segmented and divided into two sets, containing non-seizure and epileptic signals, with each segment lasting 10 s. A sample of an EEG segment taken from each set of both databases is shown in Figure 10.1. In this figure, the non-seizure and epileptic seizures are difficult to segregate by visual inspection. To evaluate distinguishable features from each of the signals, the data were subjected to TQWT.

10.2.2 Decomposition of EEG Segments Using TQWT

In the TQWT technique, the input signal is decomposed into a low-pass sub-band and a high-pass sub-band. This process is implemented iteratively on the low-pass dual-channel filter banks. Analysis of oscillatory signals such as EEG requires setting a high value of the tunable factor Q so that the underlying wavelets will have more oscillations with narrower frequency responses. Hence, based on the literature and the requirements of the current analysis, the value of Q and the redundancy factor were set to 3 with 21 decomposition levels.

10.2.3 Comparison of DWT Versus TQWT for Denoising EEG

To compare the denoising capability of TQWT versus DWT, noises of different known signal-to-noise ratios (SNRs) (i.e., 5, 10, 20, 30, 40, and 50 dB) were added to EEG signals. Then, these noisy signals were filtered within a 40 Hz range by separately applying DWT and TQWT. For visualization, the reconstructed waveforms of an EEG segment after removing 5 dB noise using DWT and TQWT are shown in Figure 10.2a and b, respectively. Similarly, the reconstructed waveforms of an EEG segment after removing 10 dB of noise using DWT and TQWT are shown in Figure 10.2c and d, respectively. From Figure 10.2, it is evident that the DWT technique is highly susceptible to low SNRs (5 and 10 dB).

The mean square errors (MSEs) of each filtered signal were also calculated to quantify the amount of denoising achieved. Figure 10.3 shows the MSEs of the

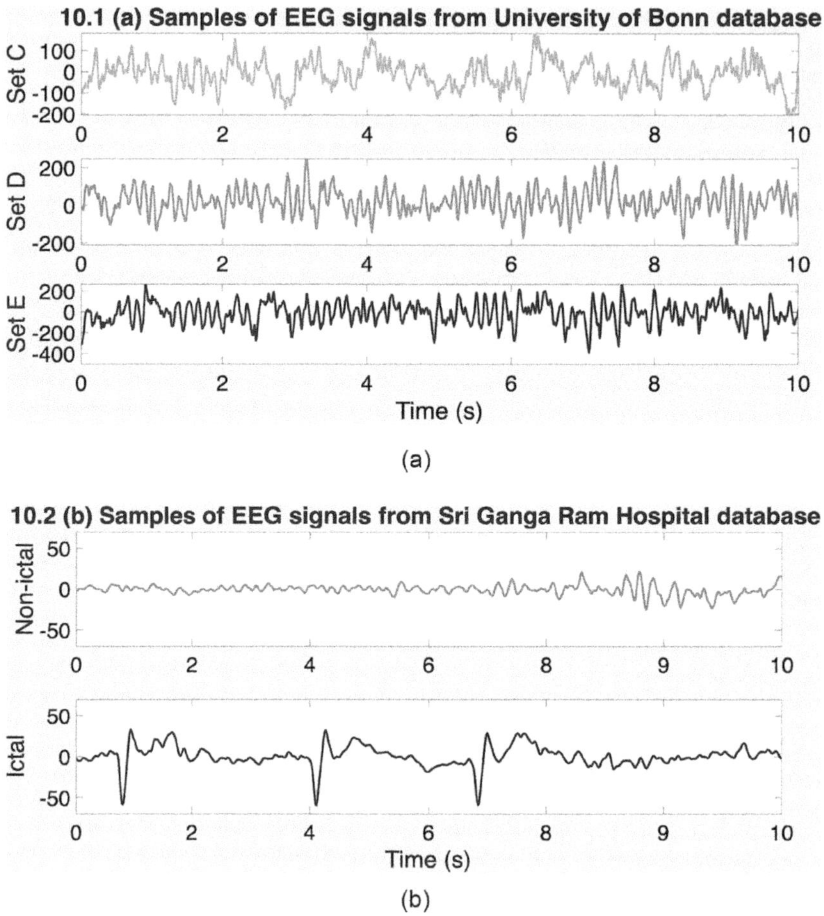

FIGURE 10.1 Sample of an EEG signal from each set obtained from (a) the University of Bonn (UoB) database and (b) our database. The *X*- and *Y*-axis in each graph represent time (in s) and amplitude (in μV), respectively.

signals shown in Figure 10.2. In Figure 10.3, it could be clearly observed that the MSEs of signals (with SNRs < 20 dB) filtered using TQWT were significantly less ($p < 0.001$). However, the MSEs of signals (with SNRs greater than 20 dB) filtered using TQWT were relatively higher than those calculated using DWT (although the difference was not statistically significant). These comparisons validate that TQWT is a more effective technique than DWT for denoising EEG signals. Therefore, the TQWT technique was used to filter the original EEG segments from both databases.

10.2.4 Feature Extraction

The features extracted for the classification of epileptic seizures included the following.

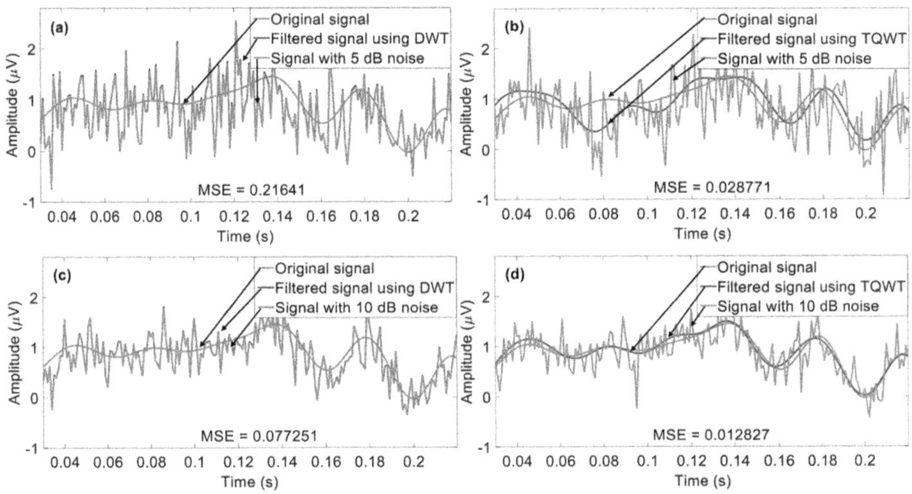

FIGURE 10.2 Filtering an EEG signal with (a) 5 dB noise, and (c) 10 dB noise using discrete wavelet transform (DWT); and filtering an EEG signal with (b) 5 dB noise, and (d) 10 dB noise using tunable-Q wavelet transform (TQWT).

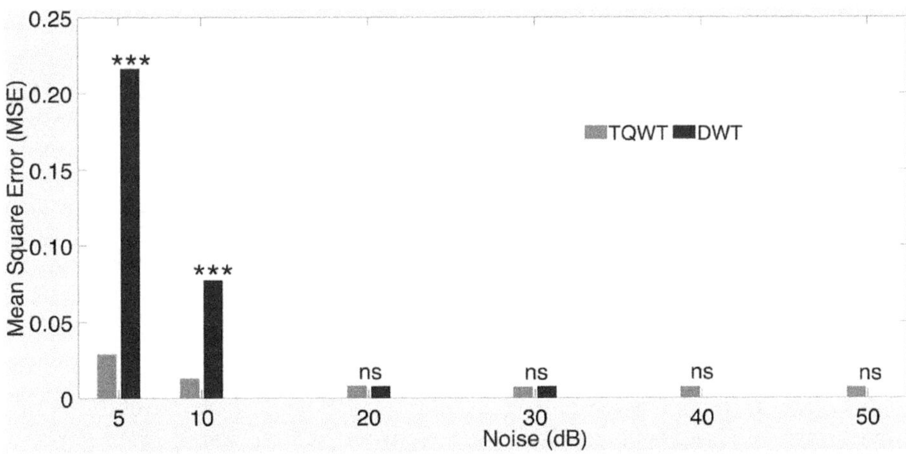

FIGURE 10.3 Comparison of mean square error (MSE) computed after filtering EEG signals using discrete wavelet transform (DWT) and tunable-Q wavelet transform (TQWT).

10.2.4.1 The First Feature Set (F_1), i.e., *SD* Evaluated from TQWT Coefficients

The statistical changes in the time–frequency-domain coefficients during the transformation of brain waves from the normal state to the epilepsy state form an important characteristic of epileptic seizures (Gajic et al. 2015; Swami et al. 2016a). Therefore, we had considered *SD* as the first feature set F_1. The *SD* for $m \times n$ number of coefficients C could be expressed by Equation 10.1:

$$F_1 \text{ or } SD_{mn} = \sqrt{(N-1)^{-1} \sum_{n=1}^{N} \left(C_{mn} - \frac{1}{K} \sum_{k=1}^{N} C_{mn} \right)^2} \tag{10.1}$$

where $n = 1, 2, 3, \ldots, N$ number of samples.

10.2.4.2 The Second Feature Set (F_2), i.e., CCE Evaluated from TQWT Coefficients

The CCE is a kernel-based measurement of correlation between multiple delayed samples of a signal (Maheshwari, Pachori, and Acharya 2017). It can be used to quantify the non-linear changes induced by epileptic seizures in EEG signals. Here, we have proposed a feature set based on the CCE derived from the coefficients of TQWT for detecting epileptic seizures. For a random process $\left[C_m(1), C_m(2), C_m(3), \ldots, C_m(N)\right]$ with delay τ and samples N, we have considered a Gaussian kernel function given in Equation 10.2 (Maheshwari, Pachori, and Acharya 2017):

$$g\left\{C_m(n) - C_m(n-\tau)\right\} = (\sigma\sqrt{2\pi})^{-1} \exp\left[-\left\{C_m(n) - C_m(n-\tau)\right\}^2 \Big/ 2\sigma^2 \right] \tag{10.2}$$

where σ is the width of the Gaussian kernel.

The CCE for $m \times n$ number of coefficients could be expressed by Equation 10.12 (Maheshwari, Pachori, and Acharya 2017):

$$F_2 \text{ or } \mathrm{CCE}_{mn} = (N - \tau + 1)^{-1} \cdot T - N^{-2} \sum_{\tau=1}^{N} T \tag{10.3}$$

$$\text{where,} \quad T = \sum_{n=\tau}^{N} g\left\{C_m(n) - C_m(n-\tau)\right\} \tag{10.4}$$

10.2.4.3 The Third Feature Set (F_3), i.e., Sliding Window Features Evaluated from Time Series

In this study, a sliding window of length $\mathcal{L}$ (fixed at 1 s) with an overlap of 0.25 s was slid over the EEG segments to extract features from the time series. An example of the application of a sliding window over an epileptic signal from Set E (UoB database) is shown in Figure 10.4. Let $x(i)$ represent an EEG signal defined on $i \in \mathbb{Z}$ with I number of samples, and $W(j)$ represent a sliding window defined on $j \in \mathbb{R}$ with J number of total windows. Then, the segments of EEG signal contained in each window can be described as

$$x_j = x_1, x_2, x_3, \ldots, x_{(I-\mathcal{L})} \tag{10.5}$$

For each shift of j windows across n signals, the following lower-order and higher-order statistical features were evaluated:

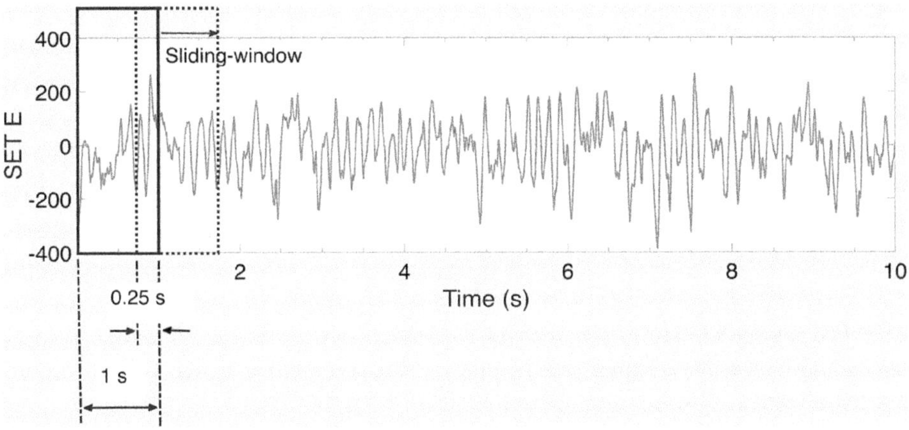

FIGURE 10.4 Application of sliding window for feature extraction over a segment from the Set of University of Bonn (UoB) database. The *X*- and *Y*-axis represent time (in s) and amplitude (in μV), respectively.

10.2.4.3.1 Mean (μ)

It is a first-order statistical parameter that represents the average of all the samples in a signal. The mean μ of $i \times n$ segments could be equated as

$$\mu_{in} = N^{-1} \cdot \sum_{i=1}^{N} x_{in} \tag{10.6}$$

10.2.4.3.2 Skewness (skw)

It is the measure of symmetry of the distribution. If a distribution is positively skewed, then the probability density function is right-tailed. Similarly, if the distribution is negatively skewed, then the probability density function is left-tailed. The skewness value is zero for a normal distribution. Here, the skewness was calculated by using Equation 10.7:

$$\text{skw}_{in} = E\left[\left(x_{in} - \mu_{in}\right)^3\right] \Big/ \sigma^3 \tag{10.7}$$

where σ is the SD, which can be expressed as

$$\sigma = \sqrt{E\left[\left(x_{in} - \mu_{in}\right)^2\right]} \tag{10.8}$$

10.2.4.3.3 Kurtosis (krt)

It is a measure to check the vulnerability of the distribution against outliers. If the kurtosis measure is *greater than 3, then* the distribution is more prone to outliers. While for a normal distribution, the kurtosis equals 3. The kurtosis was calculated by using Equation 10.9:

$$\mathrm{krt}_{in} = \frac{E\left[\left(x_{in} - \mu_{in}\right)^4\right]}{\sigma^4} \tag{10.9}$$

where σ is the SD calculated by Equation 10.8.

The entire sliding window feature matrix F_3 was constructed by the horizontal concatenation of features equated from Equations 10.6, 10.7, and 10.9. It is represented as

$$F_3 = \left[\mu_{in}\mathrm{skw}_{in}\mathrm{krt}_{in}\right] \tag{10.10}$$

The skewness and kurtosis denote third- and fourth-order statistical features, respectively, and hence obey higher-order statistics. These features further help in capturing the changes in the shape of the statistical distributions during non-seizure and epileptic stages.

10.2.5 Cross-Validation with Different Amounts of Training Data

To test the robustness of the expert system developed in this study, a cross-validation procedure was applied using different amounts of training data. Initially, we started by considering 90% of the data for training and the remaining 10% for testing. After ten folds, the percentage of training data decreased by 10%, and consequently, the percentage of testing data increased by 10%. During each iteration, the classification accuracy (CA) and statistical accuracies (sensitivity SN and specificity SP) were calculated.

10.2.6 Classifier Fusion

In this study, we have evaluated and compared the classification performance of the developed expert system using a fusion of three different types of classifiers and each of them individually. These include the following.

10.2.6.1 Decision Tree (DT) Classifier

It is a decision support tool also known as regression or classification trees (Stiglic et al. 2012). The points of segregation of the datasets are known as nodes, and it is the branch that leads to the outcome of the nodes. Further, the leaves represent the targets of the classes. The algorithm employed for the DT was as follows:

a. $A \leftarrow$ best attribute for splitting the training data
b. Decision attributes for the given node
c. For each value of A, create a child node
d. Split the training example into the child node
e. For each child node/subset:

    If the subgroup is pure
    End
    Else

Split the child node into a subset

The attributes to split were decided by the entropy (Stiglic et al. 2012) of the given set, which could be mathematically represented as

$$H(s) = -\rho \log_2(\rho) - \eta \log_2(\eta) \tag{10.11}$$

where ρ is the positive class and η is the negative class. The attribute having the least entropy is classified first. At each node, the entropies are calculated for each attribute. This process continues until we obtain either zero or at least one entropy.

10.2.6.2 Naïve Bayes (NB) Classifier

This method is based on Bayes' theorem (Efron 2013). In the NB method, we assume that the inputs (say) $X_1, X_2, X_3, \ldots, X_T$, where $t = 1, 2, 3, \ldots, T \ \forall \ T \in \mathbb{R}$ are independent of each other. This independence allows us to calculate the case-specific probabilities of a class by using Equation 10.12:

$$P(X_1, X_2, X_3, \ldots, X_T | Y_q) = \prod_{t=0}^{T} P(X_p \mid Y_q) \tag{10.12}$$

where the right-hand side terms could be calculated as the relative probabilities of individual values of X_p in class Y_q. The training set could be used to calculate the relative frequency using Equation 10.13:

$$P(X_p | Y_q) = \frac{\text{Total number of } X_p \text{ in } Y_q}{\text{Total number of cases in } Y_p} \tag{10.13}$$

If we consider the other factors to be constant, then the chances for case-specific probabilities of the class $P(X_1, X_2, \ldots\ldots\ldots X_T \mid Y_q)$ to be zero are reduced for training and independent datasets. Therefore, the NB classifier works better than the Bayes classifier.

10.2.6.3 Least Squares Support Vector Machine (LS-SVM) Classifier

The least squares formulation of SVM was used, involving the equality constraints. For binary classification, the support vector function considered in this study could be given by Equation 10.14 (Tiwari et al. 2017):

$$f(\mathbf{X}) = \text{sign}\left[\sum_{t=1}^{T} \mathcal{M}_q Y_q J(X, X_q) + \mathcal{B}\right] \tag{10.14}$$

where $\mathcal{B}$ represents the bias, $\mathcal{M}_q$ is the multiplier and $J(X, X_q)$ is the kernel function. For the given two datasets represented by $\{X_h, Y_h\}_{h=1}^{T}$, the classifier was estimated using the polynomial kernel function given by

$$\mathcal{K}(X, X_q) = (X^T X_q + 1)^{\phi} \tag{10.15}$$

where the order of the polynomial $\phi = 3$.

10.2.6.4 Classifier Fusion (CF) Design

In this study, we employed a majority voting method (Pang et al. 2012) based on the targets indicated by the three classifiers to classify the outcome. The performance parameters CA, SN, and SP were calculated across each fold.

10.3 RESULTS AND DISCUSSION

In this work, SD (F_1) and CCE (F_2) of TQWT coefficients and sliding window features (F_3) extracted from time-series data were used for the classification of EEG segments into non-seizure and epileptic signals. The results obtained from classifying the datasets taken from the UoB database and our database are shown in Tables 10.1 and 10.2, respectively. These results show the percentages of CA, SN, and SP after the application of DT, NB, LS-SVM, and classifier fusion (CF) methods individually under different training levels.

In Table 10.1, it is observed that the CA was 100% with 90% of training data by using NB and LS-SVM methods individually, while using the CF method. The CA slowly dipped to a minimum of 94.90 ± 1.75% using the NB classifier with 10% of the training data. At this training level, the CA using LS-SVM individually and the CF method was > 97%. However, the CF method outperformed the rest of the methods with SN = 98.59 ± 1.50%, SP = 97.25 ± 1.21%, and CA = 97.68 ± 0.36%. The relative SD for CA across each training level using the CF method was < 1.90%. A similar trend is observed in the classification results achieved using our database (Table 10.2). Here, the CA elicited was 100% with 90% of training data and 97.83 ± 0.50% with 10% of training data using the CF method. These measures were consistently ≥ 98% across all training levels. The SN and SP measures using the CF method with only 10% of training data were 98.76 ± 1.37% and 96.87 ± 2.06%, respectively. The overall performance trend observed in results using both databases could be summarized in the descending order as follows:

For Table 10.1, CF > LS-SVM > NB ~ DT
For Table 10.2, CF ≫ NB > DT > LS-SVM

The findings from this work were compared with a few state-of-the-art studies. To draw accurate comparisons, all state-of-the-art methods that addressed the same classification problem using the UoB database were considered (Table 10.3). The comparisons were made with the previous study conducted by our group, which utilized unifications of dual-tree complex wavelets and general regression neural network (Swami et al. 2016a). It is worthwhile to note that most existing expert systems do not discuss the performance of their expert systems under low training and high testing conditions. The present development achieved the stated objective with only 10% of the training data.

The computational cost is another vital consideration for designing an expert system. If we consider the length of the input signal as $\mathcal{N}$ and the redundancy factor as $\mathcal{R}$ then, the cost of the TQWT method is given by $O(\mathcal{R}\mathcal{N}\log_2\mathcal{N})$ (Selesnick 2011). This low computational cost again attests to the robustness of the developed expert system.

TABLE 10.1
Results Obtained Using the University of Bonn (UoB) Database (CD versus E Datasets)

Classifier	Mean Performance (in %)	Amount of Training Data (in %) for University of Bonn Datasets								
		90	80	70	60	50	40	30	20	10
Decision tree (DT)	CA ± RSD	99.17 ± 1.68	96.66 ± 0.15	96.94 ± 1.36	96.46 ± 1.91	96.66 ± 1.78	96.38 ± 0.58	95.24 ± 3.29	96.87 ± 1.43	96.38 ± 0.91
	SN ± RSD	98.08 ± 3.93	93.87 ± 2.75	94.17 ± 4.03	95.24 ± 6.37	95.24 ± 2.86	92.80 ± 2.15	90.54 ± 5.93	94.67 ± 4.93	97.70 ± 2.40
	SP ± RSD	100	98.09 ± 1.30	98.36 ± 2.23	97.25 ± 2.63	97.31 ± 2.24	98.22 ± 1.58	97.77 ± 1.90	98.08 ± 2.44	95.84 ± 1.98
Naïve Bayes (NB)	CA ± RSD	100	98.75 ± 0.85	98.34 ± 1.46	97.50 ± 2.21	98.17 ± 0.85	96.38 ± 1.10	97.14 ± 0.80	96.87 ± 0.43	94.90 ± 1.75
	SN ± RSD	100	98.75 ± 2.53	98.25 ± 1.55	98.08 ± 1.37	97.90 ± 0.09	93.58 ± 4.06	94.21 ± 5.40	94.35 ± 3.32	88.79 ± 10.66
	SP ± RSD	100	98.72 ± 1.50	98.38 ± 1.47	97.17 ± 2.95	98.29 ± 1.26	97.80± 1.08	98.34 ± 1.33	98.15 ± 1.36	98.11 ± 2.59
Least-squares support vector machine (LS-SVM)	CA ± RSD	100	98.75 ± 0.85	98.33 ± 1.39	97.71 ± 1.89	98.49 ± 0.63	98.05 ± 1.08	97.97 ± 0.83	98.64 ± 0.63	97.31 ± 1.25
	SN ± RSD	100	98.33 ± 3.39	99.17 ± 0.66	99.04 ± 0.66	99.00 ± 0.18	98.28 ± 0.15	98.37 ± 0.93	98.92 ± 1.07	99.30 ± 0.68
	SP ± RSD	100	97.73 ± 2.69	96.58 ± 3.28	95.09 ± 4.68	97.36 ± 2.04	97.85 ± 2.66	97.35 ± 2.10	98.21 ± 2.30	93.78 ± 3.70
Classifier fusion (CF)	CA ± RSD	100	98.75 ± 0.85	98.33 ± 1.39	97.71 ± 1.89	98.17 ± 0.85	97.49 ± 0.57	97.38 ± 1.52	98.02 ± 1.27	97.68 ± 0.36
	SN ± RSD	100	98.75 ± 2.53	98.25 ± 1.55	98.08 ± 1.37	97.90 ± 0.08	96.87 ± 0.31	95.56 ± 4.41	95.95 ± 4.58	98.59 ± 1.50
	SP ± RSD	100	98.72 ± 1.50	98.37 ± 1.49	97.5 ± 2.38	98.29 ± 1.26	97.80 ± 1.08	98.36 ± 0.99	99.05 ± 1.22	97.25 ± 1.21

TABLE 10.2
Results Obtained Using Our Database (Non-Seizure versus Epileptic Signals)

Classifier	Mean Performance (in %)	Amount of Training Data (in %) for Our Database 90	80	70	60	50	40	30	20	10
Decision tree (DT)	CA ± RSD	99.33 ± 1.16	99.00 ± 0.71	98.66 ± 0.48	98.10 ± 1.59	97.20 ± 0.82	97.92 ± 1.29	97.42 ± 1.99	96.94 ± 1.70	95.94 ± 1.91
	SN ± RSD	98.67 ± 2.34	97.73 ± 1.62	97.83 ± 1.32	97.05 ± 3.69	96.50 ± 2.26	97.52 ± 1.98	98.15 ± 1.34	96.75 ± 3.23	94.20 ± 3.80
	SP ± RSD	100	100	99.47 ± 0.72	99.23 ± 0.81	97.87 ± 3.11	98.28 ± 2.21	96.73 ± 3.78	97.10 ± 2.31	97.42 ± 1.54
Naïve Bayes (NB)	CA ± RSD	100	99.40 ± 0.90	98.93 ± 0.60	99.40 ± 0.42	98.64 ± 0.74	97.83 ± 1.26	97.60 ± 0.84	97.13 ± 1.54	96.39 ± 1.15
	SN ± RSD	100	98.70 ± 1.94	98.40 ± 0.59	99.14 ± 0.49	97.94 ± 1.31	96.70 ± 1.71	97.22 ± 2.35	96.41 ± 3.04	95.21 ± 1.85
	SP ± RSD	100	100	99.47 ± 0.72	99.64 ± 0.50	99.36 ± 0.67	98.98 ± 1.19	98.30 ± 0.90	97.61 ± 1.78	97.53 ± 2.17
Least-squares support vector machine (LS-SVM)	CA ± RSD	98.67 ± 1.17	99.60 ± 0.55	99.06 ± 0.90	98.40 ± 1.21	97.92 ± 1.77	97.08 ± 0.33	97.14 ± 1.70	96.25 ± 1.54	95.00 ± 1.51
	SN ± RSD	100	100	99.71 ± 0.65	100	99.48 ± 0.47	99.83 ± 0.35	99.64 ± 0.33	99.86 ± 0.28	99.75 ± 0.29
	SP ± RSD	97.52 ± 2.20	99.12 ± 1.22	98.44 ± 1.71	96.92 ± 2.05	96.55 ± 2.90	94.70 ± 0.82	94.92 ± 3.31	93.08 ± 2.77	92.35 ± 1.73
Classifier fusion (CF)	CA ± RSD	100	99.80 ± 0.45	99.46 ± 0.30	99.80 ± 0.27	98.64 ± 1.53	98.83 ± 0.85	98.10 = 1.70	98.06 ± 0.96	97.83 ± 0.50
	SN ± RSD	100	99.74 ± 0.57	99.49 ± 0.71	100	99.20 ± 0.02	99.66 ± 0.39	99.65 = 0.32	99.75 ± 0.50	98.76 ± 1.37
	SP ± RSD	100	100	99.47 ± 0.72	99.62 ± 0.52	98.05 ± 3.17	97.96 ± 2.08	96.61 = 3.62	96.36 ± 1.97	96.87 ± 2.06

TABLE 10.3
Comparison of the Present Study with a Few State-of-the-Arts

References	Methodology	Amount of Training Data (in %)	Best Mean Performance Reported (in %)
Chen (2014)	Transform(s): Dual-tree complex wavelet transform Feature(s): Fourier features Classifier(s): k-nearest neighbour	50	CA = 100
Swami et al. (2016a)	Transform(s): Dual-tree complex wavelet transform Feature(s): Energy, standard deviation, root-mean-square, Shannon entropy, mean values, maximum peaks Classifier(s): General regression neural network	10–90	CA ± SE = 95.33 ± 0.35 to 98.67 ± 0.74 SN ± SE = 95.09 ± 1.20 to 97.92 ± 1.42 SP ± SE = 95.47 ± 0.56 to 98.92 ± 0.72
Li et al. (2016)	Transform(s): DWT Feature(s): Singular values calculated from the covariance matrix of 2D and 3D phase space representations Classifier(s): ELM, SVM, back propagation neural network, Multiplicative ELM	50	CA ± SD = 98.89 ± 0.32 SN ± SD = 98.92 ± 0.01 SP ± SD = 99.93 ± 0.04
Tiwari et al. (2017)	Transform(s): Scale invariant feature transformation Feature(s): Histogram of local binary patterns Classifier(s): SVM	90	CA = 99.45 SN = 99.68 SP = 99.00
Ravi Shankar Reddy and Rao (2017)	Transform(s): TQWT Feature(s): Centred correntropy Classifier(s): Random forest, multilayer perceptron, logistic regression	90	CA = 97.70 SN = 96.00 SP = 98.50
Patidar and Panigrahi (2017)	Transform(s): TQWT Feature(s): Kraskov entropy Classifier(s): least-squares support vector machine (LS-SVM)	90	CA = 97.75 SN = 97.00 SP = 99.00

(*Continued*)

TABLE 10.3 (*Continued*)
Comparison of the Present Study with a Few State-of-the-Arts

References	Methodology	Amount of Training Data (in %)	Best Mean Performance Reported (in %)
Zhang, Chen, and Li (2018)	Transform(s): Wavelet packet transform Feature(s): Fuzzy distribution entropy Feature selection using Kruskal–Wallis test Classifier(s): k-nearest neighbour	90	CA ± SD = 99.58 ± 0.19 SN ± SD = 98.89 ± 0.37 SP ± SD = 99.92 ± 0.18
Mardini et al. (2020)	Transform(s): Discrete wavelet transform Feature(s): Mean absolute value, average power, standard deviation, variance, mean, skewness, Shannon entropy, maximum wavelet coefficients, minimum wavelet coefficients, normalized standard deviation, energy Classifier(s): Support vector machine, naive Bayes classifier, k-nearest neighbour, artificial neural network	70	CA = 99 SN = 99 SP = 100
Atlam, Aderibigbe, and Nadeem (2025)	Transform(s): Discrete wavelet transform Feature(s): principal component analysis-transformed features with selected wavelet coefficients Classifier(s): Support vector machine, random forest, decision trees, k-nearest neighbour	90	CA = 97.3 Recall = 93.08
This study	Transform(s): TQWT Feature(s): Standard deviation and centred correntropy of wavelet coefficients; mean, skewness and kurtosis values of time-series data across sliding window Classifier(s): Decision tree, naïve Bayes, LS-SVM, classifier fusion	10–90	CA ± RSD = 97.68 ± 0.36 to 100 CA ± RSD = 98.59 ± 1.50 to 100 CA ± RSD = 97.25 ± 1.21 to 100

10.4 CONCLUSIONS

The advantage of using TQWT was successfully demonstrated in this paper by comparing its denoising capability with that of the conventionally used DWT. The robust nature of TQWT is extremely beneficial for biomedical signals, such as EEG, which are often contaminated with noise (Gajic et al. 2015). The CF was performed using a simple majority voting method in this work, which not only improved the accuracy of the system but also introduced statistical coherence across different training levels and folds. This was demonstrated by the results presented in Tables 10.1 and 10.2, where all parameters using the CF method consistently maintained high performance (>97%) across every training level. Hence, the expert system developed in this work sets a high standard for classifying segments containing epileptic seizures. The proposed research could be extended in the form of a tool (Kelly et al. 2010; Chen et al. 2017) to analyse continuous EEG signals for clinical testing. This outcome will not only be helpful for the accurate diagnosis of epilepsy but will also help to leverage the acute shortage of trained healthcare professionals in many countries across the world.

Although the current work presents a highly effective binary classification scheme, it does not discuss the multi-class classification of epileptic stages. However, the proposed scheme could be extended to classify the multiple stages involved during an epileptic attack, i.e., pre-ictal versus ictal versus interictal stages (Swami et al. 2016d). In this research, all the features proved to be essential parameters for classification; however, future work could focus on reducing the feature set (Nara et al. 2016; Swami 2024a). The applicability can be further improved by its deployment in the form of hardware (Atlam, Aderibigbe, and Nadeem 2025; Swami et al. 2024b; Tirumani et al. 2018). This will help in more effective epilepsy diagnosis and would ultimately lead to better treatment. Future work can also focus on identifying the epicentre of epileptic seizures and mapping its spatiotemporal characteristics in relation to specific brain regions (Swami et al. 2016b, 2016c). This will facilitate the pre- and post-evaluation of potential surgical and/or pharmacological outcomes.

ACKNOWLEDGEMENT

The authors would like to thank Professor Ralph G. Andrzejak from the University of Bonn, Germany, for making his database publicly available and Mrs. Anu Mol, Senior Clinician at the Neurology and Sleep Centre, New Delhi, India, for her assistance in the annotation of the datasets.

REFERENCES

Acharya, U. R., S. V. Sree, G. Swapna, R. J. Martis, and J. S. Suri. "Automated EEG Analysis of Epilepsy: A Review." *Knowledge-Based Systems* 45 (2013): 147–165.

Andrzejak, R. G., K. Lehnertz, F. Mormann, C. Rieke, P. David, and C. E. Elger. "Indications of Nonlinear Deterministic and Finite-Dimensional Structures in Time Series of Brain Electrical Activity: Dependence on Recording Region and Brain State." *Physical Review E* 64, no. 6 (2001): 1–8.

Atlam, H. F., G. E. Aderibigbe, and M. S. Nadeem. "Effective Epileptic Seizure Detection with Hybrid Feature Selection and SMOTE-Based Data Balancing Using SVM Classifier." *Applied Sciences* 15, no. 9 (2025): 4690.

Chen, D., S. Wan, J. Xiang, and F. S. Bao. "A High-Performance Seizure Detection Algorithm Based on Discrete Wavelet Transform (DWT) and EEG." *PLos One* 12, no. 3 (2017): e0173138.

Chen, G. "Automatic EEG Seizure Detection Using Dual-Tree Complex Wavelet-Fourier Features." *Expert Systems with Applications* 41, no. 5 (2014): 2391–2394.

Efron, B. "Bayes' Theorem in the 21st Century." *Science* 340, no. 6137 (2013): 1177–1178.

Gajic, D., Z. Djurovic, J. Gligorijevic, S. Di Gennaro, and I. Savic-Gajic. "Detection of Epileptiform Activity in EEG Signals Based on Time-Frequency and Nonlinear Analysis." *Frontiers in Computational Neuroscience* 9 (2015): 1–16.

Gandhi, T. K., P. Chakraborty, G. G. Roy, and B. K. Panigrahi. "Discrete Harmony Search Based Expert Model for Epileptic Seizure Detection in Electroencephalography." *Expert Systems with Applications* 39, no. 4 (2012): 4055–4062.

Gandhi, T., B. K. Panigrahi, and S. Anand. "A Comparative Study of Wavelet Families for EEG Signal Classification." *Neurocomputing* 74, no. 17 (2011): 3051–3057.

Gandhi, T., B. K. Panigrahi, M. Bhatia, and S. Anand. "Expert Model for Detection of Epileptic Activity in EEG Signature." *Expert Systems with Applications* 37, no. 4 (2010): 3513–3520.

Kelly, K. M., et al. "Assessment of a Scalp EEG-Based Automated Seizure Detection System." *Clinical Neurophysiology* 121, no. 11 (2010): 1832–1843.

Li, D., Q. Xie, Q. Jin, and K. Hirasawa. "A Sequential Method Using Multiplicative Extreme Learning Machine for Epileptic Seizure Detection." *Neurocomputing* 214 (2016): 692–707.

Maheshwari, S., R. B. Pachori, and U. R. Acharya. "Automated Diagnosis of Glaucoma Using Empirical Wavelet Transform and Correntropy Features Extracted from Fundus Images." *IEEE Journal of Biomedical and Health Informatics* 21, no. 3 (2017): 803–813.

Mardini, W., M. M. B. Yassein, R. Al-Rawashdeh, S. Aljawarneh, Y. Khamayseh, and O. Meqdadi. "Enhanced Detection of Epileptic Seizure Using EEG Signals in Combination with Machine Learning Classifiers." *IEEE Access* 8 (2020): 24046–24055.

Moshé, S. L., E. Perucca, P. Ryvlin, and T. Tomson. "Epilepsy: New Advances." *Lancet* 385 (2015): 884–898.

Nara, S., P. Swami, M. Bhatia, B. K. Panigrahi, and T. K. Gandhi. "Detecting Epileptic Activities in EEG through Correlation Based Dimensionality Reduction." In *Proceedings of the IEEE International Conference on Computing for Sustainable Global Development*, New Delhi, India, pp. 2547–2550, 2016.

Pang, Y., M. Shang, Y. Yuan, and J. Pan. "Scale Invariant Image Matching Using Triplewise Constraint and Weighted Voting." *Neurocomputing* 83 (2012): 64–71.

Patidar, S., and T. Panigrahi. "Detection of Epileptic Seizure Using Kraskov Entropy Applied on Tunable-Q Wavelet Transform of EEG Signals." *Biomedical Signal Processing and Control* 34 (2017): 74–80.

Reddy, G. R. S., and R. Rao. "Automated Identification System for Seizure EEG Signals Using Tunable-Q Wavelet Transform." *Engineering Science and Technology, an International Journal* 20, no. 5 (2017): 1486–1493.

Sahu, R., T. Parija, B. Mohapatra, B. Rout, S. Sahu, R. Panda, P. Pal, and T. Gandhi. "Chaos Based Nonlinear Analysis of Epileptic Seizure." In *Proceedings of the International Conference on Emerging Trends in Engineering and Technology*, Cham, pp. 594–598, 2010.

Selesnick, I. W. "Wavelet Transform with Tunable Q-Factor." *IEEE Transactions on Signal Processing* 59, no. 8 (2011): 3560–3575.

Stiglic, G., S. Kocbek, I. Pernek, and P. Kokol. "Comprehensive Decision Tree Models in Bioinformatics." *PLoS One* 7, no. 3 (2012): e33812.

Swami, P., A. K. Godiyal, J. Santhosh, B. K. Panigrahi, M. Bhatia, and S. Anand. "Robust Expert System Design for Automated Detection of Epileptic Seizures Using SVM

Classifier." In *Proceedings of the IEEE International Conference on Parallel, Distributed and Grid Computing*, Chengdu, China, pp. 219–222, 2014.

Swami, P., B. K. Panigrahi, and T. K. Gandhi. *A Device for Automated Diagnosis of Epilepsy*. Indian Patent IN500189. Application No. 201611043166. Filed 2016, Published January 17, 2024a.

Swami, P., B. K. Panigrahi, S. Anand, M. Bhatia, and T. K. Gandhi. "Application of Machine Learning in Seizure Detection." In *Proceedings of the INCF World Congress on Neuroinformatics*, Montréal, Canada, 2018.

Swami, P., B. K. Panigrahi, S. Anand, M. Bhatia, M. Tripathi, and T. K. Gandhi. "Detection and Cortical Localization of Ictal Signatures Using Electroencephalogram Signals." In *Proceedings of the INCF World Congress on Neuroinformatics, Frontiers in Neuroinformatics*, Reading, UK, 2016c.

Swami, P., B. K. Panigrahi, S. Nara, M. Bhatia, and T. K. Gandhi. "EEG Epilepsy Datasets." 2016d. *Research Gate* 2016b. https://doi.org/10.13140/RG.2.2.14280.32006.

Swami, P., J. Maheshwari, M. Kumar, and M. Bhatia. "Evolutionary Transfer Optimization-Based Approach for Automated Ictal Pattern Recognition Using Brain Signals." *Frontiers in Human Neuroscience* 18 (2024b): 1386168.

Swami, P., M. Bhatia, M. Tripathi, P. S. Chandra, B. K. Panigrahi, and T. K. Gandhi. "Selection of Optimum Frequency Bands for Detection of Epileptiform Patterns." *Healthcare Technology Letters* 6, no. 5 (2019): 126–131.

Swami, P., M. Bhatia, S. Anand, B. K. Panigrahi, and J. Santhosh. "SVM Based Automated EEG Seizure Detection Using 'Coiflets' Wavelet Packets." In *Proceedings of the IEEE International Conference on Recent Developments in Control, Automation and Power Engineering*, Singapore, pp. 238–242, 2015b.

Swami, P., T. K. Gandhi, B. K. Panigrahi, M. Bhatia, and S. Anand. "Detection of Ictal Patterns in Electroencephalogram Signals Using 3D Phase Trajectories." In *Proceedings of the IEEE Annual India Conference (INDICON)*, Reading, UK, pp. 1–6, 2015a.

Swami, P., T. K. Gandhi, B. K. Panigrahi, M. Bhatia, and S. Anand. "Locating Ictal Activities over Human Scalp with Automated Detection Using EEG Signals." In *Proceedings of the IEEE International Conference on Signal Processing and Integrated Networks*, Noida, India, pp. 600–604, 2016b.

Swami, P., T. K. Gandhi, B. K. Panigrahi, M. Bhatia, J. Santhosh, and S. Anand. "A Comparative Account of Modelling Seizure Detection System Using Wavelet Techniques." *International Journal of System Science: Operations & Logistics* 4, no. 1 (2017): 41–52.

Swami, P., T. K. Gandhi, B. K. Panigrahi, M. Tripathi, and S. Anand. "A Novel Robust Diagnostic Model to Detect Seizures in Electroencephalography." *Expert Systems with Applications* 56 (2016a): 116–130.

Tirumani, P. V., V. K. Gudlavalleti, K. K. Mahapatra, and B. K. Panigrahi. "A Low-Noise Low-Cost EEG Amplifier for Neural Recording Applications." In *Advanced Computational and Communication Paradigms. Proceedings of ICACCP 2017*, Singapore, Vol. 1, pp. 51–59, 2018.

Tiwari, A. K., R. B. Pachori, V. Kanhangad, and B. K. Panigrahi. "Automated Diagnosis of Epilepsy Using Key-Point-Based Local Binary Pattern of EEG Signals." *IEEE Journal of Biomedical and Health Informatics* 21, no. 4 (2017): 888–896.

Tzallas, A. T., M. G. Tsipouras, D. G. Tsalikakis, E. C. Karvounis, L. Astrakas, S. Konitsiotis, and M. Tzaphlidou. "Automated Epileptic Seizure Detection Methods: A Review Study." In D. Stevanovic, ed., *Epilepsy - Histological, Electroencephalographic and Psychological Aspects*, 75–98. Rijeka, Croatia: InTech, 2012.

Zhang, T., W. Chen, and M. Li. "Fuzzy Distribution Entropy and Its Application in Automated Seizure Detection Technique." *Biomedical Signal Processing and Control* 39 (2018): 360–377.

Section V

Advanced Techniques and Machine Learning Applications

11 Arrhythmia Detection Using WPD with Bagging and Boosting Ensemble Machine Learning Methods

Abdulhamit Subasi, Muhammed Enes Subasi, and Saeed Mian Qaisar

11.1 INTRODUCTION

Arrhythmias (ARRs) are irregular heart rhythms caused by abnormal electrical signals in the heart. These can result from various factors, including changes in heart tissue, stress, imbalances in blood components such as electrolytes or hormones, side effects of medications, chronic diseases, or issues within the heart's electrical system. Cardiologists gather information about the individual's medical history through blood tests and physical examinations to evaluate a patient's cardiac health. Electrocardiography is the most commonly used non-invasive test, recording the heart's electrical activity to derive key parameters for assessing overall cardiac health. When analyzing an ECG signal, it is crucial to consider factors such as heart rate, pulse origin, pathway, and propagation speed. For instance, if an electrical impulse strays from the heart's normal conduction pathway, the ECG's morphology can be significantly altered, leading to ARR. Some cardiac conditions can be detected through short-term monitoring in a hospital setting. In contrast, others may require long-term observation using devices like the Holter monitor or modern wearable technology. These long-term monitoring methods generate large amounts of data that need to be analyzed and processed. In summary, ARRs are caused by abnormal heart signals due to irregular rhythms and from various factors. Cardiologists rely on medical history and tests to diagnose them, with ECG being a primary tool for recording electrical activity. ECG signal characteristics are crucial for identifying ARRs, which may require both short-term and long-term monitoring, leading to extensive data analysis (Montenegro et al., 2022).

Accurate methods for automatically classifying electrocardiogram (ECG) beats are essential for diagnosing heart conditions such as ARR. ARR, characterized by abnormal heart rhythms often due to heart rate discrepancies, is identified by ECG beats. These irregular rhythms, caused by chaotic electrical activity, can lead

DOI: 10.1201/9781003611448-16

to severe consequences such as cardiac arrest, hemodynamic collapse, and sudden death. Early detection of ARRs, especially life-threatening ones like ventricular tachyarrhythmias, is crucial as they can cause sudden cardiac death. Additionally, ARRs like atrial fibrillation increase the risk of thromboembolic events and stroke, highlighting the importance of computerized cardiac ARR detection and classification in diagnosing cardiovascular diseases (CVDs) (Alickovic & Subasi, 2015, 2016).

Despite advancements in automatic ECG classification, challenges remain in developing robust and efficient classifiers. Noise significantly impacts all biomedical signals during ECG recording. The primary challenge is selecting an effective denoising method, as noise elimination is crucial for ECG signal classification. ECG signals from different subjects are heterogeneous, requiring thorough preprocessing for denoising and feature extraction. Classifying ECG beats based on these features is necessary to detect various ARRs. In ambulatory ECG, unpredictable noise types such as baseline wandering, electromyogram noise, motion artifacts, powerline interference, and electrode pop or contact noise can occur. ECG preprocessing involves denoising the signal, with baseline wandering and powerline interference being the most disruptive. While powerline interference can be mitigated with hardware, baseline wandering and wideband noise require advanced software algorithms for removal. Although many studies have focused on automated ECG heartbeat classification, most algorithms struggle to achieve high accuracy. Wavelet transforms (WTs) have been used in previous research to reduce complexity and improve the signal-to-noise ratio of the ECG signal before detection (Alickovic & Subasi, 2015, 2016).

Advanced signal processing methods, such as discrete WT (DWT) denoising, can be used but are not time-invariant. To overcome this, we used a multi-scale principal component analysis (MSPCA)-based denoising method, which combines WT and principal component analysis (PCA). Alickovic and Subasi (Alickovic & Subasi, 2015) investigated the impact of MSPCA-based denoising on ECG beat classification. This study's innovation is applying an effective denoising technique to reduce noise in ECG heartbeat signals. The process involved denoising the ECG signals, extracting relevant features, and classifying the heartbeats. Accurate classification of heartbeats is essential for evaluating different types of ARRs. Automated heartbeat classification plays a significant role in detecting and diagnosing heart diseases. Several methodologies and automated systems have been developed for this purpose. The automatic classification of ECG heartbeat signals continues to be a vibrant area of research, with various data mining techniques being utilized for ECG heartbeat classification (Alickovic & Subasi, 2015, 2016). We compared the performance of different machine learning (ML) methods based on receiver operating characteristic (ROC) area, *F*-measure, Kappa, and classification accuracy.

11.2 LITERATURE REVIEW

11.2.1 Arrhythmia Detection Using Classical Machine Learning Techniques

An ECG signal is a bioelectrical signal that records the electrical activity of the heart, serving as a parameter for detecting various heart conditions. Cardiac ARR, characterized by irregular, too slow, or too fast heart rhythms, can be identified using

ECG signals. Abirami and Vincent (2020) developed a robust ML algorithm to predict cardiac ARR. The study used data from the University of California, Irvine ML repository and applied feature selection during preprocessing to enhance the performance of ML algorithms. The data was classified into two categories: normal and ARR. Features with many missing values or no variation were removed. An ensemble of ML algorithms was then used, achieving an accuracy of 85%. Ensemble prediction, which combines multiple models, generally produces more accurate results than using a single model.

Cardiac ARR is a condition where the heartbeat becomes irregular, either too slow or too fast, due to malfunctioning electrical impulses that regulate the heart's rhythm. Some severe ARRs can result in sudden cardiac death. Therefore, the main objective of electrocardiogram (ECG) analysis is to accurately detect these life-threatening ARRs to enable timely treatment and save lives. ECG signals, which reflect the heart's electrical activity, are characterized by waveforms (P, QRS, and T) whose durations, structures, and intervals between their peaks are critical in identifying heart-related issues. Taloba et al. (2021) used an autoregressive (AR) analysis to extract specific signal features represented by the parameters of the AR signal model. The AR characteristics of three different types of ECG signals are clearly distinguished in the training dataset, enabling accurate classification and diagnosis of heart conditions for each ECG signal in the dataset. A new method is introduced that combines two-event-related moving averages with fractional Fourier transform algorithms to enhance the assessment of ECG signals.

Life-threatening ventricular arrhythmias (VAs) such as ventricular tachycardia and ventricular fibrillation pose significant risks of sudden cardiac death, necessitating early detection for effective intervention. Mohanty et al. (2020) developed an automated ML system to predict VAs and potentially reduce mortality rates. They introduced a novel approach that combines variational mode decomposition for feature extraction from ECG signals with a C4.5 classifier for VA detection. Using ECG data from the PhysioNet databases, they extracted time–frequency features and ranked them using the information gain method, which calculates the degree to which the target variable's uncertainty is decreased by knowing the value of a feature. The feature provides more information about the target variable when the IG is greater. These features were classified into normal, ventricular tachycardia, and ventricular fibrillation classes using both support vector machine (SVM) and C4.5 classifiers. The C4.5 classifier outperformed the SVM, achieving 97.97% sensitivity, 99.15% specificity, and 99.18% accuracy in a 5-second window, making it an efficient tool for detecting life-threatening VAs and aiding clinicians in diagnosis.

Alickovic and Subasi (2016) introduced a random forests (RF) classifier to classify ECG heartbeat signals to diagnose heart ARR. The method uses DWT to break down ECG signals into frequency bands and extract statistical features that describe the distribution of wavelet coefficients. The study shows that the RF classifier outperforms other decision tree methods when tested with 10-fold cross-validation on ECG datasets. Evaluations on the MIT-BIH and St. Petersburg Institute of Cardiological Technics 12-lead ARR databases demonstrated that the RF classifier achieved higher accuracy—99.33% and 99.95%, respectively—compared to the C4.5 and classification and regression tree (CART) classifiers. Incorporating MSPCA denoising, DWT, and the RF classifier further enhanced performance, achieving near-perfect area under the curve (AUC) and *F*-measure values. The findings highlight the system's reliability in classifying ECG signals and accurately diagnosing CVDs.

Singh et al. (2019) presented a method for automated detection and classification of ARR. The proposed method is evaluated using the MIT-BIH Database. The performance of five ML algorithms is compared for processing the prepared feature sets. The process began with preprocessing and beat detection to isolate individual beats. Beats are isolated using a QRS detection method. Three feature sets were prepared: amplitude analysis-based feature set, area feature mining using a non-overlapping sliding window, and area feature extraction with an overlapping sliding window. Their findings revealed that the artificial neural network (ANN) using amplitude features achieved the highest accuracy of 99.59%, which is comparable to state-of-the-art counterparts.

Kumari et al. (2020) developed a method for classifying heartbeat irregularities, specifically focusing on cardiac ARR, normal sinus rhythm, and congestive heart failure. Using 162 records from the MIT-BIH Database, they divided the data into training (70%) and testing (30%) sets. The process involved four stages: (1) collecting ARR and non-ARR signals, (2) preprocessing the signals, (3) extracting features using DWT and combining them into a single feature vector, and (4) classifying the signals with an SVM classifier. The method achieved a classification accuracy of 95.92%.

Hussain et al. (2020) proposed a novel approach for ARR detection by combining hybrid features, including refined sample entropy (SampEn) and fuzzy entropy (FuzEn), with traditional features. They applied ML algorithms such as SVM, decision trees, k-nearest neighbors (k-NN), and ensemble methods. Using 10-fold cross-validation, the study found that SVM with a coarse Gaussian kernel, fine k-NN, and ensemble boosted tree achieved the best results, with an overall accuracy of 100% and an AUC of 1.00. The findings indicate that these methods outperformed traditional feature extraction techniques, particularly when using hybrid features.

Dhyani et al. (2023) used 3D DWT combined with a SVM for analyzing electrocardiogram (ECG) signals. Their method involves three stages: preprocessing the ECG signals, extracting features using 3D WT for denoising and coefficient extraction, and classifying the signals into nine heartbeat types. Analyzing around 6,400 ECG beats from the China Physiological Signal Challenge (CPSC) 2018 dataset, they found that using level 4 wavelet decomposition with Symlet-8 channels achieved the highest accuracy. The SVM classifier reached an average accuracy of 99.02%, outperforming complex SVM and weighted SVM (WSVM) methods. This approach demonstrated superior accuracy and effectiveness in classifying various ARR conditions.

Farooq et al. (2021) developed a cost-effective ECG classification system using LabVIEW. Their approach involves collecting and processing ECG signals with a sensor system and classifying them as healthy, unhealthy, or undefined. The system uses ML techniques, specifically K-means clustering, for classification. The study highlights the system's ability to detect heart anomalies and ARRs early, with a response time of under 30 s. It also suggests potential applications in the early detection of COVID-19-related breathing issues and the automatic scheduling of appointments with cardiac professionals, which can help reduce congestion in medical centers.

Tuncer et al. (2022) proposed a novel ECG signal recognition approach using a discrete wavelet concatenated mesh tree (DW-CMT) and ternary chess pattern (TCP). Their method involves four stages: preprocessing with DW-CMT, feature extraction with TCP, feature selection using neighborhood component analysis (NCA), and classification. ECG signals are first split into 15 sub-bands using DW-CMT. TCP extracts features from these sub-bands, which are combined and refined to 128 key features through NCA. The system's performance is evaluated with traditional classifiers like linear discriminant analysis (LDA), k-NN, and SVM. The method achieved 96.60% accuracy with k-NN on the MIT-BIH dataset and 97.80% with SVM on the St. Petersburg dataset, demonstrating its effectiveness in ARR diagnosis.

Mian Qaisar and Subasi (2020) introduced an innovative approach for detecting chronic heart disorders using ECG signals with an event-driven acquisition system. This system enhances real-time compression and processing efficiency, achieving a 2.6-fold improvement in compression and bandwidth usage compared to traditional methods. This advancement significantly reduces complexity and execution time in post-denoising, feature extraction, and classification. The system's effectiveness was measured by classification accuracy, *F*-measure, area under the ROC curve (AUC), and Kappa statistics, with the highest classification accuracy reaching 94.07%. The approach provides a computationally efficient, high-precision automatic diagnosis for cardiac ARR and supports cloud-based mobile health monitoring.

11.2.2 Arrhythmia Detection Using Deep Learning Techniques

In ARR classification tasks with ML algorithms, the conventional method requires manually creating features from electrocardiography signals, which is a lengthy and iterative process. Deep neural networks provide a more efficient alternative by automatically extracting features through their hidden layers, bypassing manual feature engineering. Nevertheless, having balanced datasets is essential for successfully training these algorithms.

Montenegro et al. (2022) compared the 1D-CNN and SVM for ARR classification using a combined dataset from the TNMG and CINC17 public ECG databases. The 1D-CNN achieved high performance with a precision of 93.04%, accuracy of 93.07%, recall of 93.20%, and an F1-score of 93.05%. In contrast, the SVM classifier, utilizing handcrafted features like WTs and R-peak interval features, achieved a precision of 89.04%, accuracy of 92.00%, recall of 94.20%, and an F1-score of 91.54%. Specifically, using WTs with SVM yielded a precision of 86.15%, accuracy of 85.33%, recall of 81.16%, and an F1-score of 83.58%. The study highlights that while both algorithms performed well, SVM requires more effort in feature selection, whereas CNNs automate feature extraction but demand more computational resources. Combining multiple datasets can enhance model evaluation, particularly for deep learning approaches.

ARRs are irregular heartbeat rates or rhythms that can occur sporadically, making long-term ECG data from Holter devices essential for detecting these rare events. Automating the recognition of abnormal heartbeats in extensive ECG datasets is crucial. Sannino and De Pietro (Sannino & De Pietro, 2018) developed a novel deep learning method for ECG beat classification, tested on the MIT-BIH Arrhythmia

Database. Their model surpassed current state-of-the-art methods in accuracy while maintaining strong sensitivity and specificity metrics. Ullah et al. (2021) utilized deep learning techniques to classify ARRs using the MIT-BIH Arrhythmia Database and the PTB Diagnostic ECG Database. They employed three models: CNN, CNN+LSTM, and CNN+LSTM+Attention Model, with 80% of the data used for training and 20% for testing. The results demonstrated high performance, with accuracies of 99.12% for CNN, 99.3% for CNN+LSTM, and 99.29% for CNN+LSTM+Attention Model.

The complexity and nonlinear nature of ECG signals, encompassing noise, temporal dependencies, and irregular heartbeat patterns, make manual analysis difficult. Additionally, interpretations of ECG signals can be subjective among experts. To overcome these issues, Zhou et al. (2019) developed an automated and highly accurate method for ARR classification using long short-term memory (LSTM), a type of deep learning technique. This method targets five ARR classes as defined by the Association for the Advancement of Medical Instrumentation and has been evaluated using the MIT-BIH Arrhythmia Database. The results demonstrate its effectiveness, surpassing the performance of other artificial intelligence (AI) methods previously used in similar contexts.

Yıldırım et al. (2018) developed a new deep learning approach to detect cardiac ARRs across 17 classes using long-duration ECG signals. Their method aimed to provide rapid and accurate classification by analyzing 10-second ECG signal segments rather than individual QRS complexes, thus reducing the frequency of classifications by an average factor of 13. They utilized 1,000 ECG signal fragments from the MIT-BIH Arrhythmia Database, focusing on the modified limb lead II (MLII) lead from 45 individuals. Their approach features an end-to-end deep 1D convolutional neural networks (1D-CNN) model, which combines feature extraction, selection, and classification in one stage. The method achieved an overall accuracy of 91.33% and a classification time of 0.015 s per sample, offering efficiency, real-time capability, and simplicity in implementation.

Irfan et al. (2022) proposed a new deep learning framework that combines multiple networks by stacking similar layers to create a unified model, addressing lengthy training times and manual feature selection issues. Evaluated on two public datasets for classifying five ARR micro-classes, the framework achieved high performance with a sensitivity of 98.37%, specificity of 99.59%, positive predictive value of 98.41%, and accuracy of 99.35%, outperforming state-of-the-art methods in these metrics. Swapna et al. (2018) used deep learning methods, including CNNs, LSTM, and gated recurrent units, to diagnose cardiac ARR from ECG signals with minimal preprocessing. Their CNN-LSTM hybrid model achieved a five-fold cross-validation accuracy of 0.834 in differentiating between normal and ARR ECG signals, demonstrating effective automatic detection without manual feature extraction.

Hammad et al. (2022) developed advanced deep learning models, including CNNs and convolutional long short-term memory, for automated ARR detection in Internet of Things applications. Their method converts ECG signals into a 2D representation, which these models then process to classify ARRs. This approach overcomes challenges such as overfitting and the simultaneous processing of multiple ECG leads.

Evaluated on various PhysioNet datasets, including MIT-BIH and PhysioNet 2016 and 2018, the framework achieved accuracies of 97%, 98%, 94%, and 91%, respectively. The proposed framework shows improved robustness and efficiency, especially in noisy data scenarios, compared to previous methods. Rajkumar et al. (2019) developed a deep learning-based method for classifying ECG signals using CNNs). The CNN automatically learns features from time-domain ECG signals from the MIT-BIH Database, enhancing the efficiency of diagnosing cardiac ARRs by eliminating manual feature extraction. The model, trained and tested on the MIT-BIH dataset, focuses on seven ARR types. Among various activation functions tested, the Exponential Linear Unit (ELU) achieved the best performance with an accuracy of 93.6% and a loss of 0.2%.

Patients at high risk of CVD can benefit from computerized electrocardiogram (ECG) devices designed to detect ARRs. However, these devices often require long segments of high-quality ECG data, which may lead to missed episodes. Mohonta et al. (2022) developed a deep learning model for detecting ARRs using ECG recordings' continuous WT (CWT) scalograms. The model employs 2D CNNs to classify these scalograms, focusing on five types of heartbeats: normal, left bundle branch block (LBBB), right bundle branch block (RBBB), atrial premature contraction (APC), and premature ventricular contraction. The model achieved high performance with an average sensitivity of 98.87%, specificity of 99.85%, and accuracy of 99.65%. This approach, which can effectively detect ARRs from short ECG segments, shows promise for enhancing early detection and intervention in cardiovascular health management. Hammad et al. (2021) developed a deep neural network approach to improve ECG-based diagnosis of CVDs. Their method includes robust feature extraction and uses a genetic algorithm to optimize feature extraction and classification techniques. The approach demonstrated an average accuracy improvement of 0.94 and a 0.953 increase in the F1-score compared to existing methods. This indicates that their model effectively analyzes large ECG datasets and could assist in detecting anomalies, enhancing diagnostic accuracy for medical professionals.

11.2.3 Arrhythmia Detection Using Wearables

Research in advanced health monitoring systems continues to thrive. Wearable and remote devices enable ongoing monitoring of physiological and clinical metrics like heart rate, respiration rate, and temperature. These devices often use cloud-based ML and decision support systems to predict critical health conditions. Scirè et al. (2019) introduced a solution that performs sensor data analysis and interpretation directly on wearable devices, thus eliminating the need for cloud connectivity. Their approach utilizes supervised learning to detect heartbeats and classify ARRs and is optimized for an asymmetric multicore embedded processor with a dedicated core for hardware-assisted pattern matching. Evaluations of this embedded system reveal that it achieves high detection rates and performs comparably to cutting-edge algorithms run on standard processors.

IoT technologies facilitate the digital representation of individuals, objects, or physical phenomena that can be accessed online. This capability enables remote information retrieval and data analysis by computational systems to detect patterns, make decisions, or take actions. For instance, doctors can diagnose patients based on data from an IoT system, even if the patients are in distant locations. Cañón-Clavijo et al. (2023) introduced an IoT system to monitor electrocardiogram (ECG) signals and process heart data to generate ARR alerts. The system uses a Polar H10 heart sensor and ML models for heart event classification, along with communication technology for sharing and storing patient data. They outlined the system's architecture and communication between its components. They described the experimental setup, which involved training and evaluating three classification algorithms: RF, CNN, and k-NN. The k-NN algorithm achieved the highest classification accuracy for various arrhythmias: 94% for premature ventricular contraction, 81% for fusion of ventricular beat, and 82% for supraventricular premature beat. It also effectively identified normal beats with 93% accuracy and unclassifiable beats with 97% accuracy.

In specialized healthcare settings for older people, remote ECG monitoring can provide valuable health insights through continuous analytics. Irfan et al. (2023) developed an IoT-based non-contact ECG measurement system that uses flexible fabric electrodes to measure ECG signals from the soles of the feet or the palms of the hands without direct skin contact. Each electrode is shielded with a double layer of isopotential shielding to reduce interference from electromagnetic radiation. Biosignals are stored locally on the IoT device and transmitted to the cloud for processing. An adaptive coding algorithm was implemented to optimize data storage and transmission, achieving a 72% average reduction in data volume. Full data can be reconstructed in the cloud using advanced tools on ThingSpeak for detailed analysis. The system, tested with 35 participants, including elderly individuals, adults, and children, demonstrated accurate ECG measurement with an average heart rate error of ± 1 beat per minute compared to polysomnography data. A comparison of QRS complexes detected from various body parts showed no significant differences.

11.2.4 Arrhythmia Detection Using Quantum Computing

Ozpolat and Karabatak (2023) employed a quantum-based ML method for classifying heart rhythms. They transformed ECG features into qubit structures using PCA. Subsequently, these qubits were processed to automate the identification of ARR using the quantum SVM (QSVM). The implementation and simulations were performed using the IBM Qiskit framework. A comparison between classical SVM and QSVM was performed for different dataset sizes and numbers of qubits. The classical SVM achieved an accuracy of 86.96%, while the QSVM achieved 84.64%. These findings revealed an interesting trade-off between classification performance and computational effectiveness.

11.3 MATERIALS AND METHODS

The block diagram of the devised method is shown in Figure 11.1. Details of different modules are presented below.

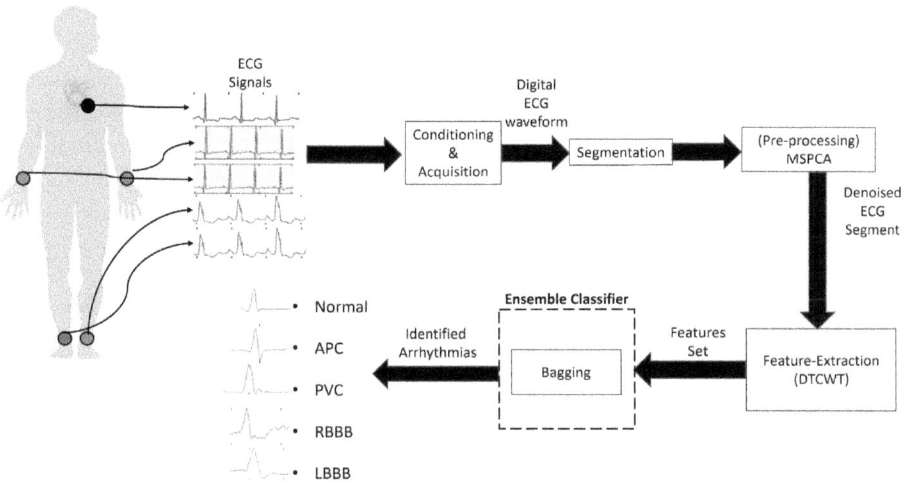

FIGURE 11.1 The proposed system block diagram.

11.3.1 Data

This study utilized the MIT-BIH Arrhythmia Database, a prominent dataset since 1980 for evaluating arrhythmia detectors and classifiers. The database includes 24-hour ECG recordings from 47 subjects, comprising 25 men and 22 women aged between 23 and 89 years. The ECG data were initially collected using Del Mar Avionics model 445 Holter recorders and digitized at 360 samples per second per channel. Each recording features two leads, typically the MLII and V1, positioned on the chest, allowing for the easy removal of 60 Hz interference with digital notch filters. The primary lead is generally MLII, while the secondary precordial lead varies among V1, V2, V4, and V5. The original annotations of the dataset include 16 rhythm classes, providing a solid foundation for analysis. The study focused on five arrhythmic beat types: normal (N), LBBB, RBBB, APC, and premature ventricular contraction (PVC). Each 30-minute recording was divided into 81-sample segments without preprocessing. These classes were extracted from 23 randomly chosen recordings, resulting in 7,500 segments. The model's performance was evaluated using 10-fold cross-validation to ensure robust and reliable results.

11.3.2 Arrhythmias

When the depolarization wave reaches the interventricular septum as expected, the PR interval (the time from the start of the P wave to the first deflection in the QRS complex) remains normal. However, abnormal conduction through the right or left bundle branches—known as a "bundle branch block"—causes a delay in the depolarization of part of the ventricular muscle. This delay lengthens the time needed for complete ventricular depolarization, resulting in a widened QRS complex. Normally, the depolarization wave travels from the interventricular septum to the edges of the ventricles in less than 120 ms, about three small squares on an ECG. A QRS complex duration exceeding 120 ms indicates abnormal and slower conduction within the ventricles (Hampton & Hampton, 2019).

A wide QRS complex can indicate a bundle branch block but may also occur if depolarization starts within the ventricular muscle itself. It is important to recognize that during sinus rhythm with bundle branch block, the P waves remain normal with a consistent PR interval, unlike rhythms originating in the ventricles. Blockage of both bundle branches has a similar effect to a His bundle block, resulting in a complete (third-degree) heart block. RBBB often suggests issues with the right side of the heart, although RBBB patterns with normal QRS durations can sometimes be seen in healthy individuals. In contrast, LBBB consistently indicates heart disease, usually affecting the left ventricle. Identifying bundle branch block is essential because LBBB complicates further ECG analysis, and RBBB makes it more challenging. Understanding the ECG patterns of RBBB and LBBB involves considering that the septum is typically depolarized from left to right, the left ventricle has a larger impact on the ECG due to its greater muscle mass, and excitation spreading toward lead results in an upward deflection on the ECG (Hampton & Hampton, 2019).

11.3.2.1 Right Bundle Branch Block (RBBB)

In RBBB, conduction is impaired in the right bundle branch, but depolarization of the septum from the left side usually proceeds. This causes an R wave in the right ventricular lead (V1) and a small Q wave in the left ventricular lead (V6). Excitation then spreads to the left ventricle, resulting in an S wave in lead V1 and an R wave in lead V6. Due to the disrupted conduction pathway, excitation reaches the right ventricle later than the left, leading to a second R wave (R1) in lead V1 and a wide, deep S wave in lead V6, which results in a widened QRS complex. An "RSR1" pattern with a QRS complex duration of less than 120 ms is sometimes referred to as "partial right bundle branch block," which is generally considered insignificant and can be regarded as a normal variant (Hampton & Hampton, 2019).

11.3.2.2 Left Bundle Branch Block (LBBB)

In LBBB, when conduction is disrupted in the left bundle branch, the septum depolarizes from right to left. This results in a small Q wave in lead V1 and an R wave in lead V6. The right ventricle depolarizes before the left, creating an R wave in lead V1 and an S wave (often just a notch) in lead V6. Any upward deflection, even if minimal, is classified as an R wave, while any downward deflection following an R wave is termed an S wave. Following this, the left ventricle depolarizes, leading to an S wave in lead V1 and another R wave in lead V6. LBBB is commonly associated with T wave inversion in the lateral leads (I, VL, and V5-V6), although this inversion may not always be present in all these leads (Hampton & Hampton, 2019).

11.3.2.3 Premature Ventricular Contractions (PVCs)

PVCs are early heartbeats originating from the ventricles. They occur when the heart's electrical system triggers an impulse from an abnormal focus within the ventricles, causing a premature and usually wide QRS complex. This abnormal beat disrupts the regular rhythm, often leading to a sensation of skipped or extra beats, which may be felt as palpitations. PVCs can occur in healthy individuals without any underlying heart disease, often triggered by stress, caffeine, alcohol, or nicotine. However, frequent PVCs or those associated with dizziness, chest pain, or shortness of breath may indicate a more serious condition requiring medical evaluation. In an electrocardiogram (ECG), PVCs are characterized by their early occurrence, wide QRS complexes, and the absence of a preceding P wave, differentiating them from normal sinus beats. Following a PVC, a compensatory pause typically occurs, allowing the next regular heartbeat to fall back into its normal rhythm. While occasional PVCs are generally benign, frequent or complex forms, such as bigeminy, trigeminy, or couplets, can be a sign of increased risk for more severe ARRs, especially in individuals with underlying heart conditions. PVC treatment varies based on frequency, underlying cause, and associated symptoms, ranging from lifestyle modifications and medications to more advanced interventions like ablation therapy (Hampton & Hampton, 2019).

11.3.2.4 Atrial Premature Contractions (APCs)

APCs, also known as premature atrial contractions, are early heartbeats originating from the atria. These contractions occur when an ectopic focus within the atria fires before the next scheduled sinus impulse, resulting in an early beat that disrupts the normal heart rhythm. On an electrocardiogram (ECG), APCs are identified by a premature P wave, which may have a different shape than normal sinus P waves, followed by a normal or slightly altered QRS complex. This early atrial impulse is typically followed by a compensatory pause, giving the heart a moment to reset before the next regular beat. While often benign, APCs can cause palpitations or a feeling of a skipped heartbeat. In many cases, APCs occur sporadically and are harmless, often triggered by factors such as stress, caffeine, alcohol, or fatigue. However, frequent or symptomatic APCs can be indicative of underlying heart conditions or an increased risk of developing more serious ARR, such as atrial fibrillation. In symptomatic individuals, APCs may cause feelings of fluttering or pounding in the chest, dizziness, or shortness of breath. Treatment for APCs focuses on managing symptoms and underlying triggers. This may include lifestyle modifications, such as reducing caffeine intake and stress management, as well as medications like beta-blockers or calcium channel blockers to help regulate the heart's rhythm. In some cases, more invasive procedures like catheter ablation may be considered if APCs are particularly troublesome and refractory to medical management (Hampton & Hampton, 2019).

11.3.3 Signal Denoising with MSPCA

MSPCA merges PCA and wavelet analysis to remove correlations and identify deterministic features in auto-correlated measurements. By performing PCA on wavelet coefficients across various scales, MSPCA integrates results from these scales to identify significant events over time and frequency, enhancing process monitoring. It filters scores and residuals, sets detection limits to spot deterministic changes, and handles auto-correlated measurements without matrix augmentation or time-series modeling due to the natural de-correlation of wavelet coefficients. This method improves the detection of deterministic changes and identifies abnormal operational characteristics, enhancing monitoring processes (Bakshi, 1998).

Utilizing multi-scale representation is crucial for effective empirical modeling from multi-scale data as it directly impacts model quality. Preprocessing through compression or filtering enhances data quality by reducing irrelevant signal features. Multi-scale methods capture contributions across multiple scales and are expected to yield superior models. Typically, each variable is preprocessed using wavelet thresholding before model development, though this approach separates filtering and modeling, often requiring heuristic and trial-and-error methods. A multi-scale, multi-variate approach improves noise removal by considering the multi-variate nature of empirical modeling problems. While latent variable regression methods reduce errors by retaining relevant variables, PCA alone cannot entirely remove errors, highlighting the need for combined methods (Malinowski, 2002).

Combining PCA with wavelet thresholding enhances noise removal and model quality by discarding noise dependent on variable relationships and eliminating less important latent variables. This dual approach leverages the relationship between wavelet thresholding and latent variable models. By decomposing variables using wavelets and applying PCA to the coefficients, Multi-scale PCA decorrelates relationships and nearly decorrelates stochastic measurements. Establishing PCA models for each scale and using scale-dependent thresholds improves noise removal, especially for non-white or scale-dependent noise. Thresholding criteria and principal component selection ensure consistency and maintain variable relationships across scales. This integrated approach, which reconstructs signals from selected coefficients and discards irrelevant components, enhances noise removal and maintains orthonormality, yielding more accurate predictive models with deeper system insights (Bakshi, 1998; Alsberg et al., 1998; Trygg & Wold, 1998; Bakshi, 1999).

11.3.4 Feature Extraction with Wavelet Packet Decomposition (WPD)

The WT enhances the time resolution of a signal's decomposition by breaking it down into basic functions. Wavelets are generated through dilations and translations of a single function (Vetterli & Herley, 1992). In the CWT for a continuous signal, $x(t)$, the signal is represented as follows (Rioul & Vetterli, 1991; Daubechies, 1990):

$$\mathrm{CWT}_x(\tau, a) = \int_{-\infty}^{\infty} x(at) \frac{1}{\sqrt{a}} \psi\left(\frac{t-\tau}{a}\right) dt, \tag{11.1}$$

Here, $\psi(t)$ denotes the mother wavelet or basic wavelet, while a is the scale factor responsible for translating the function across $x(t)$. The variable τ adjusts the time scale of the probing function ψ. The function $\psi(t)$ is a band-pass function, typically centered around specific frequencies, f_0. The scale a allows for compression or expansion of $\psi(t)$, meaning a larger scale factor compresses the function in time, whereas a smaller scale factor expands it (Rioul & Vetterli, 1991; Daubechies, 1990). When applying a larger scale factor, the same function is compressed in time, whereas a smaller scale factor results in the opposite effect, expanding the function in time (Subasi, 2013; Thakor et al., 2000).

WPD extends the DWT by decomposing low-frequency components, known as approximations, and utilizing both approximants (low-frequency components) and details (high-frequency components) (Unser & Aldroubi, 1996; Daubechies, 1990; Learned & Willsky, 1995). Unlike DWT, which only divides low-frequency components, WPD divides both low and high-frequency components into sub-levels, offering better frequency resolution for the decomposed signal. WPD is a continuous-time wavelet decomposition tested at various frequencies at each level or scale, integrating different decomposition levels to reconstruct the original signal (Kutlu & Kuntalp, 2012).

Wavelet packet decomposition (WPD) is considered an extension of DWT because it decomposes detail coefficients in addition to approximation coefficients. Although this is a subtle difference, it results in different sets of wavelet coefficients for each method. WPD generates 2^j sets of wavelet coefficients for j levels, whereas DWT produces $j+1$ sets of wavelet coefficients (Unser & Aldroubi, 1996). WPD provides superior frequency resolution compared to DWT when decomposing a signal, although it may sometimes result in the loss of important information in higher-frequency components (Kevric & Subasi, 2017). This study uses four levels of decomposition.

One approach to dimensionality reduction involves using statistical measures from the sub-bands of wavelet decomposition. This method includes first, second, third, and fourth-order statistics, resulting in a reduced feature set for ECG signal classification. Key features include the mean absolute values, average power, and standard deviation of signal coefficients in each sub-band, as well as the ratio of mean absolute values between adjacent sub-bands, and the skewness and kurtosis of the signal coefficients.

11.3.5 Ensemble Machine Learning Methods

11.3.5.1 Bagging

Models can be combined to produce a single prediction using methods like voting for classification or averaging for numerical predictions. Bagging involves creating multiple models by training on different random samples of the original dataset, with each sample drawn with replacement. This technique helps stabilize predictions by averaging results from various models, even though these models might be similar and not always independent. In contrast, boosting adjusts model weights based on past performance to focus more on effective models. While bagging creates multiple datasets from the same original data to form an ensemble, it improves performance compared to a single model trained on the original data alone. For numerical predictions, bagging averages the outputs of the individual models rather than using a voting approach (Hall et al., 2011).

11.3.5.2 AdaBoost

Boosting combines multiple models iteratively to enhance learning by focusing on instances that previous models struggled with. Similar to bagging, boosting also employs voting or averaging to combine the outputs of individual models.

However, boosting differs in that each subsequent model is influenced by the performance of those built before it, whereas bagging constructs each model independently. Boosting assigns greater weight to instances handled incorrectly by earlier models, aiming to improve performance gradually across iterations. Unlike bagging, boosting weights the impact of each model based on its performance rather than treating all models equally. The boosting algorithm initially assigns equal weight to all instances in the training data. It proceeds by training a classifier on this data and adjusting the weights of each instance based on whether it was correctly or incorrectly classified. Instances that are classified correctly receive reduced weights, while misclassified instances receive increased weights. This distinguishes between "easy" instances with lower weights and "hard" instances with higher weights. In subsequent iterations, classifiers are trained on the reweighted data, focusing more on correctly classifying the harder instances. The instances' weights are adjusted again based on the new classifier's performance. This iterative process may lead to instances originally considered hard becoming even harder, and vice versa—various outcomes are possible in practice. Throughout each iteration, the weights reflect how frequently instances have been misclassified by the classifiers developed so far. Maintaining a measure of difficulty for each instance effectively generates a sequence of classifiers that complement each other, enhancing overall predictive performance (Hall et al., 2011).

11.3.5.3 MultiBoosting

MultiBoosting is a decision committee technique that combines AdaBoost with wagging, a variant of bagging that is more suitable for specific tasks than traditional bagging. Research shows that MultiBoosting tends to be more effective at reducing errors than using either AdaBoost or wagging alone. Wagging (Bauer & Kohavi, 1999) differs from traditional bagging by requiring a base learning algorithm capable of handling training cases with varying weights. Wagging assigns random weights to instances within each training set instead of using random bootstrap samples to generate training sets. In its original formulation by Bauer and Kohavi (1999), wagging used Gaussian noise to vary the instance weights. However, this approach can reduce some instances' weights to zero, effectively removing those instances from the training set (Webb, 2000).

The observation that bagging and AdaBoost function through different mechanisms and have distinct effects, with both gaining the most benefit from the initial committee members, suggests that combining the two could be advantageous. Because bagging and AdaBoost operate differently, their combination might outperform the individual methods. Specifically, bagging primarily reduces variance, while AdaBoost reduces both bias and variance. Furthermore, evidence suggests that bagging is more effective than AdaBoost at reducing variance (Bauer & Kohavi, 1999). Therefore, combining them could retain AdaBoost's bias reduction while enhancing the variance reduction achieved by AdaBoost. MultiBoosting can be seen as akin to "wagging," a variant of bagging, where committees are formed similar to AdaBoost. While inheriting bias and variance reduction properties from each committee learning algorithm, MultiBoosting also offers potential computational advantages over AdaBoost. Notably, sub-committees in MultiBoosting can be trained in parallel, contrasting with AdaBoost's sequential process, which limits parallelization due to its ability to minimize early termination. In wagging, each classifier is independent of others, facilitating parallel computation—a characteristic MultiBoosting leverages at the sub-committee level (Webb, 2000).

11.4 RESULTS AND DISCUSSION

11.4.1 Performance Evaluation Measures

The evaluation metrics are determined by analyzing the outputs of the classification model and establishing a relationship between these results and the ground truth. In practical scenarios, the availability of datasets, particularly labeled datasets, is often limited. To address this limitation, the dataset is strategically partitioned into subsets using the cross-validation approach (Mian Qaisar, Nisar, et al., 2023). A widely adopted method for assessing the performance and generalizability of AI models is k-fold cross-validation. In this technique, the dataset is randomly divided into k equally sized subsets, referred to as folds. Given the scale of the application and the dataset size, careful consideration is required when selecting the user-defined parameter k. This method ensures that the model maintains consistent and reliable performance across different subsets of the data. In this study, k was set to 10.

The selection of an AI model for a given application is guided by the results of the chosen evaluation metrics, which aid in identifying the most suitable model from the available options. Employing multiple evaluation metrics enhances the likelihood of making an accurate selection (Cichosz, 2014; Sokolova et al., 2006). This study's evaluation metrics include accuracy, F-measure, ROC curve, AUC, and Kappa index. These metrics are calculated using confusion matrices generated through experimental analysis. A confusion matrix is typically a tabular summary that displays the outcomes of binary or multi-class classification tasks (Mian Qaisar, Nisar, et al., 2023). Figure 11.2 illustrates an example of a confusion matrix for a binary classification problem. A model with high true positives (TP) and true negatives (TN), along with low false positives (FP) and false negatives (FN), demonstrates superior performance, as indicated by higher values of the evaluation metrics.

The following equations present the process of computing the evaluation measures considered, such as accuracy, F-measure, AUC, and Kappa index.

$$\text{Accuracy} = \frac{\text{TP} + \text{TN}}{\text{TP} + \text{TN} + \text{FP} + \text{FN}} \tag{11.2}$$

Actual Values

Predicted Values

	Positive	Negative
Positive	True Positive (TP)	False Positive (FP)
Negative	True Negative (TN)	False Negative (FN)

FIGURE 11.2 Confusion matrix for a binary classification problem.

$$F - \text{measure} = \frac{2.\text{TP}}{2.\text{TP} + \text{FP} + \text{FN}} \tag{11.3}$$

$$\text{Pe} = \frac{(\text{TP} + \text{FN}) \times (\text{TP} + \text{FP}) + (\text{FP} + \text{TN}) \times (\text{FN} + \text{TN})}{(\text{TP} + \text{TN} + \text{FP} + \text{FN})^2} \tag{11.4}$$

$$\text{Kappa} = \frac{\text{Po} - \text{Pe}}{1 - \text{Pe}}. \tag{11.5}$$

In Equation 11.5, Po is the same as the accuracy in Equation 11.2.

The AUC of the ROC curve quantifies the classification performance of a model (Chawla et al., 2004). The ROC curve provides a graphical representation of the classifier's discriminative ability, with the true-positive rate (TPR) plotted on the y-axis and the false-positive rate (FPR) plotted on the x-axis. The TPR and FPR are mathematically defined in Equations 11.6 and 11.7:

$$\text{TPR} = \frac{\text{TP}}{\text{TP} + \text{FN}} \tag{11.6}$$

$$\text{FPR} = \frac{\text{FP}}{\text{FP} + \text{TN}} \tag{11.7}$$

11.4.2 Experimental Results

Several ensemble classification configurations are used to process the ECG-based extracted features to precisely identify ARR. The classification system's performance is evaluated based on the ROC area, *F*-measure, accuracy, and Kappa statistics definitions.

The experimental results of the study demonstrate that the application of these classifiers achieves high accuracy on the analyzed dataset. The complete ECG dataset is divided into training and test subsets to evaluate the performance of each classification model. The classification performance is assessed using a 10-fold cross-validation approach.

In ARR identification systems, the primary objective revolves around classifying ECG signal patterns, as the subject's cardiac states elicit specific heartbeat patterns. Notable algorithms employed in the classification process include k-nearest neighbor (k-NN), SVM, RF, ANNs, CARTs, C4.5 decision tree, reduced error pruning (REP) tree, random tree, and LDA tree.

A summary of the classification performance scores, obtained using a bagging-based ensemble configuration of the considered classifiers, is outlined in Table 11.1.

The accuracy scores achieved for the normal class are analyzed first. The highest classification accuracy, 100.00%, is achieved using the k-NNs classifier. The second-highest accuracy, 99.30%, is obtained with the SVM classifier. ANNs and RF demonstrate comparable performance, each achieving an accuracy of 99.00%. The lowest accuracy, 92.27%, is recorded for the LDA tree.

TABLE 11.1
The Obtained Performance Evaluations for the Case of Bagging

	NORMAL	APC	PVC	RBBB	LBBB	AVERAGE	*F*-Measure	AUC	KAPPA
SVM	0.993	0.97	0.977	0.987	0.993	0.984	0.984	0.998	0.980
k-NN	1	0.942	0.925	0.971	1	0.967	0.967	0.995	0.958
ANN	0.99	0.96	0.98	0.983	0.99	0.981	0.981	0.999	0.976
Random forest	0.99	0.953	0.963	0.957	0.967	0.966	0.966	0.998	0.958
CART	0.917	0.89	0.91	0.91	0.957	0.917	0.917	0.99	0.896
C4.5	0.943	0.913	0.913	0.927	0.94	0.927	0.928	0.991	0.909
REP tree	0.953	0.883	0.89	0.937	0.92	0.917	0.917	0.991	0.896
Random tree	0.967	0.943	0.927	0.923	0.94	0.94	0.94	0.993	0.925
LAD tree	0.927	0.827	0.87	0.917	0.93	0.894	0.894	0.987	0.868

For the APC class, the highest classification accuracy, 97.00%, is achieved using the SVM. The second-highest accuracy, 96.00%, is obtained with the ANN. The RF classifier ranks third with an accuracy of 95.30%, while the LDA tree records the lowest accuracy of 82.70%.

For the PVC class, the highest classification accuracy, 98.00%, is achieved by the ANN. The SVM) achieves the second-highest accuracy at 97.70%, followed by the RF classifier with an accuracy of 96.30%. The lowest accuracy, 87.00%, is obtained by the LDA tree.

For the RBBB class, the highest classification accuracy, 98.70%, is achieved using the SVM. The second-highest accuracy, 98.30%, is obtained with the ANN. The k-NN classifier ranks third with an accuracy of 97.10%, while the lowest accuracy, 91.00%, is recorded by the CART model.

For the LBBB class, the highest classification accuracy, 100.00%, is achieved using the k-NN classifier. The SVM achieves the second-highest accuracy at 99.30%, followed by the ANN with an accuracy of 99.00%. The lowest accuracy, 92.00%, is obtained by the REP tree.

The average accuracy scores across the five classes indicate that the highest overall classification accuracy, 98.40%, is achieved by the SVM. The second-highest accuracy, 98.10%, is obtained with the ANN, followed by the k-nearest neighbors (k-NN) classifier with an accuracy of 96.70%. The LDA tree records the lowest average accuracy of 89.40%. SVM demonstrates the highest overall average accuracy of 98.40% among all classifiers.

The average *F*-measure scores across the five classes reveal that the highest *F*-measure value of 0.984 is achieved by the SVM. The second-highest *F*-measure, 0.981, is obtained with the ANN, followed by the k-nearest neighbors (k-NN) classifier with a value of 0.967. The LDA tree records the lowest *F*-measure value of 0.894. Overall, SVM demonstrates the highest average *F*-measure value of 0.984.

The average AUC values across the five classes indicate that the highest AUC value of 0.999 is achieved by the ANN. The second-highest AUC value of 0.998 is

TABLE 11.2
The Obtained Performance Evaluations for the Case of AdaBoost

	NORMAL	APC	PVC	RBBB	LBBB	AVERAGE	*F*-Measure	AUC	KAPPA
SVM	1	0.997	0.99	0.987	0.99	0.989	0.989	0.997	0.986
k-NN	0.997	0.953	0.96	0.957	0.993	0.972	0.972	0.983	0.965
ANN	0.997	0.97	0.98	0.987	0.993	0.985	0.985	0.998	0.982
Random forest	0.987	0.957	0.973	0.96	0.967	0.969	0.969	0.998	0.961
CART	0.97	0.933	0.93	0.953	0.973	0.952	0.952	0.997	0.940
C4.5	0.98	0.95	0.93	0.953	0.947	0.952	0.952	0.997	0.940
REP tree	0.963	0.937	0.937	0.93	0.963	0.946	0.946	0.996	0.933
Random tree	0.87	0.8	0.77	0.813	0.86	0.823	0.823	0.889	0.778
LAD tree	0.96	0.9	0.937	0.933	0.95	0.936	0.936	0.995	0.920

attained by both the SVM and RF. The k-nearest neighbors (k-NN) classifier ranks third with an AUC value of 0.995, while the LDA tree records the lowest AUC value of 0.987.

The average Kappa values across the five classes demonstrate that the highest Kappa value of 0.9800 is achieved by the SVM. The second-highest Kappa value, 0.976, is obtained with the ANN. The k-nearest neighbors (k-NN) classifier ranks third with a Kappa value of 0.958, while the LDA tree records the lowest Kappa value of 0.868.

A summary of the classification performance scores obtained using the AdaBoost-based ensemble configuration of the considered classifiers is outlined in Table 11.2.

Initially, we examine the accuracy scores obtained for the normal class. The highest classification accuracy, 100.00%, is achieved by the SVM. The second-highest accuracy, 99.70%, is attained by both the k-nearest neighbors (k-NN) and the ANN classifiers. The RF classifier ranks third with an accuracy of 98.70%. The random tree model records the lowest accuracy score of 87.00%.

For the APC class, the highest classification accuracy, 99.70%, is achieved by the SVM. The second-highest accuracy, 97.00%, is obtained with the ANN. The k-nearest neighbors (k-NN) classifier ranks third with an accuracy of 95.30%. The random tree model records the lowest accuracy score of 80.00%.

For the PVC class, the highest classification accuracy, 99.00%, is achieved by the SVM. The second-highest accuracy, 98.00%, is obtained with the ANN. The RF classifier ranks third with an accuracy of 97.30%. The random tree model achieves the lowest accuracy score of 77.00%.

For the RBBB class, the highest classification accuracy, 98.70%, is achieved by both the SVM and the ANN. The second-highest accuracy, 96.00%, is attained by the RF. The k-nearest neighbors (k-NN) classifier ranks third with an accuracy of 95.70%, while the random tree model records the lowest accuracy score of 81.30%.

For the LBBB class, the highest classification accuracy, 99.30%, is achieved by both the k-nearest neighbors (k-NN) and ANN classifiers. The second-highest accuracy, 99.00%, is attained by the SVM. The CART ranks third with an accuracy of 97.30%, while the random tree model records the lowest accuracy score of 86.00%.

The average accuracy scores across all five classes indicate that the highest overall classification accuracy, 98.90%, is achieved by the SVM. The second-highest accuracy, 98.50%, is obtained with the ANN. The k-nearest neighbors (k-NN) classifier ranks third with an accuracy of 97.20%, while the random tree model records the lowest accuracy score of 82.30%.

The average *F*-measure scores across all five classes reveal that the highest overall *F*-measure value, 0.989, is achieved by the SVM. The second-highest *F*-measure value, 0.985, is obtained with the ANN. The k-nearest neighbors (k-NN) classifier ranks third with an *F*-measure value of 0.972, while the random tree model records the lowest *F*-measure value of 0.823.

The average AUC values across all five classes indicate that the highest overall AUC value, 0.998, is achieved by both the ANN and RF. The second-highest AUC value, 0.997, is obtained with the SVM, CART, and C4.5. The REP tree ranks third with an AUC value of 0.996, while the random tree model records the lowest AUC value of 0.889.

The average Kappa values across all five classes indicate that the highest overall Kappa value, 0.9858, is achieved by the SVM). The second-highest Kappa value, 0.9817, is obtained with the ANN. The k-nearest neighbors (k-NN) classifier ranks third with a Kappa value of 0.9650, while the random tree model records the lowest Kappa value of 0.7783. Overall, the SVM demonstrates the highest average Kappa value of 0.9858.

A summary of the classification performance scores, achieved using the MultiBoosting ensemble configuration of the considered classifiers, is presented in Table 11.3.

Starting with the accuracy scores for the normal class, the highest classification accuracy of 100.00% is achieved by the SVM. The second-highest accuracy, 99.70%,

TABLE 11.3
The Obtained Performance Evaluations for the Case of MultiBoosting

	NORMAL	APC	PVC	RBBB	LBBB	AVERAGE	*F*-Measure	AUC	KAPPA
SVM	1	0.977	0.99	0.987	0.99	0.989	0.989	0.997	0.986
k-NN	0.997	0.953	0.96	0.957	0.993	0.972	0.972	0.983	0.965
ANN	0.997	0.97	0.98	0.987	0.993	0.985	0.985	0.998	0.982
Random forest	0.987	0.957	0.973	0.96	0.967	0.969	0.969	0.998	0.961
CART	0.97	0.917	0.927	0.93	0.953	0.939	0.939	0.995	0.924
C4.5	0.95	0.913	0.953	0.95	0.973	0.948	0.948	0.996	0.935
REP tree	0.963	0.917	0.93	0.913	0.947	0.934	0.934	0.994	0.918
Random tree	0.87	0.8	0.77	0.813	0.86	0.823	0.823	0.889	0.778
LAD tree	0.94	0.893	0.9	0.91	0.923	0.913	0.914	0.991	0.892

is attained by both the k-nearest neighbors (k-NN) and ANN classifiers. The RF ranks third with an accuracy of 98.70%, while the random tree records the lowest accuracy score of 87.00%.

For the APC class, the highest classification accuracy of 97.70% is achieved by the SVM. The second-highest accuracy, 97.00%, is obtained with the ANN. The RF classifier ranks third with an accuracy of 95.70%, while the random tree model records the lowest accuracy score of 80.00%.

For the PVC class, the highest classification accuracy of 99.00% is achieved by the SVM. The second-highest accuracy, 98.00%, is obtained with the ANN. The RF classifier ranks third with an accuracy of 97.30%, while the random tree model records the lowest accuracy score of 77.00%.

For the RBBB class, the highest classification accuracy of 98.70% is achieved by both the SVM and ANN. The second-highest accuracy, 96.00%, is attained by the RF classifier. The k-nearest neighbors (k-NN) ranks third with an accuracy of 95.70%, while the random tree model records the lowest accuracy score of 81.30%.

For the LBBB class, the highest classification accuracy of 99.30% is achieved by both the k-nearest neighbors (k-NN) and ANN classifiers. The second-highest accuracy, 99.00%, is obtained with the SVM. The C4.5 classifier ranks third with an accuracy of 97.30%, while the random tree model records the lowest accuracy score of 86.00%.

The average accuracy scores across all five classes indicate that the highest classification accuracy of 98.90% is achieved by the SVM. The second-highest accuracy, 98.50%, is attained by the ANN. The k-nearest neighbors (k-NN) classifier ranks third with an accuracy of 97.20%, while the random tree model records the lowest accuracy score of 82.30%. Overall, the SVM achieves the highest average accuracy of 98.90%.

The average *F*-measure scores across all five classes reveal that the highest *F*-measure value of 0.989 is attained by the SVM. The second-highest *F*-measure value, 0.985, is achieved by the ANN. The k-NN ranks third with an *F*-measure value of 0.972, while the random tree classifier has the lowest *F*-measure value of 0.823. In terms of *F*-measure, the SVM stands out with the highest average value of 0.989.

The average AUC values across the five classes show that the highest AUC value of 0.998 is achieved by both the ANN and RF. The second-highest AUC value, 0.997, is attained by the SVM, CART, and C4.5. The C4.5 classifier ranks third with an AUC value of 0.996, while the random tree model records the lowest AUC value of 0.889. Both the ANN and RF achieve the highest average AUC value of 0.998.

The average Kappa values across the five classes indicate that the SVM attains the highest Kappa value of 0.986. The second-highest Kappa value, 0.982, is achieved by the ANN. The k-NN classifier ranks third with a Kappa value of 0.9650, while the random tree model has the lowest Kappa value of 0.778. The SVM demonstrates the highest average Kappa value of 0.986.

11.5 DISCUSSION

The results presented in Section 11.4.2 demonstrate that ensemble classification methods yield high accuracy in most cases. Among the ensemble configurations considered, the combination of MSPCA, WPD-based feature extraction, and SVM with

AdaBoost or MultiBoosting stands out as the best performer, achieving the highest average accuracy of 98.90%.

Additionally, SVM with AdaBoost or MultiBoosting also delivers the highest average *F*-measure value of 0.989, further confirming its superiority. The highest Kappa value of 0.9858 is also achieved with this configuration, highlighting the outstanding performance of SVM in terms of accuracy, *F*-measure, and Kappa index compared to other classifiers.

Although the ANN with bagging achieves the highest overall AUC value of 0.999, the SVM with AdaBoost or MultiBoosting closely follows with an AUC of 0.997, which is only 0.2% lower than the ANN's AUC. Despite this slight difference in AUC, the SVM outperforms the ANN in terms of accuracy, *F*-measure, and Kappa index.

Overall, the SVM with AdaBoost or MultiBoosting is notably more accurate than other classifiers in this study, and it also offers advantages in model development ease. While various ML techniques have been developed due to the complexity of creating practical algorithms for clinical use, alternative algorithms have also gained popularity and are suitable for many scenarios. In clinical practice, performance is the most critical factor in algorithm selection, followed by usability and interpretability. This study suggests that the most accurate ARR diagnosis can be achieved using one of these ensemble-based algorithms.

The choice of feature extraction and classification methods is the two unique considerations that determine the performance of ECG signal categorization. In fact, the performance of classification is mainly based on the preprocessing approach and feature selection method. Therefore, a tactful selection of preprocessing and feature mining pipelines should be made. The classification stage must have parameters that work well for the categorization of ARR by processing the ECG signals. The studied classifiers were chosen based on their evident performance for complex and nonlinear problems. Results have confirmed that among the intended classifiers the SVM with AdaBoost or MultiBoost is capable of more accurate ECG signal classification than other considered classifiers.

The integration of event-driven tools could further improve the performance of the proposed method, particularly in terms of real-time compression and computational efficiency (Mian Qaisar & Alsharif, 2020; Mian Qaisar, 2019, 2021). Additionally, applying metaheuristic-based offline analysis could further enhance dimension reduction (Mian Qaisar et al., 2022; Mian Qaisar, Khan, et al., 2023). Future work could explore the feasibility of incorporating these tools into the proposed approach.

11.6 CONCLUSION

This chapter explores the application of ensemble learning models combined with WPD for ARR detection based on electrocardiogram (ECG) signals. The MSPCA method is utilized to denoise the ECG signals, while WPD is employed for signal decomposition. The experimental study uses five-class ECG signals from the MIT-BIH dataset. A novel hybrid approach is proposed, integrating MSPCA, WPD, SVM, and ensemble learning for ECG ARR classification. Compared to other classifiers, the proposed ensemble model shows reduced susceptibility to overfitting.

The combination of MSPCA, WPD, and SVM with bagging achieves the highest classification accuracy of 98.90% for the five-class classification problem. This method holds the potential for enhancing automated ARR diagnosis from ECG signals.

In conclusion, this study demonstrates the potential of integrating WPD and ensemble classifiers to improve the performance of automated biomedical signal analysis, specifically in ECG-based ARR classification. The promising results highlight the need for continued research and innovation in this area to aid individuals with cardiac conditions. Further advancements in signal processing algorithms and AI-based classification models could unlock even greater potential in healthcare technology, ultimately enhancing the quality of life for those in need.

While ECG-based cardiac monitoring offers significant potential, it does have some limitations, including restricted bandwidth and spatial resolution. Additionally, noise from muscle contractions, powerline interference, and electrode artifacts can impact the ECG signals. Moreover, ECG signals can vary between individuals due to differences in cardiac anatomy and electrode placement. In the future, AI-based assistive solutions will be developed, and their performance will be evaluated while accounting for the impact of the aforementioned noise sources and interferences.

REFERENCES

Alickovic, E., & Subasi, A. (2015). Effect of multiscale PCA de-noising in ECG beat classification for diagnosis of cardiovascular diseases. *Circuits, Systems, and Signal Processing*, *34*(2), 2.

Alickovic, E., & Subasi, A. (2016). Medical decision support system for diagnosis of heart arrhythmia using dwt and random forests classifier. *Journal of Medical Systems*, *40*(4), 4.

Alsberg, B. K., Woodward, A. M., Winson, M. K., Rowland, J. J., & Kell, D. B. (1998). Variable selection in wavelet regression models. *Analytica Chimica Acta*, *368*(1), 29–44.

Bakshi, B. R. (1998). Multiscale PCA with application to multivariate statistical process monitoring. *AIChE Journal*, *44*(7), 1596–1610.

Bakshi, B. R. (1999). Multiscale analysis and modeling using wavelets. *Journal of Chemometrics*, *13*(3–4), 415–434.

Bauer, E., & Kohavi, R. (1999). An empirical comparison of voting classification algorithms: Bagging, boosting, and variants. *Machine Learning*, *36*(1), 105–139.

Cañón-Clavijo, R. E., Montenegro-Marin, C. E., Gaona-Garcia, P. A., & Ortiz-Guzmán, J. (2023). IoT based system for heart monitoring and arrhythmia detection using machine learning. *Journal of Healthcare Engineering*, *2023*(1), 6401673. https://doi.org/10.1155/2023/6401673

Chawla, N. V., Japkowicz, N., & Kotcz, A. (2004). Special issue on learning from imbalanced data sets. *ACM Sigkdd Explorations Newsletter*, *6*(1), 1–6.

Cichosz, P. (2014). *Data Mining Algorithms: Explained Using R*. John Wiley & Sons, New York.

Daubechies, I. (1990). The wavelet transform, time-frequency localization and signal analysis. *IEEE Transactions on Information Theory*, *36*(5), 961–1005.

Dhyani, S., Kumar, A., & Choudhury, S. (2023). Analysis of ECG-based arrhythmia detection system using machine learning. *MethodsX*, *10*, 102195. https://doi.org/10.1016/j.mex.2023.102195

Farooq, A., Seyed Mahmoudian, M., & Stojcevski, A. (2021). A wearable wireless sensor system using machine learning classification to detect arrhythmia. *IEEE Sensors Journal*, *21*(9), 11109–11116. https://doi.org/10.1109/JSEN.2021.3062395

Hall, M., Witten, I., & Frank, E. (2011). *Data Mining: Practical Machine Learning Tools and Techniques*. Kaufmann, Burlington.

Hammad, M., Abd El-Latif, A. A., Hussain, A., Abd El-Samie, F. E., Gupta, B. B., Ugail, H., & Sedik, A. (2022). Deep learning models for arrhythmia detection in IoT healthcare applications. *Computers and Electrical Engineering*, *100*, 108011. https://doi.org/10.1016/j.compeleceng.2022.108011

Hampton, J., & Hampton, J. (2019). *The ECG Made Easy E-Book: The ECG Made Easy E-Book*. Elsevier Health Sciences, London.

Hussain, L., Aziz, W., Saeed, S., Awan, I. A., Abbasi, A. A., & Maroof, N. (2020). Arrhythmia detection by extracting hybrid features based on refined Fuzzy entropy (FuzEn) approach and employing machine learning techniques. *Waves in Random and Complex Media*, *30*(4), 656–686. https://doi.org/10.1080/17455030.2018.1554926

Irfan, M., Shun, P., Felix, B. B., Mustafa, N., Abbasi, S. F., Nahli, A., Subasi, A., Westerlund, T., & Chen, W. (2023). An IoT-based noncontact ECG system: Sole of the feet/hands palm. *IEEE Internet of Things Journal*, *10*(21), 18718–18732.

Irfan, S., Anjum, N., Althobaiti, T., Alotaibi, A. A., Siddiqui, A. B., & Ramzan, N. (2022). Heartbeat classification and arrhythmia detection using a multi-model deep-learning technique. *Sensors*, *22*(15), 606. https://doi.org/10.3390/s22155606

Kevric, J., & Subasi, A. (2017). Comparison of signal decomposition methods in classification of EEG signals for motor-imagery BCI system. *Biomedical Signal Processing and Control*, *31*, 398–406.

Kumari, C. U., Ankita, R., Pavani, T., Vignesh, N. A., Varma, N. T., Manzar, M. A., & Reethika, A. (2020). Heart rhythm abnormality detection and classification using machine learning technique. *2020 4th International Conference on Trends in Electronics and Informatics (ICOEI)*, *48184*, 580–584. https://doi.org/10.1109/ICOEI48184.2020.9142914

Kutlu, Y., & Kuntalp, D. (2012). Feature extraction for ECG heartbeats using higher order statistics of WPD coefficients. *Computer Methods and Programs in Biomedicine*, *105*(3), 257–267.

Learned, R. E., & Willsky, A. S. (1995). A wavelet packet approach to transient signal classification. *Applied and Computational Harmonic Analysis*, *2*(3), 265–278.

M. Hammad, A. M. Iliyasu, A. Subasi, E. S. L. Ho, & A. A. A. El-Latif. (2021). A multitier deep learning model for arrhythmia detection. *IEEE Transactions on Instrumentation and Measurement*, *70*, 1–9. https://doi.org/10.1109/TIM.2020.3033072

Malinowski, E. R. (2002). *Factor Analysis in Chemistry*. Wiley, Berlin.

Mian Qaisar, S. (2019). Efficient mobile systems based on adaptive rate signal processing. *Computers & Electrical Engineering*, 79, 106462.

Mian Qaisar, S. (2021). Signal-piloted processing and machine learning based efficient power quality disturbances recognition. *PLoS One*, *16*(5), e0252104.

Mian Qaisar, S., & Alsharif, F. (2020). Signal piloted processing of the smart meter data for effective appliances recognition. *Journal of Electrical Engineering & Technology*, *15*(5), 2279–2285.

Mian Qaisar, S., Khan, S. I., Dallet, D., Tadeusiewicz, R., & Pławiak, P. (2022). Signal-piloted processing metaheuristic optimization and wavelet decomposition based elucidation of arrhythmia for mobile healthcare. *Biocybernetics and Biomedical Engineering*, *42*(2), 681–694.

Mian Qaisar, S., Khan, S. I., Srinivasan, K., & Krichen, M. (2023). Arrhythmia classification using multirate processing metaheuristic optimization and variational mode decomposition. *Journal of King Saud University-Computer and Information Sciences*, *35*(1), 26–37.

Mian Qaisar, S., Nisar, H., & Subasi, A. (2023). *Advances in Non-Invasive Biomedical Signal Sensing and Processing with Machine Learning*. Springer, Berlin.

Mian Qaisar, S., & Subasi, A. (2020). Cloud-based ECG monitoring using event-driven ECG acquisition and machine learning techniques. *Physical and Engineering Sciences in Medicine*, *43*(2), 623–634.

Mohanty, M., Biswal, P., & Sabut, S. (2020). Machine learning approach to recognize ventricular arrhythmias using VMD based features. *Multidimensional Systems and Signal Processing*, *31*(1), 49–71. https://doi.org/10.1007/s11045-019-00651-w

Mohonta, S. C., Motin, M. A., & Kumar, D. K. (2022). Electrocardiogram based arrhythmia classification using wavelet transform with deep learning model. *Sensing and Bio-Sensing Research*, *37*, 100502. https://doi.org/10.1016/j.sbsr.2022.100502

Montenegro, L., Abreu, M., Fred, A., & Machado, J. M. (2022). Human-assisted vs. deep learning feature extraction: An evaluation of ECG features extraction methods for arrhythmia classification using machine learning. *Applied Sciences*, *12*(15), 404. https://doi.org/10.3390/app12157404

Nandhini Abirami, R., & Durai Raj Vincent, P. M. (2020). Cardiac arrhythmia detection using ensemble of machine learning algorithms. In K. N. Das, J. C. Bansal, K. Deep, A. K. Nagar, P. Pathipooranam, & R. C. Naidu (Eds.), *Soft Computing for Problem Solving* (pp. 475–487). Springer, Singapore.

Ozpolat, Z., & Karabatak, M. (2023). Performance evaluation of quantum-based machine learning algorithms for cardiac arrhythmia classification. *Diagnostics*, *13*(6), 1099. https://doi.org/10.3390/diagnostics13061099

Rajkumar, A., Ganesan, M., & Lavanya, R. (2019). Arrhythmia classification on ECG using deep learning. *2019 5th International Conference on Advanced Computing & Communication Systems (ICACCS)*, Coimbatore, India, pp. 365–369. https://doi.org/10.1109/ICACCS.2019.8728362

Rioul, O., & Vetterli, M. (1991). Wavelets and signal processing. *IEEE Signal Processing Magazine*, *8*, 14–38.

Sannino, G., & De Pietro, G. (2018). A deep learning approach for ECG-based heartbeat classification for arrhythmia detection. *Future Generation Computer Systems*, *86*, 446–455. https://doi.org/10.1016/j.future.2018.03.057

Scirè, A., Tropeano, F., Anagnostopoulos, A., & Chatzigiannakis, I. (2019). Fog-computing-based heartbeat detection and arrhythmia classification using machine learning. *Algorithms*, *12*(2), 32. https://doi.org/10.3390/a12020032

Singh, V., Tewary, S., Sardana, V., & Sardana, H. K. (2019). Arrhythmia detection—A machine learning based comparative analysis with MIT-BIH ECG data. *2019 IEEE 5th International Conference for Convergence in Technology (I2CT)*, Bombay, India, pp. 1–5. https://doi.org/10.1109/I2CT45611.2019.9033665

Sokolova, M., Japkowicz, N., & Szpakowicz, S. (2006). Beyond accuracy, F-score and ROC: A family of discriminant measures for performance evaluation. *Advances in Artificial Intelligence*, *4304*, 1015–1021.

Subasi, A. (2013). Classification of EMG signals using PSO optimized SVM for diagnosis of neuromuscular disorders. *Computers in Biology and Medicine*, *43*(5), 576–586.

Swapna, G., Soman, K. P., & Vinayakumar, R. (2018). Automated detection of cardiac arrhythmia using deep learning techniques. *International Conference on Computational Intelligence and Data Science*, *132*, 1192–1201. https://doi.org/10.1016/j.procs.2018.05.034

Taloba, A. I., Alanazi, R., Shahin, O. R., Elhadad, A., Abozeid, A., & Abd El-Aziz, R. M. (2021). Machine algorithm for heartbeat monitoring and arrhythmia detection based on ECG systems. *Computational Intelligence and Neuroscience*, *2021*(1), 7677568. https://doi.org/10.1155/2021/7677568

Thakor, N. V., Gramatikov, B., Sherman, D., & Bronzino, J. (2000). Wavelet (time-scale) analysis in biomedical signal processing. *The Biomedical Engineering Handbook*, *56*, 1–56.

Trygg, J., & Wold, S. (1998). PLS regression on wavelet compressed NIR spectra. *Chemometrics and Intelligent Laboratory Systems*, *42*(1–2), 209–220. https://doi.org/10.1016/S0169-7439(98)00013-6

Tuncer, T., Dogan, S., Plawiak, P., & Subasi, A. (2022). A novel discrete wavelet-concatenated mesh tree and ternary chess pattern based ECG signal recognition method. *Biomedical Signal Processing and Control*, *72*, 103331.

Ullah, W., Siddique, I., Zulqarnain, R. M., Alam, M. M., Ahmad, I., & Raza, U. A. (2021). Classification of arrhythmia in heartbeat detection using deep learning. *Computational Intelligence and Neuroscience*, *2021*(1), 2195922. https://doi.org/10.1155/2021/2195922

Unser, M., & Aldroubi, A. (1996). A review of wavelets in biomedical applications. *Proceedings of the IEEE*, *84*(4), 626–638.

Vetterli, M., & Herley, C. (1992). Wavelets and filter banks: Theory and design. *IEEE Transactions on Signal Processing*, *40*(9), 2207–2232.

Webb, G. I. (2000). Multiboosting: A technique for combining boosting and wagging. *Machine Learning*, *40*(2), 159–196.

Yıldırım, Ö., Pławiak, P., Tan, R.-S., & Acharya, U. R. (2018). Arrhythmia detection using deep convolutional neural network with long duration ECG signals. *Computers in Biology and Medicine*, *102*, 411–420. https://doi.org/10.1016/j.compbiomed.2018.09.009

Zhou, R., Li, X., Yong, B., Shen, Z., Wang, C., Zhou, Q., Cao, Y., & Li, K.-C. (2019). Arrhythmia recognition and classification through deep learning-based approach. *International Journal of Computational Science and Engineering*, *19*(4), 506–517. https://doi.org/10.1504/IJCSE.2019.101897

12 EEG-based Biometric Authentication Using Wavelet Packet Decomposition and Ensemble Classifiers

Abdulhamit Subasi, Muhammed Enes Subasi, and Saeed Mian Qaisar

12.1 INTRODUCTION

In today's rapidly advancing technological landscape, security systems play a critical role in protecting both information and institutions. These systems are built around three core approaches: biometric-based methods (such as fingerprint or retina scanning), belonging-based methods (like magnetic cards and wristbands), and information-based methods (such as passwords). Biometric technologies, which rely on unique physical or biological characteristics for identification, have become increasingly important in addressing challenges like identity theft and fraud. With advancements in sensors and artificial intelligence-powered software, these systems have found widespread use in border control, financial security, and access management. While physical biometric traits like fingerprints remain constant throughout a person's life, bioelectrical signals like EEG (electroencephalography) are becoming more prominent due to their reliability and personalization. EEG signals stand out as stable and distinctive biometric markers, especially when enhanced through feature selection. Although some biometric features may not work for everyone, EEG-based systems offer greater accessibility, including for individuals with disabilities. Additionally, EEG data provides higher security, as it is more difficult to capture and misuse compared to other biometric types. With brain–computer interface (BCI) technology, EEG signals can also be used to control external devices, further broadening their applications in security. The potential for EEG biometrics to offer enhanced security and privacy drives ongoing research and development in this promising area (Tatar, 2023).

In computer science and cryptography, authentication is the process of confirming an entity's claimed identity, which is vital in managing access to physical and digital resources such as buildings, rooms, or computer systems. A common authentication method involves using distinct biological characteristics. Various biometric factors

DOI: 10.1201/9781003611448-17

are employed, ranging from fingerprints to palm prints, irises, and voice recognition. More recently, there has been growing interest in utilizing biometric signals like EEG, electrocardiogram, and EMG for authentication (Abo-Zahhad et al., 2015) (Blasco et al., 2016). These signals offer key advantages, particularly in their resistance to spoofing. The electroencephalogram (EEG) stands out due to its unique properties, distinguishing it from other biometric factors (Bidgoly et al., 2020).

Conventional biometric systems like fingerprints and iris scans are vulnerable to replication; for instance, attackers can forge fingerprints using photographs of fingers (Galbally et al., 2013). In contrast, EEG signals are not easily accessible or replicable by intruders (Ruiz-Blondet et al., 2016). EEG signals are influenced by the user's emotional states, meaning brainwave patterns can shift under stress or fear. As a result, EEG biometrics cannot be coerced under duress, unlike other authentication methods where an adversary could force a user to grant access. Additionally, EEG-based systems offer the advantage of ensuring that the user is alive since EEG signals rely on active brain activity, unlike other biometric methods that could be compromised using a deceased individual's body (Bidgoly et al., 2020).

EEG-based authentication is finding applications in various fields within computer science (Yang & Deravi, 2017) and is being recognized as a valuable tool for enhancing system security. Klonovs et al. (2013) developed a mobile application that increases smartphone security through two-factor authentication, combining traditional methods with EEG-based verification. Beyond traditional uses, EEG-based authentication is gaining attention in specialized areas, such as its integration with BCI technology (Lin et al., 2017; Ramzan & Shidlovskiy, 2018). BCIs create a direct communication link between the brain and electronic devices, enabling control via brainwaves. By integrating EEG-based authentication, these systems can verify a user's identity alongside executing commands. Furthermore, EEG signals are also being explored for security applications beyond authentication, such as cryptographic key generation. Each key, unique to the individual and task, offers stronger attack resistance than other encryption systems (Ravi et al., 2007; Nguyen et al., 2019).

With advancements in portable medical device technology, such as EEG headsets, and EEG's unique benefits, this biometric modality has gained considerable attention in recent years. There has been a notable increase in publications, reflecting the growing research interest, while industries are also keen to incorporate EEG into their products. For instance, companies like NeuroSky are actively researching and developing sensors for EEG-based applications. Researchers are also exploring EEG wearable systems using equipment from companies like Emotiv and NeuroSky, and even integrating EEG functionality into devices like Google Glass (Moore, 2016).

An EEG-based authentication system typically follows several stages. First, the user performs a specific task while their EEG signals are recorded. These signals then undergo preprocessing to improve their quality. After preprocessing, the enhanced signals are used to train a classifier. Traditionally, shallow classifiers like support vector machine (SVM) and k-nearest neighbors (k-NN) were widely used. However, in recent years, there has been a shift toward deep learning approaches. In shallow models, feature extraction is a separate step required before training, whereas deep learning models integrate this step, allowing the model to be trained directly on the preprocessed data (Bidgoly et al., 2020). This paper uses brain signals for biometric authentication, employing WPD and ensemble machine learning techniques.

Section 12.2 reviews the literature. Data and WPD feature extraction techniques for enhancing signals and ensemble classifiers are explained in Section 12.3. Section 12.4 presents the results and discussion. The chapter concludes in Section 12.5.

12.2 LITERATURE REVIEW

Authentication is vital for security, and various recognition systems exist. Traditional methods, like passwords, are easily stolen, while biometric systems, though more secure, have limitations, such as those found in fingerprint, facial, iris, and voice recognition. Electroencephalogram (EEG) signals have shown promise as a unique biometric method. Ong et al. (2018) analyzed EEG signals using power spectral density and classified them with a k-NN classifier. Accuracy rates for the alpha (8–13 Hz), beta (13–30 Hz), combined alpha and beta (8–30 Hz), and combined theta, alpha, beta, and gamma (4–40 Hz) frequency bands were assessed. Overall accuracy was over 80%, with the combined theta, alpha, beta, and gamma bands achieving the highest accuracy at 89.21%.

EEG brainwave signals reflect human brain activity and are confidential, sensitive, and hard to replicate, making them a promising choice for secure biometric identification and authentication. Gui et al. (2014) presented a framework for EEG-based biometric security, proposing noise reduction through ensemble averaging and low-pass filtering, feature extraction via WPD, and classification using an artificial neural network (ANN). They examined four scenarios for different authentication applications. Results show that classification rates can reach about 90% for distinguishing a subject or small group from others but drop below 11% for individual recognition from larger groups. A side-by-side method boosts overall identification rates to around 40%.

Over the past decade, interest in using brain signals for biometric recognition, particularly through EEG, has increased significantly. However, current recognition rates still fall short compared to traditional biometric traits, posing challenges for real-world applications. Maiorana (2020) examined the effectiveness of deep learning techniques—specifically convolutional and recurrent neural networks, as well as hybrid approaches—in extracting unique features from EEG data to create personal identifiers. To assess these methods' robustness, a series of challenging experimental tests were conducted to identify individuals based on their brain signals, regardless of the mental tasks performed, and to compare data collected over more than a year. The findings suggest that the proposed networks can effectively utilize the dynamic temporal characteristics of EEG signals, achieving high accuracy in brain-based biometric recognition.

The energy from brain potentials triggered by visual stimuli is being explored as a new biometric method. Palaniappan and Mandic (2007) introduced feature extraction and classification enhancements by using spatial data and sensor fusion to identify the most informative EEG electrodes with the Davies–Bouldin index. Simple black-and-white drawings of common objects are visual stimuli for the experiments.

An Elman neural network classifies the EEG energy features, with simulations on 1,600 raw EEG signals showing that 35 informative channels achieve a maximum recognition rate of 98.56 ± 1.87%. This research underscores the considerable potential of EEG biometrics, especially its robustness against fraud.

Maiorana and Campisi (2017) presented the first comprehensive study examining how aging affects the discriminative power of EEG signals across various elicitation protocols, subject counts, acquisition sessions, and time spans. They reported findings from experiments involving 45 subjects, with EEG data collected in five to six sessions over three years using four different protocols. Their analysis utilized statistical and performance-related methods, including various EEG features and hidden Markov models for classification. They also evaluated each EEG channel's uniqueness and permanence, ranking their importance for biometric applications to optimize their use. Additionally, they proposed countermeasures to mitigate aging effects on recognition performance and evaluate their effectiveness, paving the way for future cognitive recognition systems utilizing brain-based biometric traits.

Singh et al. (2015) proposed a method to identify relationships among signals from different brain regions that create unique patterns. They used magnitude squared coherence as a feature, which is robust against random amplitude fluctuations. The study optimizes the number of EEG channels for identification while maintaining accuracy, conducting experiments on various channel combinations until accuracy drops below 95%. A k-NN ($K = 1$) classifier was utilized, demonstrating that while previous studies achieved 100% accuracy with 64 channels on 108 subjects, this study reached the same accuracy on 109 subjects using just 10 channels. These findings indicate that 10 channels are sufficient for biometric identification in secure environments.

Safont et al. (2012) presented a biometric method for authorization and identification using EEG signals. The hardware features a simple setup with two signal electrodes and a reference electrode, ensuring minimal intrusion for the subject. Various features from the processed EEG signals are analyzed using multiple classifiers. The system combines all classifier and feature combinations, improving classification performance even with few observation vectors. Tested on 50 subjects and 20 intruders, the system achieved an equal error rate of 2.4% in just a few seconds, outperforming existing EEG-based systems.

Jayarathne et al. (2016) proposed a novel approach using electroencephalogram (EEG) brain signals for authentication. This method leverages distinctive EEG features to differentiate brain activities, offering greater robustness and security than traditional biometrics due to variations based on mental state. Users visualize a number while their EEG signals are recorded, which are then used to train the system for comparison during authentication. They focused on the combined alpha and beta frequency band (8–30 Hz) across all EEG channels, which is optimal for BCI. Common spatial patterns are used as training features, with linear discriminant analysis as the classification algorithm. Each user's trained model is stored in a parameter database, and users select their profile before the authentication session. Ultimately, a trained model authenticates users based on their memorization of four predetermined digits thought of during the initial process, achieving an accuracy of 96.97%.

EEG-based human recognition is becoming a popular biometric authentication method due to its liveliness and robustness against falsification. However, comprehensive studies on human authentication using EEG signals are still lacking. Meanwhile, low-cost wireless EEG recording devices are emerging, offering potential for various applications, but their feasibility requires further research. Bashar et al. (2016) proposed a method for human identification using EEG signals from these devices. The signals are first preprocessed to remove noise and artifacts using a band-pass FIR filter and then segmented. Three feature extraction techniques—multiscale shape description, multiscale wavelet packet statistics, and multiscale wavelet packet energy statistics—are applied. These features train a supervised error-correcting output code multi-class model (ECOC) with an SVM classifier, enabling individual recognition from test EEG signals. Preliminary experiments with nine EEG records show a true-positive rate (TPR) of 94.44% for the proposed method.

Mao et al. (2017) presented a novel convolutional neural network (CNN) model for biometric identification using electroencephalogram (EEG) signals, addressing the need for a straightforward and reliable approach in practical applications. The CNN operates directly on raw EEG data, eliminating manual feature engineering, and is evaluated on datasets from 100 subjects in a driving fatigue experiment. The model demonstrates fast training efficiency, requiring less than 0.5 h for over 100,000 epochs, and achieves an impressive 97% accuracy in identifying around 14,000 testing epochs, outperforming randomly sampled epochs that only reached 90% accuracy. This research underscores the potential of deep learning techniques for effective EEG-based biometric identification in real-world scenarios.

Chen et al. (2020) introduced robust models for biometric identification using electroencephalogram (EEG) signals by employing a new CNN called global spatial and local temporal filter CNN (GSLT-CNN). The GSLT-CNN processes raw EEG data without requiring feature engineering and is evaluated on datasets from 157 subjects across four experiments that measure various brain states, including driving fatigue and emotion, as well as time-locked responses like the rapid serial visual response (RSVP). The performance of the GSLT-CNN is compared to SVM, bagging tree, and linear discriminant analysis models, demonstrating its efficiency by completing over 279,000 training epochs in under 0.5 h while achieving 96% accuracy in identifying subjects. This accuracy outperforms the best SVM results by 3% on power spectral density features, 10% on autoregressive features, and 23% over a conventional CV-CNN on raw EEG data. The model also shows superior cross-session identification accuracy, achieving 99% with RSVP data compared to 97% for single-session driving fatigue data and 90% for random brain states, indicating the potential of the RSVP design for enhancing identification accuracy.

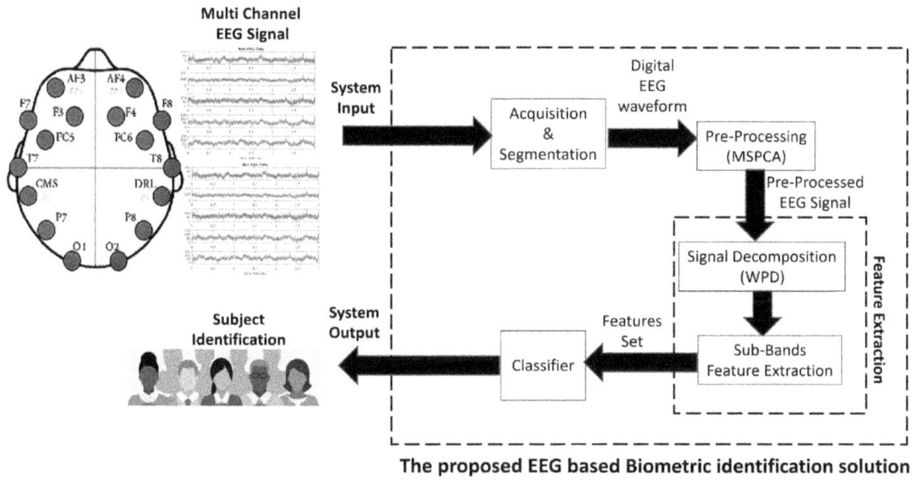

FIGURE 12.1 The proposed system block-level diagram.

12.3 MATERIALS AND METHODS

The principle of the proposed solution is depicted in Figure 12.1. A description of different system modules is provided below.

12.3.1 Dataset

Our experiment used the BCI2000 dataset to evaluate the framework's performance under various conditions. The Physionet EEG Motor Movement/Imagery Dataset provides a large number of subjects, enabling us to test the framework's robustness across a diverse user population. This dataset, recorded via the BCI2000 system, includes data from subjects either performing real movements or imagining movements such as opening and closing their fists or feet. Our experiment specifically focused on imagination tasks, where subjects were instructed to imagine opening and closing their left or right fists or both fists or feet based on visual cues displayed on the screen. The data was sampled at 160 Hz across 64 channels and included records from 109 subjects. Each subject completed six runs of 15 trials within a single session, although subjects 88 and 92 were excluded due to incomplete data. For the experiment, we captured EEG data during the imagination phase, selecting signals with a duration of 4 s per trial (Schalk et al., 2004).

12.3.2 Signal Denoising with MSPCA

Multiscale principal component analysis (MSPCA) combines principal component analysis (PCA) and wavelet analysis to effectively remove correlations and extract deterministic features from auto-correlated measurements. By applying PCA to wavelet coefficients across multiple scales, MSPCA synthesizes these results to detect significant events in both time and frequency domains, thus improving process monitoring. It filters scores and residuals, setting thresholds to identify deterministic changes while handling auto-correlated data without requiring matrix augmentation or time-series modeling, due to the natural decorrelation provided by wavelet coefficients. This approach enhances the detection of deterministic shifts and highlights abnormal operational behaviors, improving monitoring capabilities (Bakshi, 1998).

It is critical to employ multiscale representations for successful empirical modeling with multiscale data, as it directly influences model quality. Preprocessing methods like compression or filtering improve data quality by removing irrelevant signal components. Multiscale techniques, which capture contributions at various scales, typically result in better models. Each variable is often preprocessed through wavelet thresholding prior to model creation, though this process separates filtering from modeling, often relying on heuristic and trial-and-error methods. A multiscale, multivariate strategy offers better noise removal by addressing the multivariate nature of empirical modeling challenges. While latent variable regression methods mitigate errors by retaining critical variables, PCA alone is insufficient for eliminating all errors, highlighting the need for integrated approaches (Malinowski, 2002).

The combination of PCA with wavelet thresholding improves noise reduction and model accuracy by eliminating noise tied to variable relationships and discarding less significant latent variables. This approach uses the link between wavelet thresholding and latent variable models. MSPCA decomposes variables with wavelets and applies PCA to the coefficients, thereby decorrelating relationships and nearly decorrelating stochastic measurements. PCA models are created for each scale, and scale-specific thresholds further enhance noise removal, particularly for non-white or scale-dependent noise. Threshold criteria and principal component selection ensure consistency while preserving variable relationships across scales. This integrated method reconstructs signals from selected coefficients while discarding irrelevant components, enhancing noise removal, maintaining orthonormality, and yielding more accurate, predictive models with deeper insights into system behavior (Bakshi, 1998; Alsberg et al., 1998; Trygg & Wold, 1998; Bakshi, 1999).

12.3.3 Feature Extraction with Wavelet Packet Decomposition

The wavelet transform enhances the time resolution of a signal's decomposition by breaking it down into basic functions. Wavelets are generated through dilations and translations of a single function (Vetterli & Herley, 1992). In the continuous wavelet transform for a continuous signal, $x(t)$, the signal is represented as follows (Rioul & Vetterli, 1991; Daubechies, 1990):

$$CWT_x(\tau,a) = \int_{-\infty}^{\infty} x(at)\frac{1}{\sqrt{a}}\psi\left(\frac{t-\tau}{a}\right)dt, \tag{12.1}$$

Here, $\psi(t)$ denotes the mother wavelet, or basic wavelet, while a is the scale factor responsible for translating the function across $x(t)$. The variable τ adjusts the time scale of the probing function, ψ. The function $\psi(t)$ is a band-pass function, typically centered around specific frequencies, f_0. The scale a allows for compression or expansion of $\psi(t)$, meaning a larger scale factor compresses the function in time, whereas a smaller scale factor expands it (Rioul & Vetterli, 1991; Daubechies, 1990). When applying a larger scale factor, the same function is compressed in time, whereas a smaller scale factor results in the opposite effect, expanding the function in time (Subasi, 2013; Thakor et al., 2000).

WPD extends the discrete wavelet transform (DWT) by decomposing low-frequency components, known as approximations, and utilizing both approximants (low-frequency components) and details (high-frequency components) (Unser & Aldroubi, 1996; Daubechies, 1990; Learned & Willsky, 1995). Unlike DWT, which only divides low-frequency components, WPD divides low- and high-frequency components into sub-levels, offering better frequency resolution for the decomposed signal. WPD is a continuous-time wavelet decomposition tested at various frequencies at each level or scale, making it useful for integrating different decomposition levels to reconstruct the original signal (Kutlu & Kuntalp, 2012).

WPD is considered an extension of DWT because it decomposes detail coefficients in addition to approximation coefficients. Although this is a subtle difference, it results in different sets of wavelet coefficients for each method. WPD generates 2^j sets of wavelet coefficients for j levels, whereas DWT produces $j+1$ sets of wavelet coefficients (Unser & Aldroubi, 1996). WPD provides superior frequency resolution compared to DWT when decomposing a signal, although it may sometimes result in the loss of important information in higher frequency components (Kevric & Subasi, 2017). This study uses four levels of decomposition.

WPD is a powerful signal processing tool that offers a refined analysis of non-stationary signals like EEG, which exhibit complex time–frequency characteristics. The rationale for its selection in EEG-based biometric identification lies in its ability to provide enhanced time–frequency resolution, extract discriminative features, achieve notable dimension reduction, demonstrate robustness against noise, and adapt to diverse EEG characteristics. The WPD-generated sub-bands are analyzed for further dimension reduction. One method for reducing dimensions involves utilizing the first-, second-, third-, and fourth-order statistics of the sub-bands. This reduced feature set is derived from the sub-bands of the wavelet decomposition. Six statistical features are used for EEG signal classification. These features are listed in Table 12.1.

TABLE 12.1
The Extracted Features from Each Considered Sub-Band

Feature No.	Feature Definition
1	Mean absolute values of the sub-band coefficients
2	The average power of the sub-band coefficients
3	The standard deviation of sub-band coefficients
4	Skewness of the sub-band coefficients
5	Kurtosis of the sub-band coefficients
6	The ratio of the absolute mean values of sub-band coefficients in adjacent sub-bands

12.3.4 Ensemble Learning

Ensemble methods aim to enhance classification performance by combining the predictions of multiple classifiers. This approach can reduce variance and bias in classifications while diminishing reliance on any training set's characteristics. Ensemble methods are categorized into two groups based on their construction: dependent and independent methods. In dependent methods, the output of subsequent classifiers is influenced by earlier ones, while independent methods yield outputs separately before combining them for the final ensemble output. Examples of dependent methods include boosting and AdaBoost algorithms, while independent methods include bagging, boosting, and random forest (RF) algorithms (Onan et al., 2016). To assess the performance of the proposed classifier ensemble, we utilized standard ensemble learning methods such as AdaBoost, Bagging, MultiBoosting, random subspace, rotation forest (RoF) algorithms, and majority voting.

12.3.4.1 Bagging

Bagging, or bootstrap aggregating, is a resampling technique that inputs the original training dataset to create several slightly varied datasets. Denote the original training dataset as Let D_0 be the training dataset and D_i be the new training datasets obtained, where $i = 1, \ldots, m$, and m represent the number of desired classifiers. The classifiers are labeled C_i, and may utilize the same algorithm. Each D_i is generated by randomly sampling instances from D_0 using a uniform distribution with replacement, meaning that instances can be selected multiple times for each new dataset D_i. Each of these datasets is then used to train the corresponding classifier C_i, and a voting mechanism is employed to combine the outputs of all m classifiers. Regarding sampling probability, each instance has a probability $P_1 = 1 - \frac{1}{n}$ of not being chosen in a single sampling, where n is the number of instances in D_0. Consequently, the likelihood that an instance is not selected at all in a dataset of n instances can be expressed as $P_n = \left(1 - \frac{1}{n}\right)^n$. As n approaches infinity, P_n converges to $\frac{1}{e} \approx 0.368$. Therefore, when the original dataset is large, any bootstrapped set D_i typically consists of about 63% unique samples from D_0. Using bagging for classification significantly enhances accuracy compared to a single trained classifier, as the training sets D_i exhibit reduced variance, thereby minimizing the risk of overfitting. Additionally, bagging is advantageous in handling noisy data effectively (Leon et al., 2017).

12.3.4.2 AdaBoost

The AdaBoost algorithm starts by assigning equal weights to all instances in the training dataset. The learning algorithm is then used to create a classifier, after which the instances are reweighted based on the classifier's performance. Correctly classified instances have their weights reduced, while misclassified ones are given higher weights. This creates two groups: "easy" instances with low weight and "hard" ones with higher weight. In subsequent iterations, new classifiers are trained on the reweighted data, shifting the focus toward correctly classifying the harder instances. The weights are updated again after each iteration, based on the new classifier's output. As a result, some difficult instances may become even harder to classify, while others might become easier, with all possibilities occurring in practice. After each iteration, the weights reflect how frequently the instances have been misclassified. By continuously adjusting the "hardness" of each instance, AdaBoost generates a sequence of classifiers that effectively complement each other (Hall et al., 2011).

12.3.4.3 MultiBoosting

MultiBoosting is a decision committee technique that integrates AdaBoost with wagging, a variation of bagging that is considered more appropriate for specific tasks than traditional bagging. Studies show that MultiBoosting typically outperforms AdaBoost and wagging individually in reducing error rates. Wagging, introduced by Bauer and Kohavi (Bauer & Kohavi, 1999), differs from bagging by assigning random weights to instances in the training set, rather than using random bootstrap samples. The original method of wagging used Gaussian noise to modify instance weights. However, this occasionally caused some weights to drop to zero, effectively removing those instances from the training set. Researchers observed that bagging and AdaBoost operate through distinct mechanisms. This observation inspired the idea of combining the two approaches. Bagging primarily focuses on reducing variance, while AdaBoost addresses both bias and variance reduction. However, evidence (Bauer & Kohavi, 1999) suggests that bagging is more effective at minimizing variance.

MultiBoosting combines the strengths of both methods, retaining AdaBoost's ability to reduce bias while incorporating bagging's effectiveness in reducing variance. Essentially, MultiBoosting creates wagging-based subcommittees within the AdaBoost framework. Beyond combining bias and variance reduction, MultiBoosting also offers computational advantages over AdaBoost. Because AdaBoost is inherently sequential, it limits opportunities for parallel computation. In contrast, wagging trains each classifier independently, enabling parallel processing. MultiBoosting inherits this advantage, allowing subcommittees to be trained concurrently (Webb, 2000).

12.3.4.4 Random Subspace Method (RSM)

The random subspace method (RSM), introduced by Ho (1998), is a combination technique that modifies the training data by altering the feature space. A random subspace of dimensionality r is selected from the original p-dimensional feature space, creating a modified training set consisting of r-dimensional training samples. Classifiers are then built using these random subspaces, and their outputs are combined through majority voting to determine the final decision. This approach of using RSM for both creating and aggregating classifiers offers several advantages. For example, when the number of training samples is small compared to the dimensionality of the data, building classifiers within random subspaces helps address the issue of limited sample size. Additionally, if the data contains many redundant features, classifiers trained in random subspaces may perform better than those built in the original feature space. As a result, the combined decision of classifiers trained in these subspaces can outperform a single classifier trained on the full original feature set (Skurichina & Duin, 2002).

12.3.4.5 Rotation Forest (RoF)

RoF is an ensemble classifier focusing on feature extraction. The training data is generated for each base classifier by randomly dividing the feature set into K sub-sets, with PCA applied to each sub-set. To maintain the variability information within the data, all principal components are preserved. The rotation approach promotes individual accuracy and diversity within the ensemble. This diversity arises from the feature extraction process used for each base classifier. Decision trees, particularly sensitive to the rotation of the feature axes, are utilized, leading to the ensemble being referred to as a "forest." The method seeks accuracy by retaining all principal components and training each base classifier on the entire dataset (Rodriguez et al., 2006).

12.3.4.6 Weighted Majority Voting Ensemble

Ensemble learning integrates the predictions from multiple classifiers, with majority voting being a fundamental and popular technique. In this method, each classifier contributes one vote to the decision-making process, assigning the instance to the class that garners the most votes. Let n denote the number of classifiers in the ensemble, and C_t represents a classifier within the ensemble E (where $t = 1,2,\ldots,n$). The output of the tth classifier is denoted as $d_{t,j} \in \{0,1\}$, where $j = 1,2,\ldots,k$ corresponds to the number of classes. In majority voting, if the classifier selects class c_j, then $d_{t,j} = 1$; otherwise, $d_{t,j} = 0$. The overall ensemble output can be represented mathematically, though treating all classifiers equally might not yield optimal performance, as they often vary in their predictive capabilities (Dogan & Birant, 2019).

Majority voting is one of the simplest and most effective methods for combining classifiers in ensemble learning. Recent research on majority voting-based ensemble strategies shows that the robustness and accuracy of the classifier ensemble can be improved by employing weighted voting schemes (Ekbal & Saha, 2011a, 2011b). Some of these studies have focused on goal-oriented proposals for weighted voting-based ensemble classification, while others have taken a more general approach to various classification tasks.

Voting is a straightforward method for combining individual classification algorithms, with the choice of combination rule being crucial. Voting schemes can be classified as unweighted or weighted, with unweighted methods including minimum and maximum probability, majority voting, product of probabilities, and averages. Weighted voting encompasses simple, rescaled, best–worst, and quadratic best–worst approaches (Moreno-Seco et al., 2006).

Among classifier ensembles, simple majority voting is prevalent. It combines the binary outputs of k base classifiers, determining the output based on the highest vote count. Other linear combination rules include averages, products, and maxima or minima of probabilities. Research by Moreno-Seco et al. (2006) extensively analyzes weighted voting techniques in ensemble learning, which we used to evaluate our multiobjective weighted voting scheme. The techniques include simple weighted voting, rescaled weighted voting, best–worst weighted voting, and quadratic best–worst weighted voting. In simple weighted voting, each classifier's weight is determined based on its predictive performance using the training set. In rescaled voting, classifiers performing below a certain threshold are excluded, and the remaining weights are scaled proportionally. The best–worst weighted voting identifies the best and worst performing classifiers, assigning a weight of one to the best and zero to the worst, thus excluding the latter. Quadratic best–worst voting assigns higher weights to classifiers with better performance (Onan et al., 2016).

To enhance the effectiveness of the ensemble, the weighted majority voting ensemble (WMVE) approach is proposed, which assigns weights to classifiers based on their classification performance. Classifiers that exhibit greater reliability and accuracy in previous predictions receive higher weights, amplifying their influence in voting. The WMVE consists of three phases: training classifiers on a training set, determining weights using a validation set, and combining the classifiers' outputs while considering these weights. By adjusting the weights based on the classifiers' past performance on the validation set, this method seeks to optimize the final prediction for each instance, ultimately leading to improved classification results across various applications (Dogan & Birant, 2019).

12.4 RESULTS AND DISCUSSION

12.4.1 Performance Evaluation Measures

The evaluation measures are performed by analyzing the outcomes generated by the classification model and deriving a relationship of these findings with the ground truth. In real problems, dataset availability, particularly labeled datasets, is limited. To overcome this limitation of dimensionality, the available dataset is tactfully divided into sub-sets by following the cross-validation strategy (Mian Qaisar, Nisar, et al., 2023). One popular method for evaluating an AI model's performance and generalizability is the *k*-fold cross-validation. The dataset is randomly divided into *k* equally sized sub-groups, also known as the folds. Given the scale of the intended application and the available dataset, care should be taken while selecting the value of this user-specified parameter, *k*. The *k*-fold cross-validation technique ensures the model performs consistently and dependably on various data sub-sets. In this study, $k = 10$ is selected.

The selection of an AI model for the intended application is based on the outcome of the considered evaluation measures. It helps identify the most suitable model from the available choices. The use of multiple evaluation measures leads to a correct choice. The evaluation measures, used in this study are accuracy, *F*-measure, and the area under the curve (AUC). These are measurable using the confusion matrices obtained through experimentation. The confusion matrix is presented in the form of a table and it presents the outcomes of a binary or multi-class classification task (Mian Qaisar, Nisar, et al., 2023). An example of a confusion matrix for a binary classification problem is shown in Figure 12.2. A model with high TP, TN, and low FP, FN is a good model with a high value in the performance evaluation measures.

The following equations present the process of computing the considered evaluation measures such as accuracy, *F*-measure, AUC, and Kappa index:

$$\text{Accuracy} = \frac{\text{TP} + \text{TN}}{\text{TP} + \text{TN} + \text{FP} + \text{FN}} \tag{12.2}$$

Predicted Values		Actual Values	
		Positive	Negative
	Positive	True Positive (TP)	False Positive (FP)
	Negative	True Negative (TN)	False Negative (FN)

FIGURE 12.2 Confusion matrix for a binary classification problem.

$$F-\text{measure} = \frac{2.\text{TP}}{2.\text{TP}+\text{FP}+\text{FN}} \tag{12.3}$$

The area under the receiver operating characteristic (ROC) curve is called the AUC. It is a graphical representation that shows the categorization capability of a classifier (Mian Qaisar, Nisar, et al., 2023). Here, the *y*-axis represents the TPR, and the *x*-axis represents the false-positive rate. The TPR and false-positive rate are defined using Equations 12.3 and 12.4:

$$\text{TPR} = \frac{\text{TP}}{\text{TP}+\text{FN}} \tag{12.4}$$

$$\text{FPR} = \frac{\text{FP}}{\text{FP}+\text{TN}} \tag{12.5}$$

12.4.2 Experimental Results

This study uses the intended machine learning-based classifiers in ensemble configurations. The ensemble configurations considered are bagging, AdaBoost, MultiBoosting, RSM, and RoF. Moreover, a WMVE approach has also been devised. These configurations are used to process the EEG-based extracted features in order to identify a person precisely. Noise from the considered EEG signals is reduced using MSPCA. In the next step, the WPD decomposes the preprocessed EEG signals. It permits an in-depth analysis and allows effective feature mining. The sub-bands generated using WPD are processed to extract the statistical features from each sub-band, and certain relative features of the consecutive sub-bands are also mined. These extracted statistical attributes are concatenated to define each instance. The prepared feature set is further processed by the designed bagging-, AdaBoost-, MultiBoosting-, RSM-, RoF-, and WMVE-based ensemble classification configurations for automated person recognition. The classification performance assessment is performed in terms of the ROC area, *F*-measure, and accuracy.

The experimental results obtained from this study demonstrate that using the recommended classifiers offers high accuracy for the EEG dataset under consideration. The full EEG dataset is divided into training and test sub-sets so that the performance of each designed ensemble model can be assessed. The 10-fold cross-validation approach is used to assess the categorization performance.

TABLE 12.2
A Summary of the Obtained Classification Accuracy Scores

	Bagging	AdaBoost	MultiBoosting	Random Subspace	Rotation Forest	WMVE
SVM	96.44	95.44	93.24	95.67	97.67	98.89
k-NN	94.78	97.22	99.67	97.44	95.22	95.22
ANN	95.33	97.33	99.56	97.56	94.89	94.89
Random forest	94.89	97.56	99.33	97.78	94.67	95.33
CART	94.56	97.89	98.33	96.67	94.33	94.78
C4.5	95.33	98.22	98.44	93.44	94.22	93.44
REP tree	94.78	98.56	94.78	98.44	93.67	98.44
Random tree	94.67	98.67	94.67	98.56	93.56	98.44
LAD tree	94.56	98.78	94.56	98.67	93.44	98.56

Since each person generates distinct EEG patterns, the primary goal of human identity recognition systems is to classify the EEG signal patterns precisely. The k-NN, SVM, RF, ANN, classification and regression trees (CART), C4.5 decision tree, Reduced Error Pruning (REP) tree, random tree, and logical analysis of data (LAD) tree are some of the notable algorithms extensively used in the EEG signals classification process. Therefore, these classifiers are also used in this study.

A summary of the classification accuracy scores, obtained using the devised ensemble learning methods is outlined in Table 12.2.

First, let us look at the accuracy scores attained for the case of the bagging-based ensemble configuration of intended machine learning classifiers. In this case, the most significant classification accuracy of 96.44% is attained with SVM. The second highest accuracy score is 95.33%, with ANN and C4.5. RF is the third-best performer, with an accuracy score of 94.89%. CART and LAD tree attain the lowest accuracy score of 94.56%. In this case, the SVM with bagging achieves the most incredible accuracy score of 96.44%.

Now let us analyze the results obtained in the AdaBoost-based ensemble classification. In this case, the most notable classification accuracy of 98.78% is attained with LAD tree. The second highest accuracy score is 98.67%, attained with the random tree. REP tree is the third-best performer, with an accuracy score of 98.56%. SVM attains the lowest accuracy score of 95.44%. In this case, the LAD tree with AdaBoost achieves the most incredible accuracy score of 98.78%.

While analyzing the results obtained with the MultiBoosting-based ensemble classification, the highest classification accuracy of 99.67% is attained with k-NN. The second highest accuracy score is 99.56%, attained with ANN. RF follows ANN and is the third-best performer, with an accuracy score of 99.33%. SVM attains the lowest accuracy score of 93.24%. In this case, k-NN with MultiBoosting achieves the most significant accuracy score of 99.67%.

Now let us analyze the results, obtained in the case of RSM-based ensemble classification. In this case, the most notable classification accuracy of 98.67% is attained with LAD tree. The second highest accuracy score is 98.56%, achieved with the random tree. REP tree is the third-best performer, with an accuracy score of 98.44%. C4.5 attains the lowest accuracy score of 93.44%. In this case, the LAD tree with RSM achieves the most incredible accuracy score of 98.67%.

While analyzing the results obtained with the RoF-based ensemble classification, the highest classification accuracy of 97.67% is attained with SVM. The second highest accuracy score is 95.22%, attained with k-NN. ANN follows k-NN and is the third-best performer, with an accuracy score of 94.89%. LAD tree attains the lowest accuracy score of 93.44%. In this case, the SVM with RoF achieves the most significant accuracy score of 97.67%.

Now let us analyze the WMVE-based ensemble classification case results. In this case, the most notable classification accuracy of 98.89% is attained with SVM. The second highest accuracy score is 98.56%, attained with the LAD tree. REP tree and random tree are the third-best performers, with an accuracy score of 98.44%. C4.5 attains the lowest accuracy score of 93.44%. In this case, the SVM with WMVE achieves the highest accuracy score of 98.89%.

A summary of the *F*-measure values obtained using the devised ensemble learning methods is outlined in Table 12.3.

The bagging-based ensemble configuration yields the highest *F*-measure value of 97.78%, achieved by both the ANN and RF classifiers. k-NN follows closely with the second-best *F*-measure of 97.67%, while CART secures the third spot with 97.33%. The SVM records the lowest *F*-measure value at 95.44%.

Under the AdaBoost-based ensemble classification, both ANN and RF achieve the highest *F*-measure of 97.78%. CART comes in second with a score of 97.33%, while REP tree and C4.5 share the third spot with *F*-measure values of 97.22%. k-NN, on the other hand, exhibits the lowest performance, with an *F*-measure value of 93.56%.

For MultiBoosting, CART emerges as the top performer with an impressive *F*-measure value of 98.56%. C4.5 achieves the second highest score of 98.44%, followed by REP tree with 97.89%. However, k-NN shows the weakest performance in this configuration, with an *F*-measure of 92.89%.

TABLE 12.3
A Summary of the Obtained *F*-Measure Values

	Bagging	AdaBoost	MultiBoosting	Random Subspace	Rotation Forest	WMVE
SVM	95.44	95.22	97.67	90.67	98.67	95.33
k-NN	97.67	93.56	92.89	98.22	95.67	92.39
ANN	97.78	97.78	93.33	95.67	97.33	92.78
Random forest	97.78	97.78	93.56	98.44	97.56	97.22
CART	97.33	97.33	98.56	94.33	97.89	95.67
C4.5	97.22	97.22	98.44	94.22	98.22	95.56
REP tree	96.78	97.22	97.89	93.67	96.44	91.33
Random tree	96.67	95.67	97.78	93.56	96.56	92.67
LAD tree	96.56	95.56	97.67	93.44	96.67	98.89

TABLE 12.4
A Summary of the ROC Area Values

	Bagging	AdaBoost	MultiBoosting	Random Subspace	Rotation Forest	WMVE
SVM	98.33	97.67	78.56	93.44	78.56	98.78
k-NN	97.44	98.67	98.89	91.44	97.56	97.33
ANN	97.78	76.78	98.78	91.33	97.44	97.22
Random forest	90.44	67.67	98.56	91.11	97.67	97.67
CART	92.56	90.67	90.44	91.11	98.78	96.89
C4.5	92.44	90.78	90.56	96.56	98.67	96.78
REP tree	76.78	95.67	90.89	90.00	87.67	96.44
Random tree	85.89	95.78	95.56	89.67	89.89	96.33
LAD tree	96.44	95.89	90.33	89.44	98.89	97.44

In the RSM-based ensemble classification, RF delivers the best performance with an F-measure value of 98.44%. k-NN takes second place with 98.22%, while ANN secures third with a value of 95.67%. SVM records the lowest F-measure value at 90.67%.

For the RoF-based ensemble classification, SVM achieves the highest F-measure value of 98.67%, showcasing its superior performance. C4.5 follows with a value of 98.22%, while CART ranks third with 97.89%. k-NN records the lowest F-measure at 95.67%.

The WMVE-based ensemble classification shows the LAD tree as the standout performer, achieving an F-measure value of 98.89%. RF comes second with 97.22%, while CART ranks third at 95.67%. REP tree records the lowest performance, with an F-measure value of 91.33%.

A summary of the ROC area values obtained using the devised ensemble learning methods is outlined in Table 12.4.

Let us first examine the ROC area values obtained using the bagging-based ensemble configuration of machine learning classifiers. In this scenario, the highest ROC area of 98.33% is achieved by the SVM. Following closely, the ANN secures the second-best ROC area of 97.78%, while k-NN comes in third with 97.44%. On the other hand, the REP tree shows the lowest performance with an F-measure of 76.78%. Notably, the SVM with bagging stands out as the top performer with a remarkable ROC area of 98.33%.

Next, we consider the results from the AdaBoost-based ensemble classification. Here, the highest ROC area value of 98.67% is achieved by k-NN, demonstrating its strong performance. The SVM follows as the second-best performer with an ROC area of 97.67%, while the LAD tree ranks third with an ROC area of 95.89%. In contrast, RF records the lowest ROC area of 67.67%. Overall, the k-NN classifier with AdaBoost is the most effective, achieving the highest ROC area value of 98.67%.

Analyzing the MultiBoosting-based ensemble classification results, k-NN achieves the top ROC area value of 98.89%. The second highest performance is seen with the ANN, which attains an ROC area of 98.78%. The RF classifier comes third with

98.56%. In contrast, the SVM shows the lowest ROC area of 78.56%. Clearly, k-NN combined with MultiBoosting excels with the highest ROC area value of 98.89%.

The results for the RSM-based ensemble classification reveal that the C4.5 classifier achieves the highest ROC area of 96.56%. SVM secures the second position with an ROC area of 93.44%, while k-NN takes the third spot at 91.44%. At the lower end, the LAD tree records an ROC area of 89.44%. Among all, the C4.5 classifier with RSM demonstrates the strongest performance, achieving a notable ROC area of 96.56%.

In the case of RoF-based ensemble classification, the LAD tree emerges as the top performer with the highest ROC area value of 98.89%. The CART classifier follows closely, attaining an ROC area of 98.78%, while the C4.5 ranks third with 98.67%. Conversely, the SVM shows the weakest performance with an ROC area of 78.56%. Overall, the LAD tree paired with RoF achieves the most significant ROC area value of 98.89%.

Finally, the WMVE-based ensemble classification results show that SVM achieves the most remarkable ROC area of 98.78%. RF secures the second highest ROC area of 97.67%, and the LAD tree ranks third with 97.44%. The random tree records the lowest ROC area of 96.33%. In summary, the SVM with WMVE delivers the best performance, achieving the highest ROC area value of 98.78%.

12.5 DISCUSSION

In this study, we have designed various ensemble learning-based classifiers. Since the early days of computer-assisted diagnosis, alternative classification algorithms have also gained popularity and are appropriate in numerous circumstances. Performance is the most crucial factor when selecting an algorithm, followed by usability and interpretation. This study suggests that the most accurate human identity verification can be performed using an effective hybridization of EEG signal conditioning, signal analysis, feature extraction, and ensemble learning-based classifiers. The ensemble learning-based methods effectively regroup multiple machine learning classifiers and efficiently combine their findings to secure a final decision. The ensemble learning approaches are based on bagging, AdaBoost, MultiBoosting, RSM, RoF, and WMVE. For the intended base learners, these ensemble learning assemblies are designed and evaluated in this study for the automated recognition of human beings by processing EEG signals.

From the findings presented in Section 4.2, it is clear that using MultiBoosting-based ensemble classification provides the most significant accuracy for most of the considered base learners. For the considered ensemble classification configurations, the devised assembly of MSPCA, WPD-based decomposition, statistical analysis-based sub-bands feature extraction, and MultiBoosting-based ensemble classification is the best performer, securing the highest average accuracy score of 99.67% for the case of the k-NN classifier. Along with k-NN, the ANN and RF have also secured average accuracy scores of more than 99.00% for the studied dataset when effectively assembled using the MultiBoosting method. In the case of the MultiBoosting approach, the CART and C4.5 classifiers also attained more than 98.00% average classification accuracies.

The WMVE-based ensemble classifier followed the MultiBoosting-based classification approach and secured the highest average classification accuracy of 98.89% while using an assembly of SVMs as the base learners. In the case of WMVE-based approach, the REP tree, random tree, and LAD tree classifiers also attained more than 98.00% average classification accuracies.

For the other considered ensemble classifiers like bagging, AdaBoost, RSM, and RoF, the highest average accuracy scores are, respectively, 96.44%, 98.78%, 98.67%, and 97.67%. This shows a notable precise classification performance of all ensemble learning classification methods. However, among the intended methods, the best performers are the MultiBoosting- and WMVE-based approaches. Therefore, we shall mainly focus on these two methods while analyzing the *F*-measure and ROC area results.

In terms of *F*-measure values, the WMVE-based assembly has secured the highest *F*-measure value of 98.89%, while the LAD tree assembly was used as base learners. The assemblies based on MultiBoosting have also secured *F*-measure values of more than 98.00% in the case of the CART and C4.5 classifiers. This shows that in terms of the *F*-measure value, the WMVE-based approach outperforms the MultiBoosting-based assembly of the considered base classifiers.

While analyzing the performance in terms of the ROC area, the MultiBoosting-based assembly has secured the highest ROC area value of 98.89% while using an assembly of the k-NN-based learners. The assemblies based on WMVE have also secured an ROC area value of 98.78% in the case of SVM as the base learner. This shows that the MultiBoosting-based approach outperforms the WMVE-based assembly of the considered base classifiers in terms of the ROC area value.

It is evident that for the ensemble classification configurations considered, MultiBoosting and WMVE are the best performers because they secure the highest average accuracy scores, *F*-measure, and ROC area values. These models are also favorable in terms of the ease of model creation. Using a random selection of the data and characteristics for training they generate an ensemble of base deciders. This decreases the possibility of overfitting, a significant issue with individual classifiers. These classification assemblies present intricate, nonlinear decision boundaries. They can, therefore, be applied to a variety of classification problems.

The choice of input variable and classification method are the two unique considerations that determine the performance of the classifiers for EEG signal categorization. The extraction approach and feature selection strategy employed to determine which of the most valuable parameters are extracted from the EEG signal. The model's inputs must have parameters that work well for classifying EEG signals. The mentioned ensemble classifiers were chosen because they provide insight into the evolution of the EEG's regularity and complexity and are suitable for classi fying the nonlinear dynamics. Accurate EEG signal identification is necessary for biometrics purposes. It is confirmed by the results that MultiBoosting and WMVE are capable of much more accurate EEG signal-based person identification than the other considered ensemble classifiers for the targeted application.

The incorporation of event-driven tools can further enhance the performance of the devised method in terms of real-time compression and computational effectiveness (Mian Qaisar, 2019, 2021). Moreover, the metaheuristic-based offline analysis can further augment the dimension reduction (Mian Qaisar et al., 2022; Mian Qaisar, Khan, et al., 2023). The feasibility of integrating these tools into the devised approach can be investigated in the future.

12.6 CONCLUSION

This chapter introduces the application of MultiBoosting- and WMVE-based ensemble learning models combined with WPD in "electroencephalogram" (EEG)-based biometric identification. The MSPCA method removes the noise from the considered EEG signals, and signal decomposition employs WPD. Subsequently, the statistical features are extracted from the WPD-generated sub-bands. The experimental investigation involves multi-subject EEG signals from the intended dataset. A novel hybridization of the MSPCA, WPD, feature extraction, and MultiBoosting-based ensemble learning approach is proposed specifically for EEG-based biometrics. Compared with other classifiers, these two devised ensemble models demonstrate a lower susceptibility to overfitting. The devised assembly of MSPCA, WPD, and sub-bands feature mining with MultiBoosting-based ensemble classification assemblies has secured the highest average classification accuracies of more than 99.00% for the considered dataset. Consequently, these methods can potentially enhance automated biometric identification from EEG signals.

In conclusion, this study demonstrates the potential of integrating WPD and ensemble classifiers to enhance the performance of automated biomedical signal analysis, particularly EEG-based biometric identification. The promising results obtained from this study underscore the importance of continued research and innovation in the field, with the ultimate goal of improving the reliability of EEG-based biometric identification. As we move forward, further refinements and advancements in signal processing algorithms and AI-based classification models hold the potential to unlock even greater capabilities and accessibility in wearable devices and digital healthcare technology. It will ultimately improve the quality of life for those in need.

EEG-based biometric identification presents exciting opportunities. However, it has a few drawbacks, including restricted bandwidth and spatial resolution. Furthermore, various noise sources, such as muscle contractions, powerline interference, and electrode artifacts, can affect the EEG signals. Additionally, the EEG signals differ between individuals due to variations in cerebral anatomy and electrode placement. In this scenario, complementary modalities such as electrocardiogram and photoplethysmography can be employed in contemporary solutions. These multimodal approaches, with effective incorporation of AI, can lead to more robust, personalized, and humanistic biometric identification solutions. It is an approach under significant exploration and evolution.

REFERENCES

Abo-Zahhad, M., Ahmed, S. M., & Abbas, S. N. (2015). State-of-the-art methods and future perspectives for personal recognition based on electroencephalogram signals. *IET Biometrics*, *4*(3), 179–190.

Alsberg, B. K., Woodward, A. M., Winson, M. K., Rowland, J. J., & Kell, D. B. (1998). Variable selection in wavelet regression models. *Analytica Chimica Acta*, *368*(1), 29–44.

Bakshi, B. R. (1998). Multiscale PCA with application to multivariate statistical process monitoring. *AIChE Journal*, *44*(7), 1596–1610.

Bakshi, B. R. (1999). Multiscale analysis and modeling using wavelets. *Journal of Chemometrics*, *13*(3–4), 415–434.

Bashar, M. K., Chiaki, I., & Yoshida, H. (2016). Human identification from brain EEG signals using advanced machine learning method EEG-based biometrics. *2016 IEEE EMBS Conference on Biomedical Engineering and Sciences (IECBES)*, Kuala Lumpur, Malaysia, pp. 475–479. https://doi.org/10.1109/IECBES.2016.7843496

Bauer, E., & Kohavi, R. (1999). An empirical comparison of voting classification algorithms: Bagging, boosting, and variants. *Machine Learning*, *36*(1), 105–139.

Bidgoly, A. J., Bidgoly, H. J., & Arezoumand, Z. (2020). A survey on methods and challenges in EEG based authentication. *Computers & Security*, *93*, 101788.

Blasco, J., Chen, T. M., Tapiador, J., & Peris-Lopez, P. (2016). A survey of wearable biometric recognition systems. *ACM Computing Surveys (CSUR)*, *49*(3), 1–35.

Chen, J., Mao, Z., Yao, W., & Huang, Y. (2020). EEG-based biometric identification with convolutional neural network. *Multimedia Tools and Applications*, *79*, 10655–10675.

Daubechies, I. (1990). The wavelet transform, time-frequency localization and signal analysis. *IEEE Transactions on Information Theory*, *36*(5), 961–1005.

Dogan, A., & Birant, D. (2019). A weighted majority voting ensemble approach for classification. *2019 4th International Conference on Computer Science and Engineering (UBMK)*, Samsun, Turkey, pp. 1–6. https://doi.org/10.1109/UBMK.2019.8907028

Ekbal, A., & Saha, S. (2011a). A multiobjective simulated annealing approach for classifier ensemble: Named entity recognition in Indian languages as case studies. *Expert Systems with Applications*, *38*(12), 14760–14772.

Ekbal, A., & Saha, S. (2011b). Weighted vote-based classifier ensemble for named entity recognition: A genetic algorithm-based approach. *ACM Transactions on Asian Language Information Processing (TALIP)*, *10*(2), 1–37.

Galbally, J., Marcel, S., & Fierrez, J. (2013). Image quality assessment for fake biometric detection: Application to iris, fingerprint, and face recognition. *IEEE Transactions on Image Processing*, *23*(2), 710–724.

Gui, Q., Jin, Z., & Xu, W. (2014). Exploring EEG-based biometrics for user identification and authentication. *2014 IEEE Signal Processing in Medicine and Biology Symposium (SPMB)*, Philadelphia, PA, USA, pp. 1-6. https://doi.org/10.1109/SPMB.2014.7002950

Hall, M., Witten, I., & Frank, E. (2011). *Data Mining: Practical Machine Learning Tools and Techniques*. Kaufmann, Burlington.

Ho, T. K. (1998). The random subspace method for constructing decision forests. *IEEE transactions on pattern analysis and machine intelligence*, *20*(8), 832–844.

Jayarathne, I., Cohen, M., & Amarakeerthi, S. (2016). BrainID: Development of an EEG-based biometric authentication system. *2016 IEEE 7th Annual Information Technology, Electronics and Mobile Communication Conference (IEMCON)*, Vancouver, BC, Canada, pp. 1–6. https://doi.org/10.1109/IEMCON.2016.7746325

Kevric, J., & Subasi, A. (2017). Comparison of signal decomposition methods in classification of EEG signals for motor-imagery BCI system. *Biomedical Signal Processing and Control*, *31*, 398–406.

Klonovs, J., Petersen, C. K., Olesen, H., & Hammershoj, A. (2013). Id proof on the go: Development of a mobile eeg-based biometric authentication system. *IEEE Vehicular Technology Magazine*, *8*(1), 1.

Kutlu, Y., & Kuntalp, D. (2012). Feature extraction for ECG heartbeats using higher order statistics of WPD coefficients. *Computer Methods and Programs in Biomedicine*, *105*(3), 257–267.

Learned, R. E., & Willsky, A. S. (1995). A wavelet packet approach to transient signal classification. *Applied and Computational Harmonic Analysis*, *2*(3), 265–278.

Leon, F., Floria, S. -A., & Bădică, C. (2017). Evaluating the effect of voting methods on ensemble-based classification. *2017 IEEE International Conference on INnovations in Intelligent SysTems and Applications (INISTA)*, Gdynia, Poland, pp. 1–6. https://doi.org/10.1109/INISTA.2017.8001122

Lin, C.-T., Liu, Y.-T., Wu, S.-L., Cao, Z., Wang, Y.-K., Huang, C.-S., King, J.-T., Chen, S.-A., Lu, S.-W., & Chuang, C.-H. (2017). EEG-based brain-computer interfaces: A novel neurotechnology and computational intelligence method. *IEEE Systems, Man, and Cybernetics Magazine*, *3*(4), 16–26.

Maiorana, E. (2020). Deep learning for EEG-based biometric recognition. *Neurocomputing*, *410*, 374–386.

Maiorana, E., & Campisi, P. (2017). Longitudinal evaluation of EEG-based biometric recognition. *IEEE Transactions on Information Forensics and Security*, *13*(5), 1123–1138.

Malinowski, E. R. (2002). *Factor Analysis in Chemistry*. Wiley, New York.

Mao, Z., Yao, W. X., & Huang, Y. (2017). EEG-based biometric identification with deep learning. *2017 8th International IEEE/EMBS Conference on Neural Engineering (NER)*, Shanghai, China, pp. 609–612. https://doi.org/10.1109/NER.2017.8008425

Mian Qaisar, S. M. (2019). Efficient mobile systems based on adaptive rate signal processing. *Computers & Electrical Engineering*, *79*, 106462.

Mian Qaisar, S. (2021). Signal-piloted processing and machine learning based efficient power quality disturbances recognition. *PLoS One*, *16*(5), e0252104.

Mian Qaisar, S. M., Khan, S. I., Dallet, D., Tadeusiewicz, R., & Pławiak, P. (2022). Signal-piloted processing metaheuristic optimization and wavelet decomposition based elucidation of arrhythmia for mobile healthcare. *Biocybernetics and Biomedical Engineering*, *42*(2), 681–694.

Mian Qaisar, S. M., Khan, S. I., Srinivasan, K., & Krichen, M. (2023). Arrhythmia classification using multirate processing metaheuristic optimization and variational mode decomposition. *Journal of King Saud University-Computer and Information Sciences*, *35*(1), 26–37.

Mian Qaisar, S. M., Nisar, H., & Subasi, A. (2023). *Advances in Non-Invasive Biomedical Signal Sensing and Processing with Machine Learning*. Springer, Berlin.

Moore, S. K. (2016). "Brainprint" biometric id hits 100% accuracy. *IEEE Spectrum*, *53*(6), 14–14.

Moreno-Seco, F., Inesta, J. M., De León, P. J. P., & Micó, L. (2006). Comparison of classifier fusion methods for classification in pattern recognition tasks. In: Yeung, DY., Kwok, J.T., Fred, A., Roli, F., de Ridder, D. (eds) *Structural, Syntactic, and Statistical Pattern Recognition*. Springer, Berlin, Heidelberg, pp. 705–713. https://doi.org/10.1007/11815921_77

Nguyen, D., Dat, T., & Ma, W. (2019). A study on combing EEG signals and crytography for bitcoin security. *Australian Journal of Intelligent Information Processing Systems*, *34*, 1–9.

Onan, A., Korukoğlu, S., & Bulut, H. (2016). A multiobjective weighted voting ensemble classifier based on differential evolution algorithm for text sentiment classification. *Expert Systems with Applications*, *62*, 1–16. https://doi.org/10.1016/j.eswa.2016.06.005

Ong, Z. Y., Saidatul, A., & Ibrahim, Z. (2018). Power spectral density analysis for human EEG-based biometric identification. *2018 International Conference on Computational Approach in Smart Systems Design and Applications (ICASSDA)*, Kuching, Malaysia, pp. 1–6. https://doi.org/10.1109/ICASSDA.2018.8477604

Palaniappan, R., & Mandic, D. P. (2007). EEG based biometric framework for automatic identity verification. *The Journal of VLSI Signal Processing Systems for Signal, Image, and Video Technology*, *49*(2), 2.

Ramzan, Q., & Shidlovskiy, S. (2018). Evolution of the brain computing interface (BCI) and proposed electroencephalography (EEG) signals based authentication model. *MATEC Web of Conferences*, *155*, 01006.

Ravi, K., Palaniappan, R., Eswaran, C., & Phon-Amnuaisuk, S. (2007). Data encryption using event-related brain signals. *International Conference on Computational Intelligence and Multimedia Applications (ICCIMA 2007)*, Sivakasi, India, pp. 540–544. https://doi.org/10.1109/ICCIMA.2007.178

Rioul, O., & Vetterli, M. (1991). Wavelets and signal processing. *IEEE Signal Processing Magazine*, *8*, 14–38.

Rodriguez, J. J., Kuncheva, L. I., & Alonso, C. J. (2006). Rotation forest: A new classifier ensemble method. *IEEE Transactions on Pattern Analysis and Machine Intelligence*, *28*(10), 1619–1630.

Ruiz-Blondet, M. V., Jin, Z., & Laszlo, S. (2016). CEREBRE: A novel method for very high accuracy event-related potential biometric identification. *IEEE Transactions on Information Forensics and Security*, *11*(7), 1618–1629.

Safont, G., Salazar, A., Soriano, A., & Vergara, L. (2012). Combination of multiple detectors for EEG based biometric identification/authentication. *2012 IEEE International Carnahan Conference on Security Technology (ICCST)*, Newton, MA, USA, pp. 230–236. https://doi.org/10.1109/CCST.2012.6393564

Schalk, G., McFarland, D. J., Hinterberger, T., Birbaumer, N., & Wolpaw, J. R. (2004). BCI2000: A general-purpose brain-computer interface (BCI) system. *IEEE Transactions on Biomedical Engineering*, *51*(6), 6.

Singh, B., Mishra, S., & Tiwary, U. S. (2015). EEG based biometric identification with reduced number of channels. *2015 17th International Conference on Advanced Communication Technology (ICACT)*, PyeongChang, Korea (South), pp. 687–691. https://doi.org/10.1109/ICACT.2015.7224883

Skurichina, M., & Duin, R. P. (2002). Bagging, boosting and the random subspace method for linear classifiers. *Pattern Analysis & Applications*, *5*(2), 121–135.

Subasi, A. (2013). Classification of EMG signals using PSO optimized SVM for diagnosis of neuromuscular disorders. *Computers in Biology and Medicine*, *43*(5), 576–586.

Tatar, A. B. (2023). Biometric identification system using EEG signals. *Neural Computing and Applications*, *35*(1), 1009–1023.

Thakor, N. V., Gramatikov, B., Sherman, D., & Bronzino, J. (2000). Wavelet (time-scale) analysis in biomedical signal processing. *The Biomedical Engineering Handbook*, *56*, 1–56.

Trygg, J., & Wold, S. (1998). PLS regression on wavelet compressed NIR spectra. *Chemometrics and Intelligent Laboratory Systems*, *42*(1–2), 209–220. https://doi.org/10.1016/S0169-7439(98)00013-6

Unser, M., & Aldroubi, A. (1996). A review of wavelets in biomedical applications. *Proceedings of the IEEE*, *84*(4), 626–638.

Vetterli, M., & Herley, C. (1992). Wavelets and filter banks: Theory and design. *IEEE Transactions on Signal Processing*, *40*(9), 2207–2232.

Webb, G. I. (2000). Multiboosting: A technique for combining boosting and wagging. *Machine Learning*, *40*(2), 159–196.

Yang, S., & Deravi, F. (2017). On the usability of electroencephalographic signals for biometric recognition: A survey. *IEEE Transactions on Human-Machine Systems*, *47*(6), 958–969.

Index

L

M

N

O

P

Q

R

S

T

U

V

W

X

www.ingramcontent.com/pod-product-compliance
Ingram Content Group UK Ltd.
Pitfield, Milton Keynes, MK11 3LW, UK
UKHW022221190926
475702UK00003B/473
9781041007456